Exploring Climate Change Related Systems and Scenarios

Jeremy Webb draws on multiple disciplines to piece together the climate change puzzle, identifying what it would take to limit climate change and its impacts.

The book starts with a summary of the climate change problem and develops a Climate Change, National Interests, International Cooperation (CCNIIC) model of the climate response system. Webb reviews 'reverse stress testing', 'backcasting', and 'theory of change' methods, showing how they can be used to collect a large sample of possible futures. He also shows how we can explore the multiverse of futures using a new method called thematic chain analysis, finding relevant connections across scenarios. In the second half of the book, Webb explores 175 scenarios collected through 27 interviews with climate change experts. From these scenarios a signal response model is developed. Preconditions for effective social change and behaviour, political will and policy, as well as business and economic activity are synthesised. Lessons include preconditions for effective global responses to climate change, showing what it takes to limit climate change and related impacts. The book finishes with an epilogue, applying the signal response model and preconditions for effective global responses to COVID-19, demonstrating that models from this book can be applied to other global response problems – and used to quickly assess possible response strategies.

This book will be of great interest to students and scholars of climate change, environmental policy and future studies.

Jeremy Webb has a background in geology, development studies, and a doctorate in Science, Technology, Engineering and Public Policy from University College London, UK. He has worked as an international expert in environment statistics, environmental economic accounting, natural resource classification, water statistics, climate change and development, as well as minerals and development. This includes 10 years with the United Nations in New York and Addis Ababa, and another 6 years leading working groups under the United Nations Economic Commission for Europe.

Routledge Advances in Climate Change Research

Urban Planning for Climate Change
Barbara Norman

Climate Action in Southern Africa
Implications for Climate Justice and Just Transition
Edited by Philani Moyo

Climate Change Action and The Responsibility to Protect
A Common Cause
Ben L. Parr

Climate Security
The Role of Knowledge and Scientific Information in the Making of a Nexus
Matti Goldberg

COVID-19 and Climate Change in BRICS Nations
Beyond the Paris Agreement and Agenda
Edited by Ndivhuho Tshikovhi, Andréa Santos, Xiaolong Zou, Fulufhelo Netswera, Irina Zotona Yarygina and Sriram Divi

Social Transformation for Climate Change
A New Framework for Democracy
Nicholas Low

Exploring Climate Change Related Systems and Scenarios
Preconditions for Effective Global Responses
Jeremy Winston Webb

For more information about this series, please visit: www.routledge.com/Routledge-Advances-in-Climate-Change-Research/book-series/RACCR

Exploring Climate Change Related Systems and Scenarios

Preconditions for Effective Global Responses

Jeremy Winston Webb

Routledge
Taylor & Francis Group

LONDON AND NEW YORK

First published 2024
by Routledge
4 Park Square, Milton Park, Abingdon, Oxon OX14 4RN

and by Routledge
605 Third Avenue, New York, NY 10158

Routledge is an imprint of the Taylor & Francis Group, an informa business

British Library Cataloguing-in-Publication Data
A catalogue record for this book is available from the British Library

ISBN: 978-1-032-73566-5 (hbk)
ISBN: 978-1-032-73783-6 (pbk)
ISBN: 978-1-003-46591-1 (ebk)

DOI: 10.4324/9781003465911

Typeset in Times New Roman
by codeMantra

Contents

List of figures   *vii*
List of tables   *ix*
Acknowledgements   *xiii*

PART I
Introduction and background to climate responses   1

1 Introduction to the climate response problem   2

2 Climate change issues and options   9

3 Climate actors, institutions, and response system   30

4 Scenario methods and overarching themes   48

PART II
Exploring 175 global response scenarios   59

5 Climate impact, risk, and response scenarios   60

6 Actors and interests in climate scenarios   79

7 Climate change response options and scenarios   100

8 International climate cooperation scenarios   114

9 Other unusual climate-related scenarios   130

PART III
Lessons learnt   137

10   Analysing climate signals, actors, and responses   138

11   Effective global responses to climate change   170

12   How to limit climate change and its impact   196

PART IV
Epilogue   205

13   Lessons for other global response problems   206

References   *215*
List of acronyms   *234*
Glossary   *237*
Appendix A: Physical signals from IPCC assessments   *256*
Appendix B: Environment and national interests model   *258*
Appendix C: Jellyfish model of the climate regime complex   *260*
Appendix D: Methods used to explore the future   *266*
Appendix E: Respondent sentiment   *287*
Appendix F: Summary of scenario types and complexity   *289*
Appendix G: Timeliness and scale of responses   *292*
Appendix H: Effective responses and serendipity   *296*
Appendix I: Preconditions and the role of serendipity   *298*
Index   *299*

Figures

2.1 Qualitative risk categories 17
3.1 Connections between the Paris Agreement, UNFCCC, and outcome levels 35
3.2 Snapshot of interactions between climate change, national interests, and international cooperation 37
3.3 Snapshot of interactions between national interests of different sovereign states as well as climate change and international cooperation 38
3.4 The Climate Change, National Interests, International Cooperation (CCNIIC) Model 39
3.5 Options and the win set in a two-level game 41
3.6 CCNIIC Model divided into Levels I and II of a two-level game 41
3.7 A Venn diagram of preconditions for effective international cooperation 46
4.1 Semi-structured survey design 53
4.2 Overarching themes situated in relation to the CCNIIC Model 57
10.1 Climate change signal as a function of climate stress and the risk to actor interests 141
10.2 Climate stress based on the IPCC's reasons for concern and the distribution of impacts and risks to actor interests 142
10.3 Ambition levels are influenced by the options available 145
10.4 The global response to climate change depends on signals and responses 147
10.5 The global response to climate change depends on decision making including triggers and drivers, and response attitudes 149
10.6 Other factors, technologies and practices and the influences these things can have on actors and interests as well as options and actions 154
10.7 Preconditions for social change and behaviour, as well as the influences social change and behaviour might have on political will and policy, business and economic activity 156
10.8 Preconditions for climate change related political will and policy and the influence political will and policy might have on responses 158

10.9   Preconditions for business and economic activity-related
       responses to climate change as well as possible influences
       business and economic activities might have on the wider
       response to climate change   160
10.10 Venn diagram showing political will and policy as well as
       business and economic activity as a subset of social change and
       behaviour, including important interactions between these sets
       of actors and interests   161
10.11 Preconditions for GHG removals from the atmosphere   164
10.12 Preconditions for international cooperation on climate change
       and related international cooperation responses   166
10.13 The effectiveness of international cooperation depends on the
       power of coalitions and their levels of ambition   168
11.1   The extent to which climate change is a problem to society,
       applying distinctions from Kingdon (1995)   176
11.2   Climate change signals and many other factors influence
       contributions to the global response   179
11.3   The "competing angels" of individual and collective action   180
11.4   Schematic plot of effective response scenarios including climate
       stress levels and levels of serendipity required   185
11.5   Preconditions for individual and collective action on options   193
13.1   Signal response model applied to COVID-19   210
B.1   Environment National Interests (ENI) Model showing the
       interactions linking climate-related physical risks, national
       interests, and greenhouse gas emissions   258
C.1   Jellyfish model of the UNFCCC and global response
       to climate change   263
C.2   Jellyfish model of the climate regime complex   264
C.3   The climate regime is coupled with informal influences   264
D.1   Summary of methods used to collect and compile information
       about possible futures and analyse preconditions for effective
       global responses to climate change   267
D.2   Reverse stress test question order in the semi-structured interview 271
D.3   An example of a scenario compiled using theory of change,
       with four steps and seven layers, including non-essential (grey)
       and essential (dark grey) elements as well as assumptions (light
       grey). The numbers in brackets refer to rows in the database of
       transcribed responses   282
D.4   Example of a plot showing the international regime and
       organisation of international institutions at a point in time   283
E.1   Summary of survey respondent sentiment   288
F.1   Schematic illustration of a scenario plot including descriptors
       used to describe scenario complexity   291

Tables

2.1   Three conceptions of risks and their application to climate change   22

2.2   Overview of key characteristics of 1.5°C pathways   26

3.1   Success criteria based on the UNFCCC objective   34

3.2   Paris Agreement purpose and key points summarised   34

3.3   Preconditions for effective international cooperation based on the preconditions for agreement effectiveness and institutional feasibility   46

5.1   Examples of climate change hazard-related themes (in light grey rows) and related notes from the scenarios   61

5.2   Climate change risks and impact-related themes   62

5.3   No trigger responses and related definitions   69

5.4   Trigger responses and related definitions   72

6.1   Social change and behaviour themes with descriptions   80

6.2   Behavioural responses including actors, roles and actions   83

6.3   Possible behaviour-related scenarios   84

6.4   Political will and policy themes with descriptions   86

6.5   Business and economic activity themes and descriptions   94

7.1   Technology and practices definitions   101

7.2   Greenhouse gas removals and solar radiation management definitions   101

8.1   National interests that could drive risk responses for geopolitically powerful states or regions in the case of Europe   118

10.1   Summary of overarching themes and definitions organised in relation to the CCNIIC model   139

10.2   Summary of impact and risk levels on human and managed systems at 1.1°C, 1.5°C and 2°C of global warming, including generalised climate stress levels   143

10.3   Qualitative assessment of climate change signal strength showing signal strength is expected to increase in the future   144

10.4   Climate change related actions depend on triggers and drivers, strategies, justifications, and decision criteria   150

10.5   Cost-benefit analysis, cost-effectiveness analysis and social cost of carbon from the IPCC 1.5 Degree Report   152

10.6	The domestic response to climate change is made up of state and non-state actors, their interests, and actions taken	153
10.7	Responses to climate change depend on ambition levels and the availability of commercially viable options	162
10.8	Qualitative categories of risk	163
11.1	The Paris Agreement purpose includes preconditions for an effective global response to climate change	172
11.2	Key characteristics of 1.5°C pathways with related Paris Agreement preconditions for effective global responses	174
11.3	Summary of climate change related signals, responses, and contributions	180
11.4	The likely effectiveness of the global response to climate change based on signals, actors involved, their responses and contributions made	182
11.5	The characteristics of effective global responses to climate change (including effective non-responses and serendipity)	186
11.6	Preconditions for effective global responses fulfilling the UNFCCC objective	189
11.7	Preconditions for effective global responses from the Paris Agreement and IPCC's 1.5 Degree Report	190
13.1	COVID-19 triggers and drivers, strategies, justifications, decision criteria, and related actions	211
13.2	Preconditions for effective responses to COVID-19	212
A.1	Physical signals from IPCC assessments based on representative quotes	256
B.1	Relationship between vulnerability and exposure and the provision of goods, services and regulations	259
C.1	Levels used by IEGL to organise and map the climate regime	261
C.2	Types of institutions related to climate change	262
D.1	Ultimate outcomes and intermediate outcomes from the UNFCCC objective and Paris Agreement purpose	268
D.2	Degrees of success (white) and failure (dark grey) relative to the UNFCCC objective and UNFCCC legitimacy (i.e., universality of membership)	269
D.3	Sample selection criteria	275
D.4	The number of female and male invitees and respondents along with response rates	276
D.5	Primary national affiliations of invitees and respondents along with response rates	277
D.6	Primary institutional affiliations of invitees and respondents along with response rates	278
D.7	Summary of respondents including respondent numbers	278
D.8	Scenario elements	280
D.9	Fields for which notes were recorded and themes developed	284
D.10	Example of "thematic chain analysis" of the theme "social unrest"	285

F.1   Scenario type by outcome   289
F.2   Scenario complexity descriptors   290
G.1   Cynical and non-responses and implications for the timeliness
      and scale of global responses to climate change   292
G.2   Impact and risk responses and implications for the timeliness
      and scale of global responses to climate change   292
G.3   Response triggers and drivers and implications for the
      timeliness and scale of global responses to climate change   293
G.4   Response attitudes and implications for the timeliness and scale
      of global responses to climate change   295
H.1   Effective responses, including preconditions, that could
      conceivably fulfil the UNFCCC objective   296
I.1   Preconditions and the role of serendipity   298

Acknowledgements

I am immensely grateful for the work of the team at Routledge, Taylor and Francis Group, especially Senior Editor Annabelle Harris, Editorial Assistant Jyotsna Gurung, and Project Manager Saranya Megavannan from codemantra. I am also very grateful to the reviewers and their very helpful feedback on the draft manuscript.

I want to thank Professor Nick Bostrom for allowing the use of his figure showing qualitative risk categories. I want to thank Professor Michael Grubb for allowing the use of his figure showing three conceptions of risks and their application to climate change. And I want to thank the IPCC for allowing the use of the table giving an overview of key characteristics of 1.5°C pathways. The figures and table proved very helpful in understanding the climate change problem, how we might respond, and what we need to do to limit global warming levels to 1.5°C.

I am very grateful to the United Kingdom's Engineering and Physical Sciences Research Council (EPSRC) for funding the research in this book. Likewise, I am very grateful to the Department of Science, Technology, Engineering and Public Policy (STEaPP) at University College London (UCL) for the opportunity to conduct this research.

I am also in debt to the respondents involved in my research, who gave me their time, shared insights, and scenarios. I am forever grateful for the information you shared, making this book possible.

Lastly, I am extremely grateful to Professor Yacob Mulugetta, Professor Sarah Bell, and Dr Adam Cooper for their advice during the study, as well as Professor Arthur Peterson and Professor Jeremy Woods for their feedback on an earlier version of the manuscript.

Ngā mihi nui.
Jeremy Winston Webb

Part I

Introduction and background to climate responses

Part I consists of four chapters:

Chapter 1 introduces the climate response problem and the research questions this book addresses, finishing with the outline for this book.

Chapter 2 uses the analogy of a puzzle and its pieces, unpacking climate change and related issues, and the options for addressing climate change.

Chapter 3 continues with the jigsaw puzzle analogy, looking at the actors involved, the institutions established to address climate change, finishing with a model of the global response system showing how the pieces of the puzzle fit together.

Chapter 4 uses the analogy of a crystal ball, reviews foresight and related methods, then presents the methods used in this book and highlights overarching themes identified for exploration in Part II.

DOI: 10.4324/9781003465911-1

1 Introduction to the climate response problem

1.1 Introduction

"Who are the main actors?" I asked.

"We are all actors" responded the climate veteran over the noise of negotiators and observers shuffling past.

I was up to my 24th interview and a theme was emerging. Solving climate change is about you, me and everyone else, as individuals and in whatever capacities we have in society, business or government. Why? Because all of our actions contribute to the global response.

So, how can we solve climate change?

This is a big question. Answering this question is like answering a riddle. An important part of any riddle is working out what type of puzzle, or game, we are playing before responding.

From game theory,[1] there are finite games and infinite games. Finite games have winners and losers. If we lose, we can come back to the game and play again much like a game of football or chess. Infinite games are very different. It is not about winning or losing, it is about staying in the game i.e., "continuing the play" (Carse 1986). Solving climate change just means staying in the game. The defining characteristic of climate change is not its costs or benefits, although these are important, it is the uncertain but potentially catastrophic risk posed to people, communities, cultures, businesses, and even some countries.

Given that climate change is an uncertain but potentially catastrophic risk (Bostrom and Cirkovic 2008), what does it mean to solve climate change? It turns out our governments have already agreed on what solving climate change looks like. It means: stabilising atmospheric concentrations of greenhouse gases (GHGs) at levels that allow ecosystems to adapt naturally, food production is not threatened, and economic development proceeds in a sustainable manner. These are internationally agreed "objectives" from the United Nations Framework Convention on Climate Change (UNFCCC). If we achieve these things, climate change alone won't knock us out of the game.

So, we know what it means to "solve" climate change, that just leaves the "how to" part of the puzzle. Luckily, we have books on how to solve climate change. We even have global assessments of options by the Intergovernmental Panel on

DOI: 10.4324/9781003465911-2

Climate Change (IPCC). However, there is a large gap in our knowledge when comes to understanding who would act on these options, and the conditions under which we would act. These "conditions" are the "preconditions" for effective global responses to climate change.

1.2   The climate change problem

The climate change problem starts with the ordinary world, the world we were familiar with. It was a world with a relatively stable climate in which civilisations were built and fell. And while we went about our lives, climate researchers detected a change. This change started out subtly. Weather continued to vary, but long-term averages, and ranges, were changing. This included average air temperatures (IPCC 2014a), hence the expression "global warming" (see Glossary). Weather-related measurements showed subtle, and abrupt, shifts. Over the last couple of decades these shifts in long-term weather conditions have impacted our physical surroundings (i.e., environment) threatening us and our interests.

In 2007, Arctic sea ice extent collapsed surprising researchers from around the world (Zhang et al. 2008, Wadhams 2012). In 2023 record low Antarctic sea ice levels were recorded with indications that a new sea ice state had been established (Purich and Doddridge 2023). Meanwhile, the risk of prolonged Antarctic Ice Sheet collapse has been a focus of research with implications for long-term sea-level rise (Kennicutt et al. 2014, Tollefson 2016, Martin et al. 2019, Naughten et al. 2023). Polar regions may seem far away, but mountains, seas, and changes in our physical surroundings bring the impacts of climate change much closer to home including our farms, towns, and cities (i.e., human and managed systems). That is why the IPCC sees climate change as a threat that increasingly impacts people, property, and livelihoods (IPCC 2022a).

Unfortunately, risks and impacts to human and managed systems are expected to become moderate to high by 2035 and even greater mid-century (IPCC 2022a, 2023a). So, while we might have been able to ignore the impacts of climate change in the past, we will not be able to ignore them going forward (see Section 10.2.1). In short, climate change is affecting our lives and these impacts will become much more obvious over the next decades. How this influences the global response to climate change is deeply uncertain. For example, it is not known to what extent increased climate change impacts might be a problem, the extent to which these impacts might influence our individual and collective ambition levels, or, the actions you, me and everyone else might take in response to climate change.

One thing we do know is there will be no "new normal" until atmospheric concentrations of GHGs have stabilised and earth systems have reached equilibrium. Reaching equilibrium will take many generations for example, sea level rise. Climate change will impact our descendants for generations to come but the scale of these impacts depends on our actions (IPCC 2023a).

1.3   Decisions, actions, and global response scenarios

The most important part of the global response to climate change are our actions, inaction included. Each of our actions, and inactions, contributes to the global response – in terms of GHG emissions or levels of resilience. It is important to note that international cooperation is an influence on the global response, but it is only one influence among many. Most of the global response happens far away from climate negotiations or international cooperation; when we each make our own decisions for example, in the supermarket, choosing our next holiday destination or investment. In fact, the thing that links the entire global response to climate change are decisions. These decisions are made by you, me and everyone else, each day as individuals or as part of groups. They can be cynical or enlightened, impulsive or considered, based on a cost-benefit analysis or some other criteria.

Every individual and group can be thought of as an actor in the global response. With this in mind, this book takes a global perspective and focuses on a gap in the literature, consisting of the actors involved in the global response to climate change; our decision-making criteria; and, the conditions under which you, me, and everyone else would do what it takes to limit climate change and its impacts.

Given that climate change is an unprecedented challenge, with many actors, ambition levels, and possible responses, it would be nice to have something a bit like a crystal ball allowing us to ask questions, look into the future, and explore not just one possible future, but the range of possible scenarios. Instead of using a crystal ball, the research presented in this book is based on interviews with 27 people from around the world involved in climate change and related issues, and from these interviews a searchable sample of 175 possible futures is formed.

Metaphorically speaking, the searchable sample of possible futures is a bit like having a crystal ball. For any question, a keyword can be typed in and relevant scenarios are identified, including success, failure, and other scenarios. It is possible to look at, and explore, the range of possible conditions that might lead to effective global responses to climate change including the actors involved and the actions taken.

1.4   Book scope and approach

The climate puzzle has many elements. As a collective action problem (IPCC 2014a) this book considers climate change, related risks and impacts, the actors that might be affected by climate change, the actors and actions contributing to the global response (GHG emissions included), and the "institutions" that influence the global response including international agreements.

Effective responses to climate change can only be assessed at the global level, so this book takes a global perspective treating the global response to climate change as a single system with many elements. The most relevant literature, upon which this book builds, consists of IPCC assessments, other United Nations documents, as well as research articles regarding global mitigation and adaptation pathways or relevant technologies and practices.

This book draws upon, and develops, concepts and methods from a range of disciplines including international relations, finance, economics, risk management, and futures studies. In some ways each discipline is like the metaphorical blind person feeling the elephant. Individually, each person feels something different because they are exploring different parts of the same system, but taken together, they describe the elephant (Go and Carroll 2004, Evans 2006). This book explores the whole of the global response system and takes an interdisciplinarity approach drawing upon whatever concepts and methods help understand and describe the system.

Another important aspect of this book is stepping back, so the entire system can be observed and described. The parable of the blind people feeling the elephant works because those people fortunate enough to be sighted can stand back from the situation and see how each part the elephant fits to form the whole (Go and Carroll 2004, Evans 2006, Kleineberg 2013). This is analogous to the use of remote sensing, for example satellite imagery, where large features can be observed and identified that would otherwise be very difficult to identify from observations much closer to the ground (Wessman 1992, Aplin 2006). This book steps back and focuses on large-scale features of the global response system and broad response scenarios.

It is important to note that the term "global response to climate change" does not appear to have previously been defined, for example it does not appear in IPCC glossaries. To ensure the definition is consistent with the use of the term and existing practices for assessing the effectiveness of the global response, the "global response to climate change" is defined as "all human actions and inaction that influence fulfilment of the UNFCCC objective." Implicit within this definition is the assumption that there is such widespread awareness of climate change, that it can be considered a factor in all our decisions, even if the decision involves ignoring climate change. Likewise, a "response to climate change" is "any human action or inaction that influences fulfilment of the UNFCCC objective."

1.5 Five big questions

What does it take to solve climate change?

This is the "big question" that drove the research presented in this book. However, it is too big a question to be answered in one step. It needs to be unpacked if we want to answer it. So, we start with the question of:

What are the preconditions for effective global responses to climate changes?

To identify these preconditions, we need to know:

Under what conditions would actors act on effective response options?

Making the task of identifying preconditions a bit easier, there are some very likely changes in conditions going forward. For example, global warming is expected to continue along with other climate changes and related impacts, raising a subsequent question:

What influence might climate change have on actors and the global response to climate change?

Given that climate change is a collective action problem requiring some degree of international cooperation (IPCC 2014a), this raises the question:

What are the preconditions for effective international cooperation on climate change?

It is also very likely that more GHGs will be emitted into the atmosphere than a GHG budget would allow for (IPCC 2023a, UNEP 2023), raising the question:

What are the preconditions for actors to remove GHGs from the atmosphere at a scale required to limit climate change to safe levels?

1.6   Book outline

To answer these questions, scenarios were collected and analysed, new conceptual models were developed, along with new research methods, all of which shed light on the "climate crisis" and help us understand what it takes to solve climate change. Cataloguing this journey, the book has 13 chapters organised into four parts, consisting of:

- Part I: Introduction and background to climate responses
- Part II: Exploring 175 global response scenarios
- Part III: Lessons learnt
- Part IV: Epilogue

Part I: Introduction and background to climate responses

Part I includes this introduction to the climate response problem (Chapter 1) and three other background chapters (Chapters 2–4).

Chapter 2 sets out the pieces of the climate response puzzle, looking at what we know about the climate change as an issue, and the options for addressing climate change.

Chapter 3 unpacks the climate response puzzle addressing the actors that could act on the options from Chapter 2 and relevant institutions. Chapter 3 finishes with the CCNIIC model which shows how the pieces of the puzzle fit together. The model links climate change, national interests, and international cooperation into a single model of the global response system.

Chapter 4 reviews foresight and related methods, then sets out the methods used to explore the future and identify preconditions for effective global responses to climate change. This includes describing survey data collection methods and the use of backcasting and reverse stress test methodologies in the design of the semi-structured interviews; data processing methods including scenario identification using theory of change; thematic analysis of scenarios; and, the analysis of scenario themes including a new method called thematic chain analysis. The chapter finishes with a summary of scenarios collected (i.e., the sample of possible futures) and the overarching themes identified. These themes are explored in Part II, and analysed in Part III, of this book.

Part II: Exploring 175 global response scenarios

Chapter 5 looks deep into the crystal ball i.e., "sample of possible futures," exploring scenarios that address climate change impacts, risks, and responses by actors including response triggers, drivers, and attitudes. The chapter also highlights preconditions for effective responses in terms of timeliness and scale.

Chapter 6 explores future actors, interests, and actions including social change and behaviour, political will and policy, as well as business and economic activity.

Chapter 7 explores future response options including technologies and practices, GHG removals, and other factors influencing the responses of actors.

Chapter 8 explores international cooperation scenarios including factors affecting the international regime complex such as globalism versus nationalism. The chapter also presents scenarios and themes related to geopolitical power and influence, and stringent enforced climate agreements.

Chapter 9 explores other scenarios that could either influence climate or the global response to climate change, including low-probability high-impact events.

It is important to note here that Chapters 5–9 include many sections and subsections. Some of these chapters, sections, and subsections are short while others include more information including multiple possible scenarios related to a single theme. While it is normal practice to join short sections, this has not been done because all scenarios are treated as possible futures and joining sections with different themes implicitly links these themes.

When reading Chapters 5–9, it is best to think of each scenario as being a short story, that needs to be read first and then compared and contrasted with other related scenarios later. It is not until the analysis in Chapter 10 (in Part III of this book) that themes are linked, models are developed and multiple possible pathways are mapped based on the scenarios from Chapters 5 to 9.

Part III: Lessons learnt

Chapter 10 takes the scenarios and themes discovered in Chapters 5–9 and looks at what these themes mean for our future. This includes developing a climate change signal response model and mapping possible pathways for actors, interests and actions; GHG removals; and, international cooperation. Other possible scenarios are also analysed.

Chapter 11 discusses the results from Chapter 10, explores preconditions for effective global responses to climate change addressing the key questions from Chapter 1. This includes discussing preconditions from the Paris Agreement and IPCC's 1.5 Degree Report, climate change signals, possible responses, and the actors and interests involved. Preconditions for coalitions and international cooperation are discussed along with preconditions for GHG removals. Effective global responses are identified along with preconditions for effective responses and the chapter finishes with short discussions on preconditions for action and the existential risk associated with climate change.

Chapter 12 brings us back to where we are today, with recommendations on how you can respond to climate change, form coalitions and take a leading role as an individual and in whatever other capacities you have in life. Chapter 12 also answers the key questions from Chapter 1 including the question of what it means to "solve climate change." And while this book is hopefully a useful contribution to our knowledge, there is much more we need to do and understand. Hence, Chapter 12 points out areas for future research and exploration for you and others to lead.

Part IV: Epilogue

Chapter 13 reflects on how you might apply the lessons learnt from this book in the real world, helping solve climate change and other global response problems using COVID-19 as an example. Chapter 13 starts by characterising global response problems as a category, looks at how to address COVID-19 using the lessons learnt in this book, notes we are playing an infinite game when it comes to global response problems, and revisits the concepts of success and failure.

1.7   Guidance for readers

As you can see from the previous section, there is a lot to cover. So, what is the best way to read this book, and get to the information most important to you?

Part I of the book provides important background and context. If you are already familiar with IPCC assessments and the climate change literature, you may want to skip Part I and go straight to Part II. Part II may be challenging to read because there are so many possible scenarios. Skimming reading the scenarios and themes in Part II is probably adequate for most readers.

Part III of this book is the most important part, and I would recommend focusing on Chapters 10–12, rather than attempting to read this book chronologically. There is cross-referencing throughout the document, so you should be able to see which sections in previous chapters are relevant and might need to be read if you want to know more. Likewise, there are references to relevant appendices. If you want definitions for the terms used in this book, there is a comprehensive glossary.

Lastly, Part IV is the epilogue showing lessons from the global response to climate change (including related models and frameworks) are applicable to other global response problems such as COVID-19. These models and frameworks provide a basis for quickly scanning issues and options when new global response problems arise, helping us understand what we can do, how other actors might respond, and the likely effectiveness of our collective response.

Note

1 Note: This is not a book on game theory, but the research presented includes insights and methods from multiple disciplines. See Section 1.4 for more on the scope and approach used in this book.

2 Climate change issues and options

2.1 Introduction

What does it take to solve climate change? If we want to solve the climate change problem, we need to understand the problem and what we can do about it, drawing from the best available, and most salient, evidence. This chapter is the first step on that journey, reviewing the literature and looking at why climate change is an issue and what we can do about it.

A couple of useful analogies to think about while reading this chapter are the "jigsaw puzzle" and the "picture" on the front of the jigsaw puzzle box. Each part of the climate change system is like a jigsaw puzzle piece. We know what each piece looks like because we have research describing climate change in detail and international assessments unpacking the climate change problem and what we can do about it. Between the literature and the international assessments, we have a large pile of puzzle pieces. This book breaks down the puzzle into four piles of pieces, consisting of issues and options (addressed in this chapter) as well as actors and institutions (addressed in Chapter 3).

So, why is climate change an issue? Section 2.2 unpacks climate change issues. We start with physical climate change (Section 2.2.1), what this means for people in terms of impacts to date (Section 2.2.2), future risks (Section 2.2.3), and the global response (Section 2.2.4).

What can we do about climate change? Options for addressing climate change make up the second heap of puzzle pieces. These options (Section 2.3) include mitigation and adaptation of climate change (Section 2.3.1), greenhouse gas (GHG) removals from the atmosphere (Section 2.3.2), and the controversial possibility of solar radiation management (Section 2.3.3).

One final note, to ensure rigor in the use of terms and definitions, and conceptual coherence across the many fields of research addressed in this book, substantive terms are defined when first used.

2.2 Issues

Why is climate change an issue? The starting point for answering this question is to unpack what is an "issue", with some quick theory. We will then look at the characteristics of climate change and related "issues".

DOI: 10.4324/9781003465911-3

An issue can be defined as "a vital concern or unsettled problem"[1]. This definition leads to the question: how do we know if something is a "vital concern" or "unsettled problem"? Kingdon (1995) helps answer this question, when he noted

> There is a difference between a condition and a problem.[2] We put up with all manner of conditions every day: bad weather, unavoidable and untreatable illnesses, pestilence, poverty, fanaticism… Conditions become defined as problems when we come to believe that we should do something about them.
>
> (Kingdon 1995, p. 109)

Kingdon also noted, "Problems are not simply the conditions or external events themselves; there is also a perceptual, interpretive element." (Kingdon 1995, pp. 109–110). And

> There are great political stakes in problem definition. Some are helped and others are hurt, depending on how problems get defined. If things are going basically your way, for instance, you want to convince others that there are no problems out there.
>
> (Kingdon 1995, p. 110)

In the sections below, the case is made for why climate change is an issue and a problem that we need to do something about.

With this in mind, the sections below unpack the characteristics of climate change (Section 2.2.1) and related issues including impacts (Section 2.2.2), risks (Section 2.2.3), and the global response (Section 2.2.4).

2.2.1   Climate change

The IPCC defines climate change as "a change in the state of the climate that can be identified (e.g., by using statistical tests) by changes in the mean and/or the variability of its properties, and that persists for an extended period, typically decades or longer." (IPCC 2013). Climate characteristics (e.g., temperatures and precipitation) are typically assessed for periods of 30 years or more (Cubasch et al. 2013).

Climate extremes are important as they constitute hazards to people, their property and productive activities. Climate change related hazards include rapid onset events such as extreme heat or extreme precipitation events, as well as changes in storm frequency or intensity. Other hazards include slow onset events such as a warming trend, a drying trend, sea level rise, ocean acidification, and carbon dioxide fertilisation (Field et al. 2014). Internationally there is concern that climate change is a threat to development and this is reflected in the UNFCCC objective (Section 3.3.1) as well as the Paris Agreement purpose (Section 3.3.2).

The climate change problem has been defined in a number of ways, for example in agreements between states such as the UNFCCC as well as by IPCC assessments. "Climate change" is defined by the UNFCCC as "a change of climate which is attributed directly or indirectly to human activity that alters the composition of the global atmosphere and which is in addition to natural climate variability observed over comparable time periods." (Article 1, UNFCCC 1992). It is important to note that the UNFCCC uses the term climate change to refer to anthropogenic climate change while the IPCC (and this book) uses the term climate change to refer to both natural and anthropogenic changes. "Adverse effects of climate change" are defined by the UNFCCC as

> changes in the physical environment or biota resulting from climate change which have significant deleterious effects on the composition, resilience or productivity of natural and managed ecosystems or on the operation of socio-economic systems or on human health and welfare.
>
> (Article 1, UNFCCC 1992)

In response to the above framing of climate change and its adverse effects, the UNFCCC's objective is "stabilization of greenhouse gas concentrations in the atmosphere at a level that would prevent dangerous anthropogenic interference with the climate system." (Article 2, UNFCCC 1992). Article 2 goes on to state that "Such a level should be achieved within a time-frame sufficient to allow ecosystems to adapt naturally to climate change, to ensure that food production is not threatened and to enable economic development to proceed in a sustainable manner." (Article 2 UNFCCC 1992).

While it is tempting to assume anthropogenic climate change is driven exclusively by GHG emissions, this would not be accurate. There are multiple forcing agents that contribute to radiative forcing, GHGs being the most important, influencing earth's atmospheric energy balance and climate change. Any of these forcing agents, and their sources, could become a policy focus (e.g., Prins et al. 2010). Forcing agents include well-mixed GHGs, ozone, stratospheric water vapour from CH_4, surface albedo, contrails, aerosol–radiation interactions, aerosol–cloud interactions as well as natural solar irradiance. Some forcing agents are long-lived GHGs that "have lifetimes of approximately eight years or more" and other are short-lived climate pollutants (SLCPs) that "have lifetimes of approximately 20 years or less" (UNEP 2017, p. 48).

Sources of forcing agents can be natural or due to human activities. These sources may include particles and gases from volcanic eruptions, solar variations, orbital cycles, energy and industrial processes. Land use change also generates particle and GHG emissions as well as albedo changes. The combined effect of climate forcers has been to change earth's energy balance, creating a slight imbalance with the accumulation of energy mainly in the atmosphere and hydrosphere, including oceans (Cubasch et al. 2013).

2.2.2　Impacts

Attributable physical impacts of climate change on human and managed systems only emerged in the mid 2010s but impacts on natural systems were evident before this. From a review of the IPCC's first six assessment reports from 1990 to

2023, three types of physical signal can be identified, consisting of: climate science; physical risks; and physical impacts. Climate science regards "the study of relatively long-term weather conditions, typically spanning decades to centuries but extending to geological timescales." (Nature 2019). Climate risks regard "The potential for consequences where something of value is at stake" (IPCC 2014b, p. 127) and climate impacts regard "Effects on natural and human systems." (IPCC 2014b, p. 124). Representative quotes from the IPCC summaries from 1990 to 2023 are presented in Appendix A.

The climate science signal mainly regarded the detection of climate change and whether it was anthropogenic. Climate science showed an improvement in understanding of the physical system over time and was the most environmentally focused of all the signals. Climate risk was addressed in each of the summaries, but in most cases regarded long timeframes, decades in many cases and over 100 years in some cases. The ways in which physical risk were described progressively became more refined, for example with the identification of the five reasons for concern. Assessments from 1990 to 2001 included positive and negative implications of climate but assessments from 2007 to 2023 indicated risk as being negative overall.

When it came to climate change impacts on systems outside of the atmosphere, the signal went from no signal in 1990 to a clear signal in 2023. No impacts were identified in the Overview from 1990 and the Summary from 1995 stated that "Unambiguous detection of climate-induced changes in most ecological and social systems will prove extremely difficult in the coming decades." (IPCC 1995, p. 6) suggesting that limited or no signal could be expected for decades. However, by 2001 there appeared to be some observation of changes particularly in physical and biological systems. In 2007, there was medium confidence of impacts while in 2014 these had become attributable including "some impacts on human systems" (IPCC 2014a, p. 6). The IPCC Fifth Assessment Report even included an illustration summarising impacts around the world. The range of observed impacts had grown in the Six Assessment Report along with the confidence levels in many instances (O'Neill et al. 2022).

2.2.3   Risks

Climate change is a problem with long time horizons extending beyond business and political planning periods (Carney 2015) and includes the risk of feedbacks and other impacts. Long-lived GHGs such as methane[3] (CH_4) or sulphur dioxide (SO_2) remain in the atmosphere for decades (Myhre et al. 2013). Carbon dioxide (CO_2) remains in the atmosphere for over 100 years (Myhre et al. 2013). As such, emissions today have long-lasting temperature effects. However, the ultimate level of global warming and types of climate change manifested depend on feedback mechanisms that take time to reach equilibrium even after GHG concentrations stabilise in the atmosphere (Myhre et al. 2013). These feedbacks can either reinforce or limit changes in climate. Feedbacks include clouds and water vapour, emissions of GHGs other than carbon dioxide from the decomposition of peat and permafrost, the emissions of aerosols from biochemical processes such as wildfires, and ocean circulation for example. Each of these feedbacks are effective over different periods of time, from hours for longwave radiation and lapse rate, to as long as centuries for air-sea carbon dioxide exchange, permafrost, land ice and ocean circulation feedbacks.

A complicating factor when it comes to understanding and assessing climate change and related issues are uncertainties. For example, there are large uncertainties related to forcing agents and despite limited uncertainty around the radiative forcing of GHGs there remain large uncertainties around the negative radiative forcing related to aerosols resulting in large uncertainties around total anthropogenic radiative forcing. IPCC Working Group I noted that there are large uncertainties related to the longevity of GHG emissions and related radiative forcing (Myhre et al. 2013, Forster et al. 2021). Large uncertainties can also be found in climate change projections, including levels of global warming and warming over land and sea areas. The possible effects of global warming on precipitation, storms, and even sea level rise are not certain and as such this makes understanding the hazard climate change poses to human and managed systems difficult to assess.

To help get an understanding of possible futures, the IPCC has used models and scenarios since its first report. Scenarios from the IPCC's Sixth Assessment Report included a mix of Representative Concentration Pathways (RCPs) (which focused on atmospheric concentrations of GHGs and related radiative forcing) as well as Shared Socio-economic Pathways (SSPs) (which focused on socio-economic characteristics of scenarios). Scenarios were also classified according to global warming pathways and by the policies involved.

Global warming pathways, between now and 2100, were classified into the following categories:

- C1: Limit warming to 1.5°C (with greater than 50% likelihood) with no or limited overshoot
- C2: Return warming to 1.5°C (with greater than 50% likelihood) after a high overshoot
- C3: Limit warming to 2°C (with greater than 67% likelihood)
- C4: Limit warming to 2°C (with greater than 50% likelihood)
- C5: Limit warming to 2.5°C (with greater than 50% likelihood)
- C6: Limit warming to 3°C (with greater than 50% likelihood)
- C7: Limit warming to 4°C (with greater than 50% likelihood)
- C8: Exceed warming of 4°C (with greater than or equal to 50% likelihood)

According to the IPCC, global surface temperature reached 1.1°C above 1850–1900 in the period 2011–2020 (IPCC 2023a). The UNEP emissions gap report for 2023 noted "Under current policies, the carbon budget for a 50 per cent chance of limiting warming to 1.5°C is expected to clearly be exceeded by 2030." (UNEP 2023 p. 27). This is consistent with the IPCC's Sixth Assessment Report which indicated 20-year running average temperatures are expected to show 1.5°C of global warming sometime between 2030 and 2035. If global warming is to be limited below 1.5°C by 2100, then global warming overshoot, driven by an emissions budget overshoot is likely. The challenging part is how to remove atmospheric GHGs and bring temperatures back down. It is unclear how these removals would be achieved (UNEP 2017, van Vuuren et al. 2018) although there are removal options available (Section 2.3.2).

Complicating matters, global warming pathways and related scenarios include considerable uncertainties. For example, there is a very small likelihood that a C5 pathway (see above) could result in peak global warming staying below 1.5°C (a median likelihood of 4%). There is also an extremely small likelihood that a C7 pathway (see above) could still result in peak global warming staying below 2°C (a median likelihood of 0% but a 2% likelihood at the 95th percentile) (Riahi et al. 2022). Based on the physical science and global warming scenarios, the IPCC has assessed risks to the environment, biological systems, and human systems. For many risks there are considerable uncertainties.

Complicating the issue of scenarios and related studies is uncertainty around other events that might happen affecting climate. For example, in any given period of time, there is a chance of volcanic activity and the longer the period of time the greater the likelihood of volcanic eruptions capable of perturbing the climate locally, regionally, or globally (Robock 2000, Self 2006). However, most volcanic eruptions are not expected to influence long-term global warming although many eruptions influence climate variability (Bethke et al. 2017). The IPCC acknowledges that models informing its assessments struggle with low-likelihood high-impact events (IPCC 2023b) also known as black swan events (see Section 4.3.2).

The IPCC Sixth Assessment Report Summary for Policy Makers notes that "The likelihood of abrupt and/or irreversible changes increases with higher global warming levels. Similarly, the probability of low-likelihood outcomes associated with potentially very large adverse impacts increases with higher global warming levels." (IPCC 2023a, p. 18). Abrupt irreversible changes were defined in the Fifth Assessment Report as "a large-scale change in the climate system that takes place over a few decades or less, persists (or is anticipated to persist) for at least a few decades, and causes substantial disruptions in human and natural systems" (Collins et al. 2013, p. 1114). Irreversibility is defined as "A perturbed state of a dynamical system... [where] ...the recovery timescale from this state due to natural processes is significantly longer than the time it takes for the system to reach this perturbed state." (adapted from Collins et al. 2013).

Collins et al. (2013), in Chapter 12 of Working Group II's contribution to the IPCC's Fifth Assessment Report, listed potential abrupt or irreversible changes that could affect the climate. These include the loss of carbon from forests to the atmosphere due to die back, methane releases from the seafloor sediments, losses of ice affecting albedo, and in the case of Greenland or Antarctic ice sheets, sea levels. There is also a possibility of climate changes resulting in long-term droughts and changes in monsoonal circulation. A shutting down of the Atlantic meridional overturning circulation would result in cooling of Europe.

There is also concern related to ocean acidification (IPCC 2023a). This is due to elevated levels of carbon dioxide in the atmosphere causing more carbon dioxide to be absorbed by water in the sea or entering the sea, which in turn acidifies the water with carbonic acid (IPCC 2014a). Even without climate change, increased atmospheric concentrations of carbon dioxide would be an ocean acidification issue threatening marine life.

At the aggregate level, five reasons for concern have been reported since the Third Assessment Report and even have their own acronym (i.e., the RFCs). These are also referred to as the "burning embers" due to the graphical representation transitioning from white to yellow, then red, and eventually purple with increasing risk or impacts. The five reasons for concern (i.e., burning embers) consist of:

- RFC1: Unique and threatened systems
- RFC2: Extreme weather events
- RFC3: Distribution of impacts
- RFC4: Global aggregate impacts
- RFC5: Large scale singular events

These reasons for concern provide a synthetic indicator of anticipated risk and impacts by global warming level. In addition to updating the RFCs, the IPCC's Sixth Assessment Report also introduced eight "Representative Key Risks". Like the RFCs, Representative Key Risks have their own acronym (RKR). RKRs represent a condensed set of more than 120 key risks, where a key risk is "a potentially severe risk" that is "especially relevant to the interpretation of dangerous anthropogenic interference (DAI) with the climate system" and "the prevention of which is the ultimate objective of the UNFCCC as stated in its Article 2." (O'Neill et al. 2022, p. 2450).

The eight RKRs consist of:

- RKR-A: Risk to low-lying coastal socio-ecological systems
- RKR-B: Risk to terrestrial and ocean ecosystems
- RKR-C: Risks associated with critical physical infrastructure, networks and services
- RKR-D: Risk to living standards
- RKR-E: Risk to human health
- RKR-F: Risk to food security
- RKR-G: Risk to water security
- RKR-H: Risks to peace and to human mobility

RKRs were determined based on the: magnitude of adverse consequences; likelihood of adverse consequences; temporal characteristics of the risk (i.e., timing of the risk); and ability to respond to the risk. The magnitude of adverse consequences consisted of the: irreversibility of consequences; potential for impact thresholds or tipping points; and, potential for cascading effects beyond system boundaries. Each of these things is an important consideration when it comes to being able to fulfil the UNFCCC objective (for more on the UNFCCC objective see Section 3.3.1).

Of special concern when it comes to RFC "unique and threatened systems" and RKR "risk to terrestrial and ocean ecosystems" are coral reefs. At 1.1°C global warming, there is very high risk of loss and degradation of coral reefs in Australia, and warm water coral reefs around the world are at a high risk of impacts around the world (IPCC 2022a). At 1.5°C, the risk to warm water corals around the world is expected to be somewhere between high to very high.

For the RFCs in general, the risk and impacts at 1.1°C range from barely detectable through to high. At 1.5°C, risks and impacts are expected to range from moderate-high, and at 2°C, risks and impacts are expected to range from moderate to very high (IPCC 2022a). As such, the next period of global warming looks set to have much more noticeable impacts and risks, while up until this point, there appears to have been limited impacts and risks, and in some cases, barely detectable levels of impact and risk.

Reflecting concerns around climate change impacts and risks, the United Nations (UN 2020), United States Congress (US Congress 2019) and others (e.g., Crist 2007, Davidson et al. 2020) have each described climate change as a "climate crisis". According to the Oxford Dictionary, a crisis is "a time of intense difficulty or danger", meanwhile, danger refers to "the possibility of suffering harm or injury" i.e., risk. It is unclear whether the UN or US Congress are referring to immediate impacts or the risk of climate change. Stating there is a "climate crisis" might give the impression of intense difficulty (i.e., immediate impacts) or could refer to the risk of dangerous climate change. The extent to which climate change is a crisis, and the extent to which the climate crisis regards impacts versus risks, or the global response to climate change, is an important question which is addressed in Section 12.2.

According to Xu and Ramanathan (2017), without an adequate global response, climate change constitutes a dangerous to catastrophic risk. While the term catastrophic is used to describe anticipated impacts of unmitigated climate change, it is not altogether clear what is meant by the term. It would have been ideal if Xu and Ramanathan (2017) had used Bostrom and Cirkovic's (2008) taxonomy of risks, later refined by Bostrom (2013), where the scale of risks (i.e., scope) can range from personal to pan-generational, and the intensity of these risks can range from imperceptible to crushing (Figure 2.1) or terminal.

The framework is useful for distinguishing different climate change impacts and risks. For example, David Attenborough's concern that climate change is a risk to civilisation (Attenborough 2018) suggests climate change could be classified as being an endurable catastrophic global and transgenerational risk, rather than a terminal existential risk. Some climate change impacts and risks will be personal including those impacts and risks that affect individuals and their property, meanwhile other climate change impacts and risks may be global or transgenerational. While evidence shows climate change is a risk, the taxonomies from Bostrom and Cirkovic (2008) and Bostrom (2013), are a reminder that there are other risks in addition to climate change related risks.

Physical risks are unpacked in Chapter 19 of Working Group II's contribution to the IPCC's Fifth Assessment Report (Oppenheimer et al. 2014) in a conceptual model for understanding climate change related risks. Chapter 19 noted "... the term risk is used primarily to refer to the risks of climate-change impacts."

SCOPE

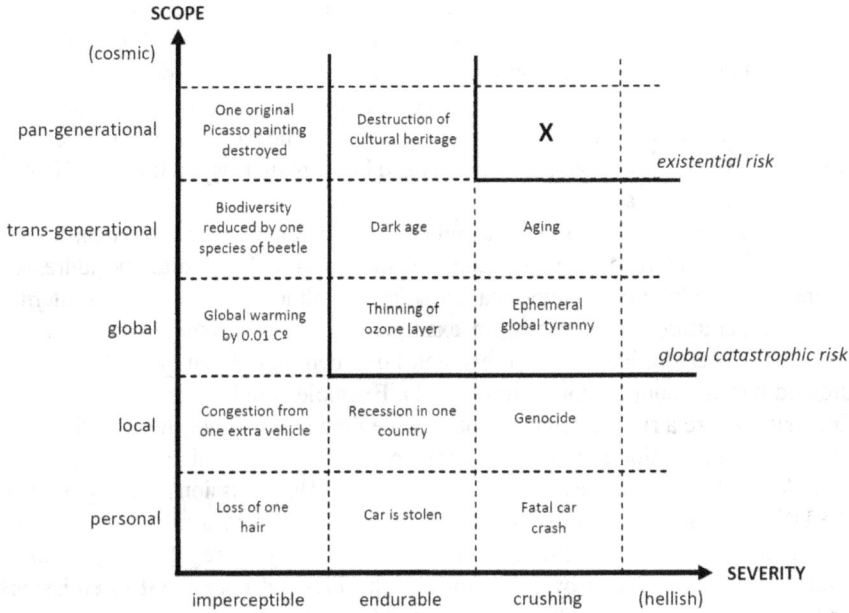

	imperceptible	endurable	crushing	(hellish)
(cosmic)				
pan-generational	One original Picasso painting destroyed	Destruction of cultural heritage	X	*existential risk*
trans-generational	Biodiversity reduced by one species of beetle	Dark age	Aging	
global	Global warming by 0.01 C°	Thinning of ozone layer	Ephemeral global tyranny	*global catastrophic risk*
local	Congestion from one extra vehicle	Recession in one country	Genocide	
personal	Loss of one hair	Car is stolen	Fatal car crash	

SEVERITY

Figure 2.1 Qualitative risk categories.
Source: Bostrom (2013).

(Oppenheimer et al. 2014, p. 1048). Risk was defined generically as "The potential for consequences where something of value is at stake and where the outcome is uncertain" (Oppenheimer et al. 2014, p. 1048). It was noted that "Risk is often represented as probability of occurrence of hazardous events or trends multiplied by the impacts if these events or trends occur." and "Risk results from the interaction of vulnerability, exposure, and hazard." (Oppenheimer et al. 2014, p. 1048). Vulnerability was defined as "The... predisposition to be adversely affected" (Oppenheimer et al. 2014, p. 1048) for example due to the quality of infrastructure or housing construction. Exposure was defined as "The presence of [things] that could be adversely affected" (Oppenheimer et al. 2014, p. 1048) and in the context of physical risk, this regards the physical location of people, property, and livelihoods including infrastructure, housing, assets, and productive activities. A hazard was defined as "The potential occurrence of a... physical event... trend or physical impact" (Oppenheimer et al. 2014, p. 1048). As such, physical risk regards the potential for, and scale of, possible climate change and related physical impacts based on the likelihood of occurrence, the presence of things of value and their predisposition to be adversely affected.

In the model, natural climate variability and anthropogenic climate change contribute to hazards, meanwhile socio-economic processes are shown to influence vulnerability and exposure to these hazards. Hazards, vulnerability, and exposure together create climate risk. Feedbacks include impacts, GHG emissions and land use change. Impacts are defined as "effects on natural and human systems" (Oppenheimer et al. 2014, p. 1048) i.e., risk that has become a reality. Emissions and land use change are shown as being a result of socio-economic processes and contributing to anthropogenic climate change. Importantly, hazards and impacts can include "climatic" hazards and impacts (e.g., extreme heat or changes in precipitation) as well as "related" physical hazards and impacts such as sea level rise, flooding, or ocean acidification.

The risk model can be applied at subnational, national, or international levels. Physical risks and impacts are location specific and as such will often be addressed by individual subnational or national jurisdictions, although in some cases international cooperation may be useful for example where the actions of one sovereign state can exacerbate hazards for other states or a common threat might be best addressed through cooperation (Farber 2011). Examples might include downstream flood risk where a river catchment spans more than one state, or the establishment of the European Union Solidarity Fund (Hochrainer-Stigler et al. 2017) which limits risk by sharing the cost of climate impacts. GHG emissions are reported to the UNFCCC at the national level but may be regulated by a mix of national or subnational authorities. Similarly, land use change may be regulated by national authorities or more often by subnational authorities. Of course, GHG emissions and land use change may also be influenced by wider economic trends including for example issues of behaviour, preferences and demand, technology and prices as well as international trade (Meyfroidt et al. 2013).

Note: For a more detailed analysis regarding the concepts of hazards, exposure, vulnerability, and interactions with actors within a state constituting a socio-economic system, see the Environment and National Interests Model in Appendix B.

It is important to note, there are other risks beyond the physical risk of climate change. In a speech titled "Breaking the tragedy of the horizon – climate change and financial stability", delivered at a dinner for insurers just weeks before COP21, the Governor of the Bank of England and Chairman of the Financial Stability Board, Mark Carney (2015) highlighted three types of climate risk consisting of physical risk; liability risk; and transition risk.

Carney's (2015) classification of climate change risks, and definitions from the Bank of England Prudential Regulatory Authority, were framed in the context of finance and insurance. These definitions can be generalised to ensure the classification is comprehensive and more widely applicable. For example, physical risks can be generalised to: Potential physical impact of climate change. Liability risks can be generalised and defined as the potential for legal compensation claims related to climate change. Carney's (2015), and the BOE's (2015), definitions of transition risks only referred to risks related to mitigation, but it is conceivable that there may be transition risks related to adaptation as well. As such, transition risk could be generalised to: Potential disruption due to adjustment towards a climate resilient low GHG emissions economy.

In 2017, the Task Force on Climate-related Financial Disclosures (TCFD) Chaired by Michael Bloomberg, published recommendations (TCFD 2017) that identified two overarching categories of climate change related risk, specifically physical and transition risks. Liability risks were combined with policy risks to create a category titled "policy and legal risk" under transition risks. Other transition risks identified included technology, reputation and market. Physical risks were divided into acute and chronic.

From the work of the TCFD (2017), BOE (2015), Carney (2015), Grubb et al. (2014), Bostrom and Cirkovic (2008), and Oppenheimer et al. (2014), climate change and related responses pose direct and indirect risks to people, businesses, government, and their interests.

2.2.4 *Global response*

An important part of the climate change problem is the lack of an adequate global response. The global response to climate change is tracked by the United Nations Environment Programme (UNEP) and its Emissions Gap Report series. The UNEP Emissions Gap Report has stated that

> The overarching conclusions of the report are that there is an urgent need for accelerated short-term action and enhanced longer-term national ambition, if the goals of the Paris Agreement are to remain achievable — and that practical and cost-effective options are available to make this possible.
>
> (UNEP 2017, p. XIV)

Figueres et al. highlighted the same risk in their commentary titled "Three years to safeguard our climate" published in *Nature* (Figueres et al. 2017). They noted that "When it comes to climate change, timing is everything" and "should emissions continue to rise beyond 2020, or even remain level, the temperature goals set in Paris become almost unattainable." (Figueres et al. 2017, p. 593) due to the steep reduction in GHG emissions that would be required. Figueres et al. (2017) also highlighted goals and steps that could be taken to address climate change (Section 2.3).

Emissions budgets help give a sense of scale to the climate change problem and response required. Figueres et al. (2017) noted that "After subtracting past emissions, humanity is left with a 'carbon credit' of between 150 and 1,050 gigatonnes" (Figueres et al. 2017, p. 594) reflecting different ways of calculating the carbon budget and uncertainty. The UNEP Emissions Gap Report series also uses a budget to assess the gap in emissions reductions required to limit global warming to well below 2°C from pre-industrial times (UNEP 2017). This includes an assessment of current progress as well as what progress would be made if Parties fulfilled their unconditional Nationally Determined Contributions (NDCs) and conditional NDCs[4]. The UNEP Emissions Gap Report (2017) found that

> The NDCs that form the foundation of the Paris Agreement cover only approximately one third of the emissions reductions needed to be on a least cost pathway for the goal of staying well below 2°C. The gap between the reductions needed and the national pledges made in Paris is alarmingly high.
>
> (UNEP 2017, p. XIV)

UNEP goes on to state, "it is clear that if the emissions gap is not closed by 2030, it is extremely unlikely that the goal of holding global warming to well below 2°C can still be reached." (UNEP 2017, p. XIV). Furthermore, "Given currently available carbon budget estimates, the available global carbon budget for 1.5°C will already be well depleted by 2030." (UNEP 2017, p. XIV). As such, the current policy trajectory is diverging from the pathways required to limit global warming to well below 2°C. The UNEP Gap Reporting (UNEP 217) shows that conditional and unconditional NDCs are better than the current policy trajectory but are far from sufficient to put the global response to climate change on a path towards achieving the UNFCCC objective or Paris Agreement purpose.

Further complicating the situation, most scenarios that limit global warming to well below 2°C also require carbon dioxide removals from the atmosphere (UNEP 2017, van Vuuren et al. 2018, IPCC 2018a). It is not clear which technological options could become cost effective or would be practical "at scale"[5] as there is limited experience in this area. Section 2.3 discusses these options further. Other organisations have conducted similar assessments and come up with similar findings (e.g., Shell 2018).

Likewise, the 1.5 Degree Report found carbon dioxide removals need to be at scale by 2050, if global warming is to be limited to 1.5°C of global warming. Furthermore, the 1.5 Degree Report indicated overshoot scenarios are a real possibility where global warming levels have to be brought down after exceeding 1.5°C.

The measurement of progress using carbon, carbon dioxide, carbon dioxide equivalent and any other type of budget tends to simplify issues. As noted by Figueres et al. (2017), the actual budget is uncertain and as noted in Section 2.2.3, there are large uncertainties around the level of global warming that might happen given a particular concentration of GHGs in the atmosphere in combination with other climate forcers and feedbacks.

Further complicating the assessment of the global response to climate change is the fact that emissions are recorded by territory from which they are emitted (i.e., sovereign states and their territorial jurisdictions) (Goodwin et al. 2019), however, only a fraction of emissions are from the government activities. Emissions inventories also include emissions from the activities of economic and social actors within a territory. National governments can influence GHG emissions and other climate forcers through policies for example, levies, legislation, regulations, investments, and plans as well as the extent to which these interventions are applied and enforced. In many cases there will be sub-national government bodies and governance processes that also have an influence on social and economic activities (Farber 2011). There are many other influences on social and economic activities including, for example, the availability of technology, knowledge and practices, culture and history, prices, costs and benefits as well as legal, security and other considerations (Dryzek et al. 2011). Furthermore, international trade and issues of supply and demand may also influence economic activities and their emissions within a sovereign state (Keohane and Victor 2011). Hence, the entity for which emissions are measured, i.e., the sovereign state, has limited direct and indirect influence on the GHG emissions being emitted from its territory (Farber 2011). There are many actors with competing interests that have an influence on the social and economic activities within a sovereign state as well as the policies and plans that the government of a sovereign state may adopt (Putnam 1988, Kingdon 1995, Farber 2011).

To help with understanding how social and economic factors might influence the global response to climate change, O'Neill and others have developed SSPs that "describe plausible alternative changes in aspects of society such as demographic, economic, technological, social, governance and environmental factors" (O'Neill et al. 2017, p. 170). The first iteration of the five SSPs emerged from a workshop held in 2012 (O'Neill 2012), consisting of SSP1 sustainability, SSP2 middle of the road, SSP3 fragmentation, SSP4 inequality, and SSP5 conventional development. In 2017, O'Neill et al. published updated "narratives for Shared Socioeconomic Pathways describing world futures in the 21st century". The five SSPs in O'Neill et al. (2017) were similar to the five published by O'Neill et al. (2012) and consisted of: SSP 1 – sustainability, SSP 2 – middle of the road, SSP 3 – regional rivalry, SSP 4 – inequality, and SSP 5 – fossil-fuelled development. For each SSP, short and detailed narratives were produced, addressing: population and human resources; economic development; human development; technology; lifestyles; environment and natural resources; and, policies and institutions.

O'Neill et al. (2017) mapped the SSPs relative to the socio-economic challenges for mitigation as well as the socio-economic challenges for adaptation. SSP 3 regional rivalry was indicated as being the greatest challenge for mitigation and adaptation while SSP1 sustainability had the least challenge for mitigation and adaptation.

While Bostrom and Cirkovic (2008) addressed how bad a physical risk might be and Mark Carney (2015) highlighted transition risks to actors and their interests (Section 2.2.3), Grubb et al. (2014) addressed how a risk might be conceived and related response strategies. How actors respond to climate change depends in part, on how actors conceive risk and the strategies they are willing to employ (Grubb et al. 2014). Grubb et al. (2014) identified climate change risk conceptions and related national interests, societal processes and climate change timescales assuming progressively greater impacts. If a sovereign state and its leader's risk conception is one of indifference or disempowerment, then any options with a cost associated will likely be ignored at the international level, and the state is likely to freeride. Grubb et al. (2014) noted that climate change risk conceptions based on tangible and attributable costs follow the so-called rational approach. If the risk conception is one of securitisation, then the sovereign state and its leaders are expected to mitigate as much as practical while adapting to changes that can not be mitigated. Grubb et al. (2014) did not address the extent to which states or other actors might be willing to coerce other actors into participating in the global response to climate change (Table 2.1).

Table 2.1 Three conceptions of risks and their application to climate change

Risk conception	Basic belief	Typical strategy	Societal process	Timescale of climate change
Indifferent or disempowered	Not proven, or "What you don't know can't hurt you"	"Ignorance is bliss"	Environmental group campaigns vs resistance lobbying	First few decades of climate change
Tangible and attributed costs	Weigh up costs and benefits	Act at costs up to "social cost of carbon"	Technocratic valuation and politics of pricing	As impacts rise above the noise – next few decades
Disruption and securitisation	Personal or collective security at risk, climate change as a "threat multiplier"	"Containment and defence"	Mitigate as much as practical and adapt to the rest	Ultimately, for all (systemic and global risk), most vulnerable, sooner, with international spill over

Source: Grubb et al. (2014).

In each of the models, from Grubb et al. (2014), Bostrom and Cirkovic (2008) and Oppenheimer et al. (2014), climate change and related physical risks can impact people, their property and livelihoods. Luckily, there are options for limiting the impacts of climate change on people.

2.3   Options

Why are options important? Because we have a climate change problem, and related set of issues, that need addressing. With regards to climate change, there is no new normal until atmospheric GHG emissions are stabilised and reduced. Even then there will be long-term changes for example sea level rises and we have to hope we don't pass any tipping points or thresholds leading to positive feedbacks. It is in this context that we need to know what options exist and what we might do to limit climate change and its impacts. Options include mitigation and adaptation options (Section 2.3.1), GHG removal options (Section 2.3.2) as well as controversial solar radiation management options (Section 2.3.3).

2.3.1   Mitigation and adaptation

Mitigation regards options and actions for avoiding climate change meanwhile adaptation regards options and actions in anticipation of, or response to, physical climate change (see Glossary for definitions). Global assessments of mitigation options include IPCC Working Group III's Fifth Assessment Report, the 2017 edition of the UNEP Emissions Gap Report, a cooperative assessment of options by researchers around the world culminating in a book titled Drawdown, a scenario from Shell, as well as studies by researchers (e.g., Blok et al. 2018 or van Vuuren et al. 2018) and more recently the IPCC's 1.5 Degree Report. Likewise, there is a growing body of literature on adaptation including IPCC Working Group II's contribution to the Fifth Assessment Report.

IPCC Working Group II addressed adaptation along with climate change impacts and risks. In its Fifth Assessment Report the IPCC (2014c) stated that "Adaptation options exist in all sectors, but their context for implementation and potential to reduce climate-related risks differ across sectors and regions." (IPCC 2014c, p. 95). As such, there is no one adaptation solution, but rather there are many possible options, and appropriate options depend on the local situations. The IPCC also noted that

poor planning or implementation, overemphasizing short-term outcomes or failing to sufficiently anticipate consequences can result in maladaptation, increasing the vulnerability or exposure of the target group in the future or the vulnerability of other people, places or sectors (medium evidence, high agreement).

(IPCC 2014a, p. 20)

The IPCC Fifth Assessment Report summarised adaption options. Broad approaches included: options that reduce vulnerability and exposure to climate change related hazards, through development and planning related options; adaptation through incremental or transformative changes; and, transformation related to institutions and wider societal changes. Furthermore, the IPCC (IPCC 2014c, p. 95), noted that broad options include:

- Social, ecological asset and infrastructure development
- Technological process optimisation
- Integrated natural resources management
- Institutional, educational and behavioural change or reinforcement
- Financial services, including risk transfer
- Information systems to support early warning and proactive planning

It should be noted that adaption options need to be taken together with mitigation. The IPCC noted that "Adaptation can reduce the risks of climate change impacts, but there are limits to its effectiveness, especially with greater magnitudes and rates of climate change." (IPCC 2014a, p. 19). As such, mitigation is essential if adaptation is to be effective (Pers. Comm. Youba Sokona 2011). Furthermore, climate change interventions may help with both adaptation and mitigation, for example, the installation of solar electricity systems can mitigate GHG emissions while at the same time powering air-conditioning and refrigeration during extreme heat events (Anand et al. 2015).

IPCC Working Group III addressed mitigation and noted that "Mitigation options are available in every major sector." (IPCC 2014c, p. 98). These sectors include energy, transport, buildings, industry (i.e., manufacturing) as well as human settlements and infrastructure. The IPCC also identified policy options including economic instruments (e.g., taxes and subsidies as well as tradable allowances), regulations, information, and the provision of goods and services by government institutions. The IPCC Fifth Assessment Report also showed how these policy options can be applied to various sectors. The IPCC did not provide a definitive list of options or the emissions that could be mitigated, but rather demonstrated that there are a variety of options available to governments.

Project Drawdown provided a comprehensive inventory of mitigation options, including the amount of mitigation that could be achieved, the costs and savings. Project Drawdown identified 100 options, 80 of which were ranked and another 20 were identified as being "coming attractions" (i.e., other possible solutions too early to assess). The options identified as "coming attractions" demonstrated that new climate change options may emerge over time, especially when it comes to technology, management systems and behaviours. Project Drawdown not only addressed technical options but also social options, educating girls, family planning or indigenous peoples' land management requiring investment in people, culture and social systems.

Project Drawdown succeeded in its aim of creating an inventory, which serves as a useful resource for anyone interested in mitigation options, but it did not specify how these options might be brought together to actually address climate change. As noted by the IPCC,

> Mitigation can be more cost-effective if using an integrated approach that combines measures to reduce energy use and the greenhouse gas intensity of end-use sectors, decarbonize energy supply, reduce net emissions and enhance carbon sinks in land-based sectors.
>
> (IPCC 2014c, p. 98)

Fortunately, Figueres et al. (2017) provide guidance on which options to pursue, meanwhile van Vuuren et al. (2018), Blok et al. (2018) and Shell (2018) have identified scenarios in which global warming would likely be limited to 2° or less.

In addition to noting the need to go "further, faster, together" involving the use of science to guide decisions, scaling up solutions rapidly, and encouraging optimism, Figueres et al. (2017) identified specific interventions that together would likely limit global warming to well below 2°C. This included interventions in the sectors of energy (e.g., renewables), infrastructure (including cities), transport (e.g., electric vehicles), land (e.g., reforestation), industry (e.g., efficiency and emissions reductions) and finance (e.g., private sector and new instruments).

Ecofys (Blok et al. 2018) published a report titled "Energy transition within 1.5°C" with the subtitle: "A disruptive approach to 100% decarbonisation of the global energy system by 2050". Blok et al. (2018) prepared a scenario taking into account population and economic growth and the need for additional goods and services to meet development expectations. The scenario focused on energy options and found:

> Despite the global energy system's rapid reduction of CO_2 emissions in our disruptive decarbonisation scenario, cumulative CO_2 emissions beyond 2014 are calculated to be 680 billion tonnes, likely exceeding the carbon budget.

However, combined with options such as afforestation and agricultural carbon sequestration, it looks possible to stay within a carbon budget compatible with a maximum temperature increase of 1.5°C.

(Blok et al. 2018, p. ii of summary)

As such, focusing on energy alone is insufficient to limit climate change. Importantly, Blok et al. (2018) noted "Most of the required technologies are already available and developments in some sectors go so fast that transitions become cost-competitive." (Blok et al. 2018, p. 11). The paper also noted the need for "massive investments" and the fact that there is "no time to lose" (Blok et al. 2018, p. 11).

In 2018, Shell, the oil and gas company, released its Sky scenario which limited global warming to less than 2°C in 2100, complementing its Oceans and Mountains scenarios, both of which exceeded 2°C of global warming. The Sky scenario was backcast, and as such differed in approach from the development of Mountains and Oceans scenarios which were developed using engagement, feedback and forward-looking processes. Sky was normative. The Sky scenario recognises peoples' need for development and that "in the context of the UN Sustainable Development Goals, several billion people are still pursuing a better life through much-needed access to clean water, sanitation, nutrition, health care, and education. Energy is a key enabler for these basic needs." (Shell 2018, p. 12). Shell also addressed issues of competing interests and the need for government policies and leadership. A large role is played by carbon capture and storage, carbon capture and use, as well as reforestation, with the net effect of allowing more oil and gas extraction. The scenario also covered a range of sectors, GHGs, and related mitigation actions but adaptation was not addressed.

van Vuuren et al. (2018) published a paper in *Nature* titled "Alternative pathways to the 1.5°C target reduce the need for negative emission technologies". van Vuuren et al. (2018) prepared a series of scenarios and assessed which scenarios, or combination of scenarios, would be likely to limit global warming to 1.5°C in 2100. Scenarios included a focus on efficiency, renewable electricity, agricultural intensification, non-CO_2 emissions reductions, lifestyle changes, and low population. Additionally, a combination of all these climate-related actions was assessed and compared with SSP2 as a baseline, as well as RCP 2.6 and a 1.9 W m^{-2} (i.e., 1.5°C) scenario. van Vuuren et al. (2018) noted that GHG removals are problematic and as such wanted to know if it was possible to limit global warming to 1.5°C following "deep mitigation pathways" limiting the need for carbon dioxide removals. The scenarios were "found to significantly reduce the need for CDR, but not fully eliminate it." (van Vuuren et al. 2018, p. 391).

The UNEP Gap Report is a collaboration that involves many authors including Blok and van Vuuren for example, forming an assessment of the current state of knowledge of the emissions gap. The UNEP Gap Report 2017 addressed sectoral options for limiting GHG emissions. These sectors consisted of energy, industry, forestry, transport, agriculture, buildings, and others. According to the UNEP Emissions Gap Report 2017, the largest potential contributions to the global response to climate change came from energy, followed by industry and forestry. The UNEP

Gap Report also provided a breakdown of the emissions reductions by category, many of which are related to the mitigation actions discussed above.

With regards to limiting climate change to 1.5°C, the IPCC's 1.5 Degree Report included options and their feasibility. This included options related to green infrastructure, industrial mitigation options, adaptation options, and behavioural options related to mitigation and adaptation (de Coninck et al. 2018). In a section aptly titled "Disentangling the whole-system transformation", the IPCC also noted "There is a diversity of potential pathways consistent with 1.5°C, yet they share some key characteristics" (Rogelj et al. 2018, p. 129). These key characteristics (Table 2.2) are important as they represent actions (i.e., options) that need to be fulfilled if the global response is to be effective.

Table 2.2 Overview of key characteristics of 1.5°C pathways[6]

1.5°C Pathway characteristic	Supporting information
Rapid and profound near-term decarbonisation of energy supply	Strong upscaling of renewables and sustainable biomass and reduction of unabated (no CCS) fossil fuels, along with the rapid deployment of CCS, lead to a zero-emission energy supply system by mid-century.
Greater mitigation efforts on the demand side	All end-use sectors show marked demand reductions beyond the reductions projected for 2°C pathways. Demand reductions from IAMs for 2030 and 2050 lie within the potential assessed by detailed sectoral bottom-up assessments.
Switching from fossil fuels to electricity in end-use sectors	Both in the transport and the residential sector, electricity covers markedly larger shares of total demand by mid-century.
Comprehensive emission reductions will be implemented in the coming decade	Virtually all 1.5°C-consistent pathways decline net annual CO_2 emissions between 2020 and 2030, reaching carbon neutrality around mid-century. In 2030, below-1.5°C and 1.5°C-low-OS pathways show maximum net CO_2 emissions of 18 and 28 $GtCO_2$ yr^{-1}, respectively. GHG emissions in these scenarios are not higher than 34 $GtCO_2$ e yr^{-1} in 2030.
Additional reductions, on top of reductions from both CO_2 and non-CO_2 required for 2°C, are mainly from CO_2	Both CO_2 and the non-CO_2 GHGs and aerosols will be strongly reduced by 2030 and until 2050 in 1.5°C pathways. The greatest difference to 2°C pathways, however, lies in additional reductions of CO_2, as the non-CO_2 mitigation potential that is currently included in integrated pathways is mostly already fully deployed for reaching a 2°C pathway.
Considerable shifts in investment patterns	Low-carbon investments in the energy supply side (energy production and refineries) are projected to average 1.6–3.8 trillion 2010 USD yr^{-1} globally to 2050. Investments in fossil fuels decline, with investments in unabated coal halted by 2030 in most available 1.5°C-consistent projections, while the literature is less conclusive for investments in unabated gas and oil. Energy demand investments are a critical factor for which total estimates are uncertain.

Options are available to align 1.5°C pathways with sustainable development	Synergies can be maximised, and risks of trade-offs limited or avoided through an informed choice of mitigation strategies. Particularly pathways that focus on a lowering of demand show many synergies and few trade-offs.
CDR at scale before mid-century	By 2050, 1.5°C pathways project deployment of BECCS at a scale of 3–7 $GtCO_2$ yr^{-1} (range of medians across 1.5°C pathway classes), depending on the level of energy demand reductions and mitigation in other sectors. Some 1.5°C pathways are available that do not use BECCS, but only focus on terrestrial CDR in the AFOLU sector.

Source: Rogelj et al. (2018).

The feasibility of options for addressing climate change was also addressed in the 1.5 Degree Report. It was stated that "Feasibility depends on geophysical, ecological, technological, economic, social and institutional conditions for change." (Allen et al. 2018, p. 71).

2.3.2   *Greenhouse gas removals*

Beyond conventional mitigation and GHG emissions abatement, there is likely to be a need for atmospheric GHG removals (Sections 2.2.4 and 2.3.1), in particular carbon dioxide removals. Carbon dioxide removals can be characterised as being "restoration" of atmospheric concentrations of carbon dioxide to safe levels. The UNEP Gap Report 2017 notes that this can be done using natural or technological means. Natural means include afforestation and reforestation, biochar, soil carbon sequestration, and other land use initiatives. The options are relatively low cost and can be deployed currently but are vulnerable to reversal. For example, a forest fire could eliminate a stock of carbon in days that might have taken decades to accumulate. Technological options for GHG removals include accelerated weathering, direct air capture, ocean alkalinity enhancement, and carbon dioxide being converted to durable carbon materials. The benefits of technological approaches include being much less vulnerable to reversal, however the technologies require much more research and development before they can be deployed at scale or at reasonable cost (UNEP 2017).

Natural and technological options can be combined for example to form Bioenergy with Carbon Capture and Storage (BECCS). While conceptually BECCS is possible, practical details are much less compelling, for example the large land areas that would be required (Smith et al. 2016), the possibility that forests might not be sustainably managed (Fajardy et al. 2019) or the possibility of cheating resulting in ecologically sensitive timber being harvested instead of new growth.

As noted in Section 2.2.1, it is not just long-lived GHGs that are climate forcers. Short-Lived Climate Pollutants SLCPs are also important. Key messages emerging from the UNEP Emissions Gap Report 2017 included "Early reductions in SLCP emissions would provide substantial health benefits, limit the short-term rate of climate change, slow self-amplifying feedbacks, and facilitate the achievement of the Paris Agreement's long-term temperature target." (UNEP 2017, p. 57). Furthermore, the UNEP Gap Report stated that "Over the period 2018–2050, stringent SLCP reductions based on existing, demonstrated technical measures could reduce warming by between 0.3°C and 0.9°C relative to current emissions projections." and "Roughly half of the mitigation potential is associated with methane, one third with black carbon, and the remainder with hydrofluorocarbons." (UNEP 2017, p. 57). Some GHG mitigation actions would also have the effect of limiting SLCPS, but for the maximum effect to be achieved, other measures would need to be taken beyond those under the Paris Agreement. Furthermore, the accounting of SLCPs needs to be conducted separately due to the differing properties of these pollutants (UNEP 2017).

The UNEP Emissions Gap Report included a chapter on coal and estimates the carbon dioxide emissions that would be locked in if announced coal-fired power plants, pre-permitted plants, permitted plants, plants under construction, and operating plants were to progress as expected. Once finance is committed and a coal-fired power plant is under construction it is anticipated that there will be lock-in of the carbon dioxide emissions (UNEP 2017).

2.3.3   Solar radiation management

In addition to GHG removals, there is the possibility of solar radiation management. Solar radiation management is "the intentional modification of the Earth's shortwave radiative budget with the aim to reduce climate change according to a given metric (e.g., surface temperature, precipitation, regional impacts, etc.)." (IPCC 2013). Working Group I, in its glossary, goes on to note that

> Artificial injection of stratospheric aerosols and cloud brightening are two examples of SRM techniques. Methods to modify some fast-responding elements of the long wave radiative budget (such as cirrus clouds), although not strictly speaking SRM, can be related to SRM. SRM techniques do not fall within the usual definitions of mitigation and adaptation.
>
> (IPCC 2013)

In its Summary for Policy Makers, the IPCC noted that "SRM is untested and is not included in any of the mitigation scenarios. If it were deployed, SRM would entail numerous uncertainties, side effects, risks and shortcomings and has particular governance and ethical implications." (IPCC 2014a, pp. 25–26). Furthermore, solar radiation management does nothing to limit ocean acidification and if efforts to limit solar radiation are stopped, temperatures would rebound quickly (IPCC 2014a).

2.4   Conclusions

The global response to climate change has many elements to it. Issues including climate science, climate change impacts and risks as well as the state of the global response. Options include mitigation, adaptation, GHG removals or even solar radiation management. Options on their own are of little use. We need actors to act on these options, and institutions to support implementation of these options. That is why actors, institutions and the global response system are addressed in the next chapter (Chapter 3).

Notes

1 Adapted from Merriam-Webster (2019).
2 In the quote Kingdon thanks James Q. Wilson for highlighting this point.
3 Methane is classified as being both a long lived greenhouse gas and short lived climate pollutant.
4 Nationally Determined Contributions (NDCs) to the global response to climate change are efforts volunteered by states, towards fulfilling the Paris Agreement purpose, including unconditional NDCs as well as conditional NDCs requiring the support of other states. NDCs are at the heart of the Paris Agreement and are periodically reviewed in global stocktakes and new NDCs submitted every five years by parties to the Paris Agreement (UNFCCC 2019).
5 "At scale" means "at the required size to solve the problem" (PCMag Encyclopedia 2020), typically involving large volumes.
6 Use of IPCC data is at the User's sole risk. Under no circumstances shall the IPCC, WMO or UNEP be liable for any loss, damage, liability or expense incurred or suffered that is claimed to have resulted from the use of any IPCC data, without limitation, any fault, error, omission, interruption or delay with respect thereto. Nothing herein shall constitute or be considered to be a limitation upon or a waiver of the privileges and immunities of WMO or UNEP, which are specifically reserved.

3 Climate actors, institutions, and response system

3.1 Introduction

There are many pieces to the climate change puzzle, including the issues and options which we looked at in the previous chapter. This chapter continues with the analogy of the jigsaw puzzle, including the pieces that need to be put together.

An important bunch of puzzle pieces are the actors (you, me, and everyone else) who might act on the options from Chapter 2. Section 3.2 looks at actors, including state actors (Section 3.2.1) and non-state actors (Section 3.2.2).

When putting together, and solving, a puzzle, people start forming rules, for example who gets to decide where a piece goes. These "rules of the game" are informal when putting together a jigsaw puzzle but have been formalised when it comes to the global response to climate change. Section 3.3 looks at the institutions and the "rules of the game", including the United Nations Framework Convention for Climate Change (UNFCCC) (Section 3.3.1), Paris Agreement (Section 3.3.2), other institutions (Section 3.3.3), and how these institutions interact forming a climate regime complex (Section 3.3.4).

Between the issues and options from Chapter 2 and the actors and institutions discussed in this chapter, we have many paragraphs. Each of these paragraphs are like puzzle piece that need to be put together. This chapter finishes with the big "picture", akin to the image on the front of the jigsaw puzzle box, showing what the global response system should look like when all the pieces are put together (Section 2.6). This picture shows how climate change, national interests, and international cooperation are related (Section 3.4.1) and fit together to form the Climate Change, National Interests, International Cooperation (CCNIIC) Model (Section 3.4.2). Section 3.4 also addresses criteria for effective international cooperation (Section 3.4.3).

The literature reviewed in Sections 3.2 and 3.3 provides a basis for critically engaging with the scenarios presented in Chapters 5–9. If you are already familiar with the literature on climate change actors and institutions you might want to skip ahead to Section 3.4 and the CCNIIC Model.

3.2 Actors

Why are actors important? Actors are affected by climate change and actors make options happen. The extent to which the global response is effective depends on

DOI: 10.4324/9781003465911-4

actors and all their actions combined. In this section actors are discussed including their roles in relation to climate change and the global response. Section 3.2.1 addresses sovereign states and central government and Section 3.2.2 looks at domestic non-state actors. Given that climate change is a global problem, international institutions such as the UNFCCC are important too, and these are addressed in Section 3.3.

3.2.1   State actors

In the context of the UNFCCC and the global response to climate change, sovereign states are an important unit of analysis, because the global response to climate change is set out by states in their nationally determined contributions (NDCs).

Each state has its own national interests, but in many cases these interests will be shared with other sovereign states and as such, sovereign states that are parties to the UNFCCC form groups that cooperate on climate negotiations and climate-related issues. In some cases, common interests have led to the formation of Party Groupings that are recognised by the UNFCCC and others, for example Small Island Developing States (SIDS), Least Developed Countries (LDCs), Environmental Integrity Group (EIG), or Arab Group (UNFCCC 2020).

Physical characteristics are the defining feature of SIDs and the Coalition for Rainforest Nations (CfRN). Development and economic characteristics define the G-77, LDCs and Brazil, South Africa, India and China (BASIC) groupings and their common interests. The Organization of the Petroleum Exporting Countries (OPEC) is defined by a common interest in energy, in particular petroleum production. Other groups such as the European Union (EU), Central Asia, Caucasus, Albania and Moldova (CACAM) Group, or the Independent Association of Latin America and the Caribbean (AILAC) are defined by geographic proximity and varying levels of international cooperation and integration. Other interests linking countries include trade and security.

In a report titled "A New World: The Geopolitics of the Energy Transformation" the Global Commission on the Geopolitics of Energy Transformation stated, "Fundamental changes are taking place in the global energy system which will affect almost all countries and will have wide-ranging geopolitical consequences." (IRENA 2019, p. 12). Countries that depend on oil, gas, or coal for energy or revenues will face challenges (IRENA 2019). Meanwhile, "The majority of countries can hope to increase their energy independence significantly, and fewer economies will be at risk from vulnerable energy supply lines and volatile prices." (IRENA 2019, p. 15). Furthermore, countries that have mineral resources needed as an input to the production of technologies driving the transformation, stand to benefit from increased demand (IRENA 2019). According to the report, the transformation which includes increased energy efficiency, the adoption of renewables and electrification, will "redraw the geopolitical map of the 21st century." (IRENA 2019,

p. 12). It is an interesting question as to how different states and other actors might respond to these changing circumstances?

3.2.2   *Non-state actors*

Non-state actors include individuals, civil society, businesses, and local government for example. Non-state actors can be domestic or international (see Section 3.3.3 for more on international non-state actors in the form of international institutions). When it comes to non-state actors and actions outside NDCs, the UNEP Emissions Gap Report 2017 stated that "There is still limited evidence that non-state action will fill a significant part of the emissions gap, although there is significant potential for it to do so." (UNEP 2017, p. XIV). In many cases contributions made by non-state actors will be counted as part of an NDC due to the way NDCs are formulated and emissions reductions are accounted for using national inventories. The UNEP Gap Report 2017 did note that "subnational and non-state actions could possibly make a significant contribution to narrowing the gap" and "The aggregate impact of the initiatives could be in the order of a few $GtCO_2e$ in 2030 beyond the current NDCs, if the initiatives reach their stated goals and if these reductions do not displace actions elsewhere." (UNEP 2017, p. 25). Graichen et al. (2016) in a paper titled "Climate initiatives, national contributions, and the Paris Agreement" identified non-state mitigation actions that would be additional and would make meaningful contributions to the global response to climate change in the order of gigatonnes.

3.3   Institutions

Why are institutions important? Institutions are "systems of established and embedded social rules that structure social interactions" (Hodgson 2006, p. 18). Given that climate change is a collective action problem at a global scale (IPCC 2014a) and collective action depends on institutions (Ostrom 1990), institutions are going to play an important role in the global response to climate change, including international institutions. According to Mearsheimer (1994) international institutions are "sets of rules that stipulate the ways in which states should cooperate and compete with each other" (p. 8, in Simmons and Martin 2002) however some definitions also include agreements and initiatives by non-state actors as part of the international system[1]. This book includes agreements and initiatives by non-state actors.

The United Nations Framework on Climate Change (Section 3.3.1) and the Paris Agreement (Section 3.3.2) are influential institutions at the core of the climate policy architecture. However, there are other institutions that influence the global response to climate change (Section 3.3.3) and as such, climate change is part of a regime complex (Section 3.3.4) with multiple influences.

3.3.1 UNFCCC

According to the UNFCCC "The 1992 United Nations Framework Convention on Climate Change... ...provides the foundation for multilateral action to combat climate change and its impacts on humanity and ecosystems." (UNFCCC 2018a). As a framework convention, the UNFCCC simply "serves as an umbrella document which lays down the principles, objectives and the rules of governance of the treaty regime" (UNECE 2011) and leaves "more detailed rules and the setting of specific targets either to subsequent agreements between the parties, usually referred to as protocols, or to national legislation." (OPIL 2011). As such, the UNFCCC forms a framework for the global response to climate change, consisting of (UNFCCC 2018a):

- An objective
- Broad principles
- General obligations
- Basic institutional arrangements
- An intergovernmental process for agreeing to specific actions over time.

Essentially, the UNFCCC can be divided into two parts, the first is the objective that the international community aspires to achieve. The second part consists of the principles, obligations, institutional arrangements, and intergovernmental processes which together form the "rules of the game". However, it is important to note there is a mismatch between the universal goal (i.e., UNFCCC objective) which depends on the global response to climate change, and the rules of the game which are not universally accepted or applied, for example when it comes to issues of trade, technology, or land use.

The ultimate objective of the Convention is to achieve stabilisation of atmospheric greenhouse gas (GHG) concentrations at safe levels (UNFCCC 1992). Safe levels mean ecosystems are able to adapt naturally to climate change, food production is not threatened and economic development is able to proceed in a sustainable manner. As such, there are four success criteria that can be taken from the UNFCCC objective (Table 3.1).

The UNFCCC has many bodies and the most important of these bodies consist of: the Conference of Parties (COP) to the UNFCCC; the Conference of the Parties serving as the meeting of the Parties to the Kyoto Protocol (CMP); the Conference of the Parties serving as the meeting of the Parties to the Paris Agreement (CMA); and the Bureau for the COP, CMP, and CMA. Beneath these bodies are permanent subsidiary bodies that support the COP and CMP on technical issues related to science and technology as well as implementation. Then there are convention bodies, bodies related to the technology mechanism under the Convention, expert groups under the Convention, Kyoto Protocol bodies, bodies related to financial mechanisms, and bodies related to other financial arrangements. Lastly, there is the UNFCCC Secretariat that supports these bodies along with organisation and record keeping under the UNFCCC (see the Jellyfish model in Appendix C).

Table 3.1 Success criteria based on the UNFCCC objective

UNFCCC objective (Article 2)	Key points (i.e., success criteria)
…stabilization of greenhouse gas concentrations in the atmosphere at a level that would prevent dangerous anthropogenic interference with the climate system.	Stabilise atmospheric GHG concentrations at safe levels
Such a level should be achieved within a time-frame sufficient to allow ecosystems to adapt naturally to climate change, to ensure that food production is not threatened and to enable economic development to proceed in a sustainable manner.	Ecosystems adapt naturally Food production not threatened Economic development to proceed in a sustainable manner

Source: Compiled from UNFCCC (1992).

Table 3.2 Paris Agreement purpose and key points summarised

Paris Agreement purpose (Article 2)	Key points
Strengthen the global response to the threat of climate change, in the context of sustainable development and efforts to eradicate poverty	Strengthen the global response to climate change
Holding the increase in the global average temperature to well below 2°C above pre-industrial levels and to pursue efforts to limit the temperature increase to 1.5°C above pre-industrial levels, recognizing that this would significantly reduce the risks and impacts of climate change	Limit global warming to well below 2°C Efforts to limit global warming to 1.5°C (from pre-industrial levels)
Increasing the ability to adapt to the adverse impacts of climate change and foster climate resilience and low greenhouse gas emissions development, in a manner that does not threaten food production	Improved adaptive capacities Climate resilient low greenhouse gas emissions development Food production not threatened by the response*
Making finance flows consistent with a pathway towards low greenhouse gas emissions and climate-resilient development	Align finance with climate resilient low carbon development
This Agreement will be implemented to reflect equity and the principle of common but differentiated responsibilities and respective capabilities, in the light of different national circumstances.	Equitable responses from sovereign states

Source: Compiled from UNFCCC (2015).

*Note: The Paris Agreement, like the UNFCCC, refers to food production not being threatened, but the Paris Agreement focuses on limiting the threat from the global response to climate change, rather than the threat of climate change itself.

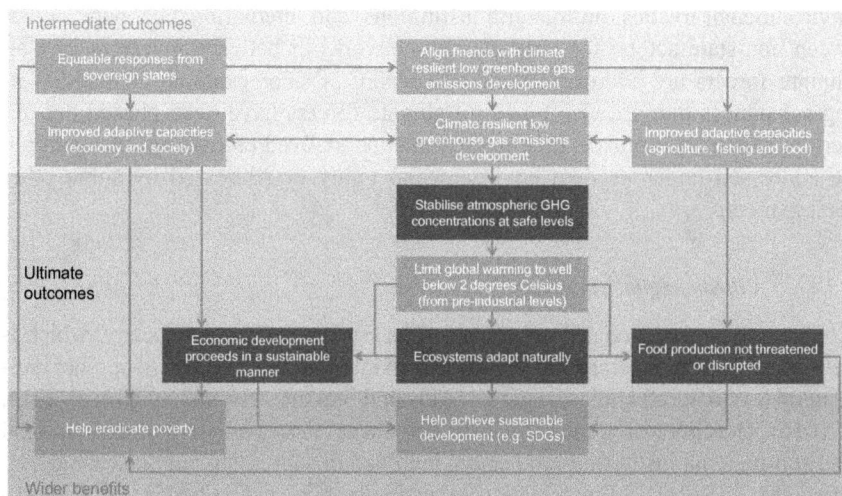

Figure 3.1 Connections between the Paris Agreement, UNFCCC, and outcome levels.

3.3.2 Paris Agreement

According to the IPCC, "The provisions of the [UNFCCC] are pursued and implemented by two treaties: the Kyoto Protocol and the Paris Agreement." (IPCC 2018b, p. 560). While the Kyoto Protocol represents an important step in the global response to climate change, it is effectively superseded by the Paris Agreement (Savaresi 2016). Article 2 of the Paris Agreement sets out its purpose and this is summarised in Table 3.2.

Paris Agreement supports the UNFCCC objectives (Figure 3.1). For example, finance for climate resilient low GHG emissions development supports the achievement of climate resilient low GHG emissions development (as specified in the Paris Agreement purpose), which in turn helps stabilise atmospheric GHG concentrations at safe levels (as specified in the UNFCCC objective). Similarly, limiting global warming to well below 2°C (as specified in the Paris Agreement purpose) is expected to help allow ecosystems to adapt naturally to climate change (as specified in the UNFCCC objective) which along with improved adaptive capacity (as specified in the Paris Agreement purpose) helps avoid disruption to food production systems, while at the same time helping economic development to proceed in a sustainable manner (as specified in the UNFCCC objective).

3.3.3 Other institutions

Related to the UNFCCC and the Paris Agreement are many other institutions at international, regional, national, and subnational levels. This includes other environmental treaties, multilateral institutions, and international cooperation between non-state actors, for example city networks or between investors. Not all climate forcers are addressed under the UNFCCC. For example, ozone-depleting substances that are also powerful climate forcers have been phased out under the Montreal Protocol. It is possible that short-lived climate pollutants might be addressed under its own environmental treaty or addressed by some other arrangement.

3.3.4   Climate regime complex

Collectively, international institutions form a climate policy architecture, which is defined as "the basic nature and structure of an international agreement or other multilateral (or bilateral) climate regime" (Aldy and Stavins 2010 in Stavins et al. 2014, p. 1016). The climate regime can also be defined as "the set of international, national and sub-national institutions and actors involved in addressing climate change" by Moncel et al. (2011). Climate policy architecture options identified by IPCC Working Group III prior to the Paris Agreement included strong multilateralism-based architectures, harmonised national policies, or decentralised architectures and coordinated national policies (Stavins et al. 2014). The UNFCCC and Paris Agreement are central to the climate policy architecture and climate regime.

Importantly, there are other international institutions and regimes that also affect climate change, such as the World Trade Organisation, multilateral development banks, and international corporations for example. As such, the UNFCCC is part of a wider climate regime complex defined as "A loosely coupled set of specific regimes" (Keohane and Victor 2011, p. 7). Importantly, "Regime complexes are marked by connections between the specific and relatively narrow regimes but the absence of an overall architecture or hierarchy that structures the whole set" (Keohane and Victor 2011, p. 8).

To explore preconditions for effective global responses to climate change, it is important to address the climate regime complex (i.e., all the agreements and rules influencing climate and the global response). Focusing on the climate regime (i.e., agreements and rules formed as part of the global response to climate change) means other influences on the global response to climate change would be ignored. Section 3.4 defines the global response system taking into account the climate regime complex. For more information on the climate regime complex see Appendix C.

3.4   The global response system

Why is the global response system important to understand and define? Everything we do to limit climate change and its impacts is a contribution to a wider global response. However, here in lies the challenge. What you or I do may be undermined or enhanced by what others do. In the section below, the global response to system is pieced together, so we can better understand where we fit in the system and how each of us relates to others in the system. The model developed in this section will be an important part of the framing and analyses in Chapters 4, 8, 10, and 11.

The global response system and scope of this book are defined in the sections below. Climate change, national interests, and international cooperation are recognised as important parts of the global response system (Section 3.4.1). Based on these elements, Section 3.4.2 develops the CCNIIC Model by looking at the relationships between climate change and national interests, the sovereign state and national interests, national interests and international cooperation, and other issues. Given that climate change is a global problem, requiring a global response, criteria for international cooperation are addressed in Section 3.4.3.

Importantly, each of us fits somewhere in the CCNIIC Model. Many of us will be non-state actors, but some of us will have roles in central government or even in international organisations such as the UNFCCC. I personally have had roles as a non-state actor, in central government and within international organisations. Each of us has influence, and interests, according to where we are situated.

3.4.1 *Climate change, national interests, and international cooperation*

Interactions between climate change and socio-economic systems, including sovereign states, were addressed in Section 2.2.3 as well as Appendix B. These interactions consist of climate and related physical hazards to socio-economic systems; of which sovereign states are especially important. Meanwhile socio-economic systems, such as sovereign states include national interests that generate GHG emissions and other changes to the environment, driving climate change (Figure 3.2). Central governments determine negotiations mandates and negotiating positions for climate negotiators (Hermwille 2018). As such, Figure 3.2 shows the national interests of a sovereign state informing positions on international cooperation and engagement with international institutions. The UNFCCC creates obligations on central governments which affect national interests (Hermwille 2018). As such, Figure 3.2 shows international cooperation creating options and norms that sovereign states and their central governments may act on or ignore depending on their national interests (Figure 3.2).

Figure 3.2 is a simple model with a central focus on the national interests of a state. However, the international community as represented by the United Nations has 193 members (UN 2018a) and the UNFCCC has 197 Parties (UN 2018b). International cooperation and institutions such as the UNFCCC depend upon

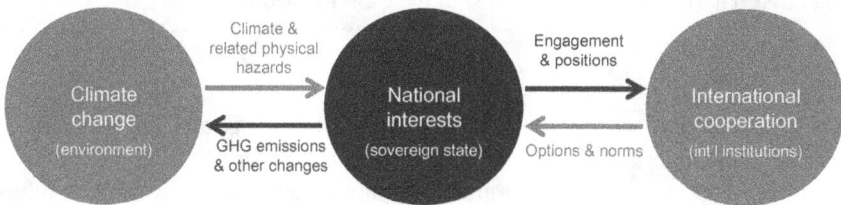

Figure 3.2 Snapshot of interactions between climate change, national interests, and international cooperation.

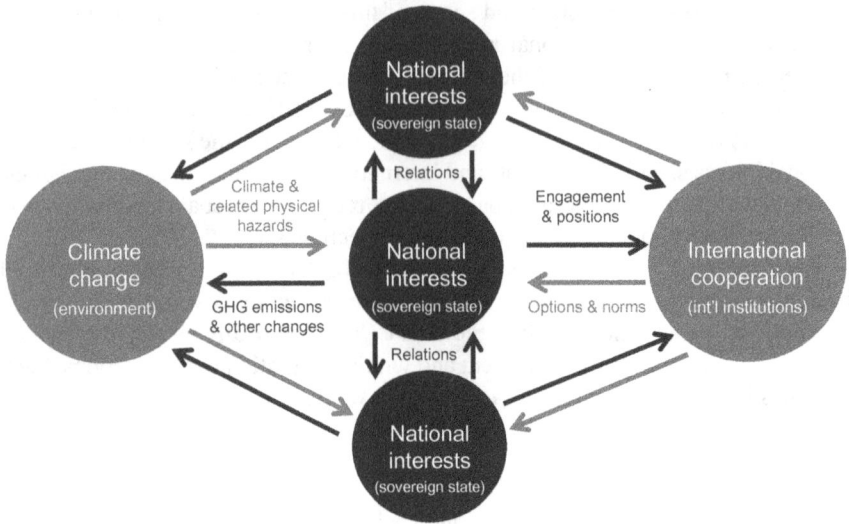

Figure 3.3 Snapshot of interactions between national interests of different sovereign states as well as climate change and international cooperation.

the national interests of many states and are the result of many positions coming together. Figure 3.3 represents a case with three states for simplicity's sake (rather than attempting to represent 197 Parties to the UNFCCC). Figure 3.3 also includes relations between the states which can influence national interests and positions for example at climate negotiations. Importantly, international sanctions can be used to enforce international agreements if applied by member states.

While Figure 3.3 includes multiple sovereign states, it lacks detail when it comes to international cooperation. For example, Keohane and Victor (2011), Stavins et al. (2014), and IEGL (2017) each refer to international cooperation by state actors as well as non-state actors such as civil society, local government, and businesses. Furthermore, international cooperation includes the UNFCCC as well as other international institutions such as the World Trade Organisation, multilateral development banks, and others. These other international institutions can provide options or set norms that may, or may not, be aligned with the UNFCCC objective. These things are addressed in the CCNIIC Model.

3.4.2   CCNIIC Model

From the previous section, climate change, national interests, and international cooperation are important parts of the global response system. These three elements can be are unpacked below to form the Climate Change, National Interest, International Cooperation (CCNIIC) Model (Figure 3.4). Note: CCNIIC can be pronounced "scenic". The elements of the CCNIIC Model, and linkages between these elements, are detailed below in terms of climate change and national interests, the

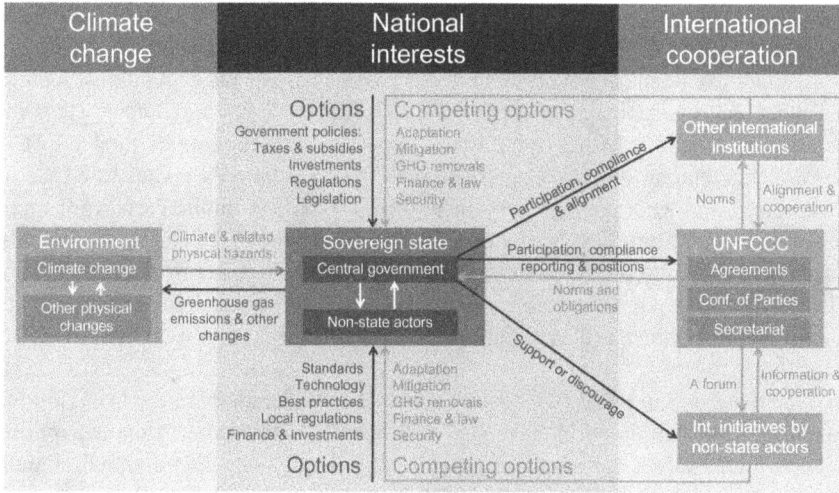

Figure 3.4 The Climate Change, National Interests, International Cooperation (CCNIIC) Model.

sovereign state and national interests, national interests and international cooperation, and other issues.

Climate change and national interests

The CCNIIC Model (Figure 3.4) starts on the left with climate change. Following the discussion in Sections 2.2 and 3.4.1, climate change includes climate change and other related physical changes such as sea level rise, flooding, or ocean acidification. Meanwhile, actors in states, through their activities create GHG emissions and other changes to the environment.

Note: Interactions between climate change and national interests can be broken down in more detail, but for the purposes of this book and defining the climate change and global response system, the level of detail is restricted in the CCNIIC Model. For a detailed description of climate change and related physical changes, or GHG emissions and other changes to the environment, see Appendix B and the Environment, National Interests Model.

The sovereign state and national interests

Within the state, two actors are identified consisting of central government and non-state actors (Figure 3.4). Central government leaders have a special role in deciding what is in the national interest of a state (Putnam 1988). However, central government leaders, to varying degrees depending on their governance regime, need to take into account the interests of domestic non-state actors. Domestic non-state actors can include local government, business, civil society, researchers, the media, religious organisations, communities, and individuals.

Leaders from central government decide what is in the national interest including whether to participate in, or comply with, international agreements. This includes decisions on whether or not to participate in, or comply with, the UNFCCC or Paris Agreement, what is reported to the UNFCCC when it comes to NDCs or inventories, and what positions to take at the UNFCCC or on the Paris Agreement for example regarding levels of ambition. However, the UNFCCC is not the only international institution that a state might choose to participate in or comply with. States will be a member of other international institutions and can take positions influencing whether these other international institutions align with the UNFCCC objective.

It is noted by the IPCC that "When it comes to understanding the behaviour of sovereign states, national interests are important." Furthermore, "domestic political conditions affect participation in, and compliance with, international climate policies." and "This has been addressed in the literature on 'two-level' games" (Stavins et al. 2014, p. 1010).

The term "two-level games" was coined by Robert Putnam, who stated that:

> At the national level, domestic groups pursue their interests by pressuring the government to adopt favourable policies, and politicians seek power by constructing coalitions among those groups. At the international level, national governments seek to maximize their own ability to satisfy domestic pressures, while minimizing the adverse consequences of foreign developments. Neither of the two games can be ignored by central decision-makers... Each national political leader appears at both game boards.
>
> (Putnam 1988, p. 434)

Putnam (1988) referred to international negotiations as "Level I", and domestic negotiations as "Level II". The options available to a negotiator at Level I depend upon the options acceptable domestically at Level II (Figure 3.5). The domestic acceptability of options depends upon the preferences and coalitions within society as well as institutional arrangements within the sovereign state (Putnam 1988). The set of options available to a negotiator is referred to as a "win set". It is generally easier to reach an international agreement if a win set is larger (Putnam 1988).

In addition to societal preferences and coalitions, institutional arrangements can limit the size of a win set. For example, in the United States two thirds of the Senate need to approve international treaties, which in turn limits "institutionally viable options" and the size of the win set available to negotiators. For many other states, the win set is determined by central government leaders based on the prospect of upcoming elections, simple majorities or coalition agreements rather than a ratification process. In other states, the win set may be determined by a single party and may involve opaque coalitions of domestic interests (Putnam 1988).

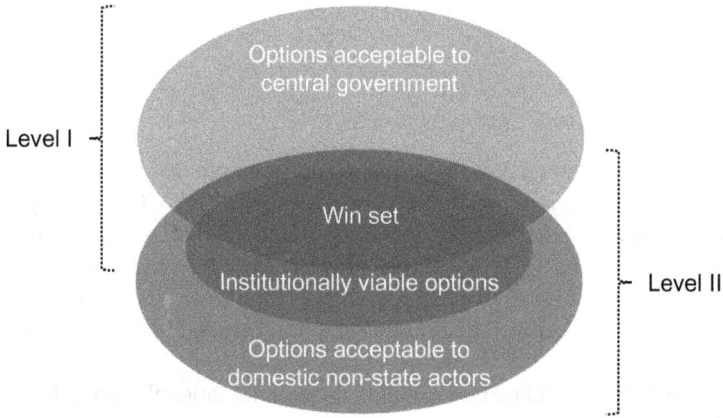

Figure 3.5 Options and the win set in a two-level game.

Source: Author representation of Putnam (1988).

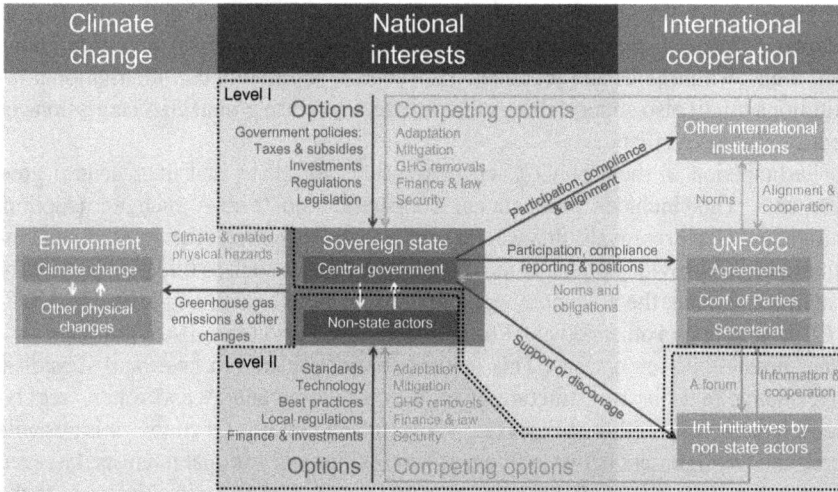

Figure 3.6 CCNIIC Model divided into Levels I and II of a two-level game.

From Putnam (1988), national interests and win sets reflect the interests of central government leaders and domestic non-state actors (Figure 3.6). Given the many interests, preferences, and coalitions of non-state actors it can be difficult to judge Level II interests. Further complicating the understanding of national interests, it was noted by Putnam that "on nearly all important issues 'central decision-makers' disagree about what the national interest and the international context demand." Furthermore, "Even if we arbitrarily exclude the legislature from 'the state'… it is wrong to assume that the executive is unified in its views." (Putnam 1988, p. 432).

As such, national interests are difficult to ascertain as there is uncertainty around interests at Level I and Level II.

National interests and international cooperation

As discussed in Section 3.3.4, the global response to climate change can be characterised as a climate regime complex including the UNFCCC, Paris Agreement, state actors, and non-state actors including other international institutions. The right-hand side of the CCNIIC Model presents the "International Cooperation" part of the global response system, consisting of the UNFCCC (Paris Agreement included), other international institutions and international initiatives by non-state actors.

The UNFCCC and related agreements create international norms, for example the UNFCCC objective and Paris Agreement purpose indicate that efforts should be made to mitigate GHG emissions and adapt to climate change. These norms have an influence on other international institutions, for example within the UN system and to a lesser extent international institutions outside the UN system (Keohane and Victor 2011). Other international institutions cooperate with the UNFCCC to varying degrees, for example attending the UNFCCC COP, support mitigation or adaptation through their own activities, or including climate change considerations into their own policies and practices. However, it is possible that international institutions might also support things that in effect increase global GHG emissions or reduce climate resilience.

Participation in the UNFCCC creates options for states and their central governments. This includes international cooperation on themes such as adaption, mitigation, GHG removals, finance, law, or security. However, other international institutions also provide options in these areas or on issues such as trade or security for example. As such, there are competing options, some of which align with the UNFCCC objective and others do not. Furthermore, these options are also competing with other possible policy options. This includes options related to taxes and subsidies, investments (e.g., in infrastructure or utilities providing energy and water), regulations, legislation, or other policies (e.g., government targets). Of course, it is possible there can be alignment between international options and national interests, but even in such situations, there will be competing interests and options when it comes to the design, modalities, and oversight of international cooperation initiatives.

There are a range of possible international initiatives by non-state actors. From IPCC Working Group III, these include: (1) Public–private partnerships; (2) Private sector governance initiatives; (3) Non-governmental organisation transnational initiatives; and (4) Sub-national transnational initiatives (Stavins et al. 2014).

International initiatives by non-state actors provide options for domestic non-state actors related to standards, technology, best practices, local regulations, as well as finance and investments. The activities of domestic non-state actors will in many cases count towards the state's contribution to the global response to climate change. However, it should be noted that options arising from international initiatives are competing with other options in the same areas. Options include options based on domestic innovation and knowledge, or competing options arising from technology transfer from other non-state actors from other states. All options, regardless of source, compete and require coalitions of actors and interests to be successful. These coalitions, at the level of non-state actors, can come in the form of businesses, local government policies, or wider societal changes.

UNFCCC COP side events provide a forum for international initiatives by non-state actors, where they can meet and discuss issues, as well as meet with delegates (e.g., negotiators) from member states, UN observers, and other influencers (i.e., so-called "track 2 diplomacy"). At the same time, non-state actors share information including valuable experiences that might be replicated or avoided in the future. Furthermore, some non-state actors can help the UNFCCC with ideas and cooperation towards the UNFCCC's objective through their projects and activities.

Importantly, central governments can have a supportive or discouraging influence on international initiatives by non-state actors (Figure 3.6). For example, a supportive central government might include non-state actors as delegates to the COP improving access to influential people and facilitating interactions with other non-state actors (Böhmelt et al. 2014).

In Figures 3.4 and 3.6, the UNFCCC is represented with three parts consisting of "agreements" including the Convention as well as the Kyoto Protocol and Paris Agreement. The "Conference of Parties" includes other conferences under the UNFCCC as well as permanent subsidiary bodies that report to these conferences. The "Secretariat" is included as it supports many of the processes related to the UNFCCC including reporting. For a more detailed representation of the UNFCCC in the context of the international regime complex, see Appendix C and the Jellyfish Model.

Other issues

It is important to acknowledge that there are many influences that are not explicitly included in Figure 3.4 and the CCNIIC Model. For example, behaviour or development goals are not highlighted in the Model. Other national interests related to security or trade are not directly addressed either. All of these can have an influence on national interests and the perceived advantages and disadvantages of addressing climate change.

Figure 3.4 highlights options but does not show how they emerge. For example, technology is widely regarded as a key issue, and while technology is included in the options in Figure 3.4, the model does not provide any insight on how technological or other options might come into existence or change incentive structures. Time series information, including scenarios, is better suited to looking at changes over time than system snapshot models.

3.4.3   *Criteria for effective international cooperation*

As a global collective action problem, international cooperation is widely regarded as an essential part of the global response to climate change (e.g., Stavins et al. 2014). From the literature, two sets of related criteria are identified for effective international cooperation, one for an international agreement to be effective and another for an international institution to be effective.

Barrett (2005) identified issues of compliance and participation as being central to the effectiveness of international agreements along with the issues of incentives and enforcement. Barrett (2005) noted that states are sovereign and as such, can decide whether to participate or not in international agreements and whether to comply or not with an agreement. In many cases, a state will not participate in an international agreement if they think they will not be able to comply (Barrett 2005), or in other cases will not comply unless there is an incentive to do so. Bodansky (2012) referred to Barrett (2005) and likewise identified participation and compliance as being preconditions for effective international agreements along with stringency[2] (Bodansky 2012, Bodansky and Diringer 2010). In an independent piece of work, Wilson (2015) looked at the limits of international energy cooperation through multilateral organisations. Wilson (2015) found the effectiveness of multilateral energy organisations was limited by issues of membership (i.e., a lack of participation), design (including issues of stringency) or commitment (i.e., compliance). Specifically, Wilson (2015) found that multilateral organisations either lacked membership (i.e., participation) by key actors, the design was focused on soft law approaches and therefore lacked stringency, or there was a lack of commitment (i.e., compliance) to agreements that were stringent.

Neither Bodansky (2012) nor Bodansky and Diringer (2010) define stringency. While the stringency of environmental agreements has been addressed at domestic levels (i.e., within a single jurisdiction, few authors have addressed the concept of stringency at the international level across multiple jurisdictions. Fortunately, Hofmann (2019) has reviewed concepts and definitions related to stringency as part of his research developing a "stringency index" for international environmental regulations. According to Hofmann, stringency in the context of international agreements and regulations is a "function of formal tightness and substantive ambition" (Hofmann 2019, p. 219). "The tightness of regulation depends on its legality, precision, and compliance system." (Hofmann 2019, p. 219). An "international regulation is tight when it is legally binding, highly precise, and endowed with strong compliance mechanisms." (Hofmann 2019, p. 220). Meanwhile, substantive ambition "results from the scope of the regulation and its requirement levels in relation to the external effect and compared with other international regulations." (Hofmann 2019, p. 220). A substantively "ambitious international regulation possesses a large scope as well as high requirement levels in relation to the external effect and other international regulation." (Hofmann 2019, p. 221).

The UNFCCC and Paris Agreement have a large scope but don't have "high requirements" instead relying on Parties to make "nationally determined contributions" to the global response to climate change. In short, the UNFCCC and Paris Agreement lack "substantive ambition" in themselves, and instead seek "voluntary ambition" from Parties. Hofmann (2019) also notes that "In case of full compliance, [substantively ambitious] regulation would virtually eliminate the external effect addressed." (Hofmann 2019, p. 221). In the case of the UNFCCC and Paris Agreement the "external effect" is climate change, meanwhile, full compliance would mean all parties complying (i.e., full participation and compliance – see Table 3.3). So, while there has been near universal membership and participation in the UNFCCC and Paris Agreement, stringency (i.e., tightness and substantive ambition) of these agreements is low. Instead, the voluntary ambition of Parties to the Paris Agreement is an important precondition for success.

In a blog for the World Resources Institute, Cameron and DeAngelis (2012) defined ambition in the context of climate change as "countries' collective will—through both domestic action and international initiatives—to cut global GHG emissions enough to meet the 2°C goal." Cameron and DeAngelis (2012) also stated "Ambition further represents the actual steps countries are taking to meet that temperature goal." As such, Cameron and DeAngelis (2012) consider ambition in terms of "will" (i.e., determination) as well as "action" (i.e., responses).

With regards to ambition in the Paris Agreement, the concept involves Parties voluntarily making greater contributions towards the global response through their NDCs, compensating for a lack of stringency in the Paris Agreement. This is referred to as ratcheting (e.g., Meltzer 2016, Rockström et al. 2017).

The IPCC's chapter on international cooperation in the Fifth Assessment Report identified criteria for "institutional feasibility". These criteria consisted of participation, compliance, legitimacy, and flexibility (Stavins et al. 2014). According to Stavins et al. (2014), "participation in an international climate agreement might refer to the number of parties, geographical coverage, or the share of global GHG emissions covered." (Stavins et al. 2014, p. 1010). Compliance refers to the fulfilment of an agreement's provisions (Stavins et al. 2014). Legitimacy refers to the belief by sovereign states and others, that the institution is credible with some degree of authority. Stavins et al. (2014) noted

legitimacy of substantive rules is typically based on whether parties evaluate positively the results of an authority's policies, while procedural legitimacy is typically based on the existence of proper input mechanisms of participation and consultation for the parties participating in an agreement.

(Stavins et al. 2014, p. 1010)

Flexibility refers to the mechanisms under the institution specifically the ability to adjust or adapt as required to new information or changes in economic situations or politics (Stavins et al. 2014).

Institutional feasibility is different from agreement effectiveness. Institutional feasibility in its most basic form is the ability of an institution to perpetuate itself (i.e., perpetuate the rules of the game). Meanwhile, agreement effectiveness is the ability of an agreement to achieve its goals. In the context of the UNFCCC, both sets of criteria are relevant. This is because the UNFCCC has the characteristics of being both an international agreement and an international institution. As a framework convention, the UNFCCC depends upon additional international agreements such as the Kyoto Protocol or Paris Agreement, for its objective to be achieved (Section 3.3.1). Figure 3.7 presents the criteria for institutional feasibility along with the criteria for agreement effectiveness. Notably, participation and

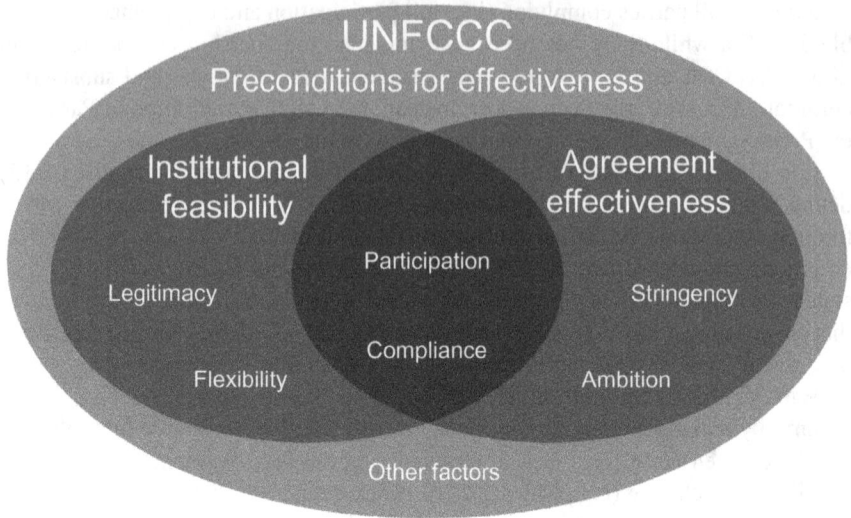

Figure 3.7 A Venn diagram of preconditions for effective international cooperation.

Table 3.3 Preconditions for effective international cooperation based on the preconditions for agreement effectiveness and institutional feasibility

Preconditions	Definition
Participation	When States consent to be bound by an agreement (Source: adapted from Vienna Convention 1969) "participation in an international climate agreement might refer to the number of parties, geographical coverage, or the share of global GHG emissions covered." (Stavins et al. 2014, p. 1010)
Compliance	The fulfilment of an agreement's provisions. (Source: Adapted from Stavins et al. 2014).
Stringency	The formal tightness or substantive ambition of an agreement. Formal tightness means the agreement is legally binding, highly precise, with strong compliance mechanisms. Substantive ambition means the agreement is comprehensive in scope and has obligations sufficient to fulfil the goals, objectives, or purpose of the agreement. (Source: Modified from Hofmann 2019) Note: Agreements with formal tightness or substantive ambition require participation and compliance to be effective.
Ambition	The level of determination, and action required, to fulfil an agreement and achieve its goals, objectives or purpose. (Source: Generalised from Cameron and DeAngelis 2012)
Legitimacy	The belief by sovereign states and others, that the institution is credible with some degree of authority. (Source: Adapted from Stavins et al. 2014).
Flexibility	The ability of Parties to an agreement to adjust or adapt their actions while still fulfilling their obligations. Such adjustments and adaptations are required when there are changes in information, economic situations or politics. (Source: Adapted from Stavins et al. 2014)

Source: Author, Stavins et al. (2014), Cameron and DeAngelis (2012).

compliance are included in both sets of criteria. The criteria are treated as preconditions as each of them needs to be fulfilled if the UNFCCC is to be an effective agreement and feasible institution (Table 3.3).

When it comes to collective action and the institution of the UNFCCC, one of the key issues identified by Stavins et al. (2014) was the legitimacy of the UNFCCC. Without the legitimacy the institution would likely fail and ultimately cease to be recognised. As such, legitimacy was identified as a key success or failure criterion. Meanwhile, flexibility, participation, compliance, and stringency are preconditions towards being able to deliver the objective of the UNFCCC.

Stavins et al. (2014) noted that "The UNFCCC is currently the only international climate policy venue with broad legitimacy, due in part to its virtually universal membership (robust evidence, medium agreement)." (Stavins et al. 2014, p. 1005). As such, Stavins et al. (2014) identified participation (i.e., membership) as a key indicator of legitimacy.

3.5   Conclusions

The global response to climate change depends on actors and institutions. Actors involved in the global response to climate change include state actors as well as non-state actors. Meanwhile, institutions have formed around the issue of climate change including the UNFCCC and the Paris Agreement but these institutions compete with other institutions. Together, these elements form a complex global response system made up of climate change, national interests, and international cooperation as represented in the CCNIIC Model.

Linking everything within the global response system are actors. Climate change impacts actors and creates risks for actors, meanwhile these actors make decisions on which options to take while also negotiating institutions and the "rules of the game" at local, national, and international levels. It is possible to explore preconditions for effective global responses to climate change by focusing on actors, their interests and the decisions they might make under changing conditions (see Chapters 10 and 11).

Before diving into global response scenarios (Chapters 5–9), we need to look at scenario methods and some of the overarching themes from scenarios collected (Chapter 4).

Notes

1 Duffield defined international institutions as "relatively stable sets of related constitutive, regulative, and procedural norms and rules that pertain to the international system, the actors in the system (including states as well as nonstate entities), and their activities." (Duffield 2007, pp. 7–8). It should also be noted that "There are at least as many definitions of [international] institutions as there are theoretical perspectives." (Thomas Risse 2002, p. 605 in Duffield 2007).
2 Cambridge Dictionaries (2016) defines stringency as "a situation in which a law, test, etc. is extremely severe or limiting and must be obeyed."

4 Scenario methods and overarching themes

4.1 Introduction

In the two previous chapters, the analogy of a jigsaw puzzle was used. The pieces of the puzzle were described (i.e., climate change issues, options, actors, and institutions) and Chapter 3 finished with the big "picture" on the front of the box, showing how the pieces fit together forming the global response system (linking climate change, national interests, and international cooperation). However, the question of how the actors making up the system might respond to climate change, and under what conditions they would act on effective response options, remains a mystery. To address these questions, it would be nice if we could look into the future and see how people respond to unprecedented climatic conditions and other influences. Foresight methods can help us describe possible futures (Section 4.2) and search for effective response scenarios fulfilling the UNFCCC objective (Section 4.3.1).

In this chapter, the analogy of a crystal ball is used. The value of a crystal ball is not just that you can look into the future, but you can ask questions of the future and see what happens. Taking the place of the crystal ball is a "searchable sample of possible futures". By collecting a "sample of possible futures" (Section 4.3.2) and processing these data so they are "searchable" (Section 4.3.3), it is possible to analyse the conditions that lead to effective global responses (Section 4.3.4).

Section 4.4 takes a first look into crystal ball, highlighting links between the methods used and the types of scenarios identified. Section 4.4 also summarises the overarching themes identified.

Unlike a crystal ball, where a fortune teller can inexplicably see your future, the methods used in this book are described below, and in more detail in Appendix D.

4.2 Foresight and related methods

Scenarios and models have been used extensively in environmental studies including in IPCC reports (e.g., IPCC 2018a). The IPCC has used scenarios since its first assessment report (Moss et al. 2010, IPCC 2022b). The research presented in this book used scenarios reported by the IPCC to check the plausibility of the information collected.

DOI: 10.4324/9781003465911-5

Bishop et al. (2007) noted that scenario development is concerned with creating stories about the future. Bishop et al. (2007) also noted that "A good scenario grabs us by the collar and says, 'Take a good look at this future. This could be your future. Are you going to be ready?'" (Bishop et al. 2007, p. 5). With regards to being ready for the future, Mangalagiu et al. (2011) noted that "their value is not in enhancing prediction but in revealing and testing critical assumptions about the future, in enabling more shared and systemic understanding and in providing a means for identifying more and better options." (Mangalagiu et al. 2011, p. 11).

Steenmans (2019) highlighted scenarios as being an important method for getting useful intelligence on the future to help with policy decision making. Steenmans (2019) noted that traditional decision-making processes focused on reducing uncertainty by agreeing on assumptions about current and future conditions and then selecting a preferred pathway towards achieving some objective or goal. The problem with this approach is that it gambles on predictions of future conditions, which in many cases are not correct. Furthermore, the path chosen may be sensitive to any deviations from expected conditions. An alternative approach, that is especially useful when there is deep uncertainty, is to analyse the robustness of options and pathways under multiple scenarios, and test these options for robustness (Swart et al. 2013, Steenmans 2019).

Having a large set of global response scenarios could also help with the identification of conditions that lead to effective global responses to climate change. This set of scenarios would have to include effective global response scenarios, even if these scenarios are challenging to achieve or difficult to imagine.

According to Bishop et al., one of the drawbacks of scenario development using extrapolation is that "the future then carries the 'baggage' of the past and present into the future. The baggage limits creativity and might create futures that are… not as bold as the actual future" (Bishop et al. 2007, p. 13). Bishop et al. go on to state that "An antidote to carrying too much baggage is to leap out into the future, jab a stake in the ground, and then work backward on how we might get there." (Bishop et al. 2007, p. 13). This method is called backcasting (Robinson 1990, Bishop et al. 2007, Rounsevell and Metzger 2010).

For backcast scenarios, "The storyline itself is a description of the series of events and causal relationships that lead from the current world condition to the desired future world." (Rounsevell and Metzger 2010, p. 608). Furthermore, "Inherent to this type of thinking is that very different pathways may exist that converge on the same desired outcome." (Rounsevell and Metzger 2010, p. 608). Backcasting, like many other scenario development exercises, uses intuitive logic, which "aims to provide understanding of broad trends and processes" (Rounsevell and Metzger 2010, p. 608). Intuitive logic is similar to judgement techniques that "rely primarily on the judgement of the individual or group describing the future." (Bishop et al. 2007, p. 11). Rounsevell and Metzger noted that "most scenario storylines use the expert judgement of scenario analysts" (Rounsevell and Metzger 2010, p. 610).

Bishop et al noted that backcast scenarios can be developed for the "plausible or fantastical, preferred or catastrophic" (Bishop et al. 2007, p. 13). Backcasting can be used to identify "signposts" and critical nodes, for example technological breakthroughs or the breaching of thresholds that have a bearing on options, contingencies, and how the future is likely to unfold (Bishop et al. 2007).

The backcasting approach was adopted to collect effective global response scenarios. However, scenarios and conditions that lead towards ineffective global responses are also important to understand, constraining pathways towards effective global responses. Reverse stress test methods can help elicit these failure scenarios.

A stress test is an assessment with the purpose of understanding how something responds under difficult conditions, informing risk management and decision making. Climate change stress testing emerged following the great financial crisis, with methods for assessing climate change related risks, including physical, transition, and liability risks (Carney 2015). Brown and Wilby (2012), Stern et al. (2013), Swart et al. (2013), and Carney (2015) have each made the case for using stress tests to investigate climate change related risks. Between them, they noted that history and precedent may not be a reliable guide to the future and stress tests may help with understanding future risks arising from the interaction of climate change and human systems.

However, stress tests are not without criticism. According to Borio and others, financial "stress tests failed spectacularly when they were needed most: none of them helped to detect the vulnerabilities in the financial system ahead of the recent financial crisis." (Borio et al. 2014, p. 12). Borio et al. (2014) noted an alternative approach to using "severe but plausible"[1] scenarios is to instead conduct a "reverse stress test" to see what it would take for the system to fail; an approach similar to that used in engineering.

Reverse stress testing involves backcasting from a failure state, to derive scenarios. An important part of reverse stress testing is understanding and setting the conditions that define success or failure. A reverse stress test scenario must satisfy these conditions (Kilavuka 2013). According to Füser et al. (2012), qualitative reverse stress tests take a "birds eye view" and tend to focus on actors and strategies that are generally overlooked by quantitative stress tests. In this regards, qualitative reverse stress tests appear to be a promising method for collecting information for this book, especially when it comes to system conditions and the situations actors might find themselves in or identifying possible chains of events and understanding the roles, incentives and motivations various actors might have in each scenario.

The reverse stress testing approach was adopted to collect failure scenarios. Like backcasting, reverse stress testing works backwards from a predefined outcome. The difference is reverse stress testing works backwards from failure and backcasting works backwards from success. Given that both backcasting and reverse stress tests collect information on possible sequences of conditions or events, a method for plotting these changes over time could be very helpful.

Theory of change is defined as "The description of a sequence of events that is expected to lead to a particular desired outcome" by Davies (2012) in Vogel (2012). At its core, theory of change analyses follow a logic of "if... then" (Stein and Valters 2012), using this logic to map "pathways" (Taplin and Rasic 2012) or "outcome chains" (Harries et al. 2014) where "A pathway is the sequence in which outcomes must occur to reach long-term goal." (Taplin and Rasic 2012) such as the UNFCCC objective. Importantly, it is possible to have "multiple outcomes chains" (Harries et al. 2014, p. 15) towards a specified goal and these outcomes constitute "preconditions because they are conditions that must exist... ...for the next outcome in the pathway to be achieved." (Taplin et al. 2013, p. 21). As such, theory of change analysis is uniquely suited as a method for exploring preconditions while also being akin to backcasting and reverse stress testing.

Vogel (2012) reviewed theory of change methods and identified a set of six basic elements, consisting of: context, long-term change, sequence of change, assumptions, and the use of diagram and narrative summaries. James (2011), along with Stein and Valters (2012), also highlights the use of narratives and diagrams in theory of change analyses. However, Stein and Valters (2012) suggested diagrams should only be used as part of the analysis and then removed from final documentation, due to their complexity. With regards to diagrams, James (2011) along with Taplin and Rasic (2012) present outcome chains vertically in the form of a hierarchy, meanwhile Harries et al. (2014) presented chains horizontally from left to right giving a sense of being a time series.

The quality of theory of change narratives, scenarios, and stress tests depends on the quality of analysis and understanding of how the world works. In this regard, Stein and Valters note, "At its best, ToC requires an engagement with wider social science theory and research-based evidence. Such work is ultimately an attempt to describe and understand how change happens in the world" (Stein and Valters 2012, p. 9). However, they also caution that "The extent of practitioner engagement with... social science theory and research, may well reflect whether ToC approaches ultimately reveal or oversimplify the complexity of processes of change." (Stein and Valters 2012, pp. 9–10). This is poignant in the case of climate change, where social processes and responses to climate change are not yet adequately addressed or understood especially within models (IPCC 2023b).

4.3   Methods used

The methods used to collect, process, and analyse global response scenarios in this book consisted of: backcasting and reverse stress test methods to collect scenarios (Section 4.3.2), theory of change methods to process and visually plot these scenarios (Section 4.3.3) and "thematic chain analysis" to synthesis the sequences of conditions leading to effective global responses (Section 4.3.4). But before describing these methods, we need to be clear about what constitutes an effective global response to climate change (Section 4.3.1) as both backcasting and reverse stress testing methods need criteria on what constitutes success and failure.

4.3.1   *Effective global response criteria*

To find effective global responses to climate change, we need to know what an effective global response looks like. From Section 3.3.1, the UNFCCC objective provides an internationally accepted description of what an effective global response looks like, which can be summarised as:

* stabilising atmospheric concentrations of greenhouse gases (GHGs) at levels that
    * allow ecosystems to adapt naturally,
    * food production is not threatened, and
    * economic development can proceed in a sustainable manner.

These criteria for effective global responses to climate change serve as a key reference point for data collection. Note: For more information on effective global response criteria, including a discussion of the Paris Agreement in relation to the UNFCCC objective, see Appendix D and Section D.2.2.1.

4.3.2   Data collection

Referring back to the analogy of a crystal ball, data collection can be thought of as filling the crystal ball with scenarios. The purpose of data collection was to get information on possible futures. Meanwhile, the aim was to collect as many scenarios as possible, forming a representative sample of possible futures. So long as a scenario was considered possible it was included in the sample.

The possibility of conducting structured scenario workshops, online simulations, and surveys was investigated. However, these methods involved varying levels of framing. The concern was that with more framing there would be more bias introduced into the sample of possible futures.

It was decided to develop a semi-structured survey using backcasting and reverse stress test methods as they allow respondents to form their own scenarios with little influence from the interviewer or anyone else. Respondents simply describe the system conditions and situations leading towards success or failure, including the actors involved.

Invitations to participate in the survey were sent to 44 people of which, 27 people responded and participated in the survey (8 women and 19 men). Primary regional affiliations included 3 respondents from Africa, 2 from Asia, 11 from Europe, 3 from North America, 7 from Oceania, and 1 from South America.

With regards to institutional affiliations, 1 was primarily affiliated with a United Nations related climate change institution, 4 were affiliated with other international organisations, 5 with sovereign states and their governments, and 17 were primarily affiliated with non-state actors such as research organisations or civil society. There were no business, local government or religious respondents. However, the extent to which this is a problem depends on the extent to which the "survey sample" influences the "sample of possible futures" collected. It is possible the sample of possible futures missed out on effective global response scenarios involving business, local government or religion, although some scenarios involving business, local government and religion were collected.

Of the 27 interviews, 19 interviews were conducted in person and 8 were conducted by telephone or over the internet. Interviews took between 30 and

85 minutes and were recorded for transcription purposes. Each semi-structured interview started with background information including a description of the survey process and what the information was going to be used for (Figure 4.1). The first set of questions collected attitudinal information to get a sense of the underlying beliefs and views of the respondents.

The substantive part of the survey started with a question asking respondents to describe the current climate change situation. These responses were later compared with IPCC assessments and the information in Chapters 2 and 3, to check if they were consistent with the literature.

Respondents were then asked if they thought the global response was on track towards fulfilling the UNFCCC objective or not. Every respondent said the global response was not on track. Following the reverse stress test approach, respondents were asked to describe a scenario where the global response fails to achieve the UNFCCC objective. Respondents were then asked to describe a scenario where the global response to climate change succeeds in achieving the UNFCCC objective following the backcasting approach. Some respondents provided more than one success scenario.

The scenarios literature notes the need to consider so-called "black swan" low probability high impact scenarios and other scenarios that might influence outcomes (CSF and CSC 2012, Oura and Schumacher 2012). Climate change is a long-term problem, and many historic and other events will happen while the climate responds to the accumulation of GHGs in the atmosphere. Respondents were asked what other things might happen that could influence climate change or the global response including the UNFCCC. Follow-up questions were asked clarifying details in the scenarios being provided. Lastly, respondents were asked reflective questions, including which scenarios they, or their organisations, were contributing towards.

To ensure a wide set of scenarios were collected, new interviews were initiated until there was thematic saturation i.e., very little new information was emerging from

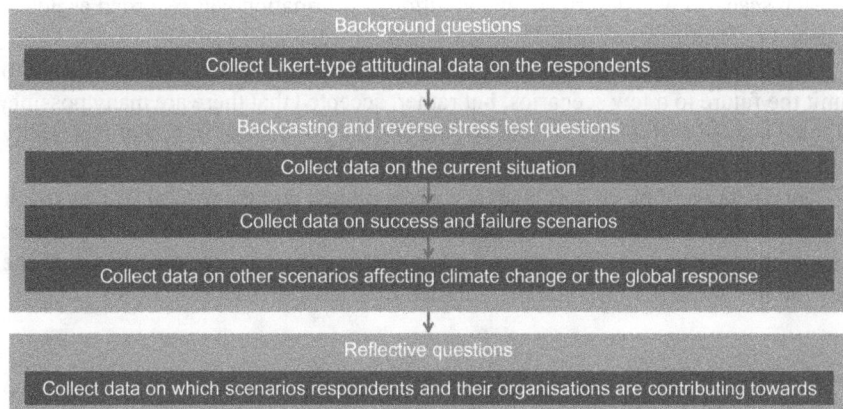

Figure 4.1 Semi-structured survey design.

respondents because so many scenarios were already collected. By having a wide range of scenarios, the aim was to form a representative sample of possible futures.

See Appendix D for more information the questions asked included follow-up questions and the reasons why these questions were asked.

4.3.3   Data processing

From the 27 survey responses, over 145,000 of the respondents' own words were transcribed and analysed in detail using a theory of change approach. This resulted in 175 scenario plots. Referring back to the metaphor of a crystal ball, the process of creating plots is akin to seeing how future might unfold over time.

Plots started with the current situation as described by respondents in their own words. The analysis of transcripts included looking for "if then" statements. Where "if then" statements were found, the respondent's own words were copied and pasted into a box and plotted. Each box was referred to as a "scenario element". Sequences of scenario elements over time were plotted left to right starting with the current situation. Scenario elements regarding local issues were plotted towards the bottom of the plot and scenario elements regarding more global issues were plotted towards the top of the plot. Direct linkages between scenario elements were shown using solid arrows and indirect linkages were shown using dotted arrows. Assumptions provided by the respondent were also recorded as scenario elements on the plot. In total, 37,000 words of respondent quotes were included in the 175 scenarios plotted.

Each scenario plot was thematically analysed and coded, following a process of familiarisation, initial codes, searching for themes, reviewing themes, naming and defining themes and reporting of themes (following steps from Braun and Clarke 2006). The thematic analysis included identifying and recording the range of assumptions, issues and options, actors and interests, and responses (i.e., actions) by actors. Outcomes were also recorded, based on the extent to which conditions at the end of a scenario fulfilled the UNFCCC objective. This consisted of success, failure, branching success and failure scenarios, and other scenarios. The complexity of each scenario was described, and additional information was recorded as notes. Lastly, key themes were coded in a short comma-separated list.

Given that no one knows how the future will unfold, the study did not attempt to limit the future to a few scenarios, but rather, accepted that there are many possible climate change and related scenarios. The challenge was how to accept this reality and analyse information from across the "multiverse" of scenarios.

Once data were processed, with scenarios plotted and thematically analysed, it was possible to search the "sample of possible futures" for keywords and by theme. Searchable text included the respondent quotes in the scenario plots, as well as the key themes and other information recorded for each scenario.

4.3.4   Data analysis

Data analysis had the aim of identifying the conditions leading to effective global responses.

With this aim in mind, the analysis included "thematic chain analysis", a method developed as part of this study, allowing sequences of conditions from multiple scenarios to be analysed together. Thematic chain analysis involved identifying a theme of interest, making a search for relevant scenarios, which were then reviewed and a table created, with the theme of interest in the central column, preconditions recorded in the left column and subsequent conditions (i.e., following conditions) recorded to the right (see Appendix D for an example). Using this process, chains of conditions leading to effective global responses were synthesised and then mapped in Sections 10.3–10.5.

The analysis also focused on a series of overarching themes, and was organised into five parts, reflected in the structure of upcoming chapters. This included an analysis of:

- the influence climate change impacts might have on actors and the global responses to climate change (Chapter 5) informing in a typology of climate change signals and responses (Section 10.2).
- actors and interests from scenarios (Chapter 6). This included social change and behaviour (Section 10.3.1), political will and policy (Section 10.3.2), business and economic activity (Section 10.3.3).
- response scenarios (Chapter 7), including an analysis of GHG removal options (Section 10.4).
- international cooperation scenarios (Chapter 8), including coalitions and issues of power and capacity (Section 10.5).
- types of response and the likelihood of these responses fulfilling the UNFCCC objective (Chapter 11) including the possibility of serendipity playing a role in effective global responses (Section 11.11), in addition to ambition.

Failure scenarios were also identified and thematically analysed but due to space constraints, these scenarios do not feature in the analyses presented in upcoming chapters.

4.4 Initial results, and overarching themes

Using the methods above, data was collected, processed, and analysed. This section provides the first look into proverbial crystal ball providing a broad summary of the types of scenarios and overarching themes identified. This section also provides some important context on respondent sentiment, the nature of responses and how methods were applied.

As noted in Section 4.3.2, respondent sentiment (i.e., attitudinal data) was sampled to help get a sense of possible respondent biases. Sentiment analysis suggests the survey sample includes respondents with mixed levels of optimism and faith in the UNFCCC and its role in the global response to climate change (Appendix E). This is encouraging as the purpose of the survey was to collect a broad sample of possible futures.

Each of the scenarios compiled is like a short story of the future. Some of these scenarios are very short, for example consisting of a single element, while other scenarios are much more complex consisting of multiple steps and layers (Appendix F). Each scenario was analysed and coded starting with outcomes (success, failure, branching success or failure, and other outcomes).

Branching scenarios could either lead towards success or failure when it comes to fulfilling the UNFCCC objective. Seven of these branch scenarios were identified, each of which had a critical node i.e., responses and conditions that strongly influence the likelihood of achieving the UNFCCC objective. These critical nodes highlight themes critical for effective or ineffective global responses to climate change.

Respondents, without exception, indicated that the current situation is not on path for achieving the UNFCCC objective, or indicated that the global response has already failed.

When asked to provide a scenario where the UNFCCC objective is achieved, some respondents struggled to respond with success scenarios or stated that they considered the success scenarios they provided unlikely. This is consistent with the UNEP Emissions Gap Report (UNEP 2017) (Section 2.2.4). Furthermore, some respondents challenged the notion of success. For example, Respondent 17 noted, "…we return to 1.5, but it's the Earth without X." where "X" could be a set of ecosystems. In an overshoot-drawdown-stabilise scenario, Respondent 23 noted "… maybe we will begin to be adapting to… 500 PPM, and that becomes a new desirable level." Some responses aligned more closely with effective response criteria than others.

During interviews, respondents were asked what other things could happen that could either affect climate or the global response to climate change. Other scenarios were identified from responses to this question that show other, often extreme, sets of conditions or events that could affect the climate or the global response to climate change. In some cases, other scenarios did not focus on responses to climate change, but rather featured other issues and options, for example, some event, crisis, catastrophe or issue not directly related to climate change.

While systematically coding and analysing the scenarios, it became apparent that some themes were related and could be grouped under overarching themes. At the same time, it also became apparent that these overarching themes fitted into different parts of the CCNIIC Model (Section 3.4.2) but could also be grouped as being "impacts and risks" or "responses". As such Figure 4.2 shows the climate change, national interests and international cooperation categories from the CCNIIC Model, but also includes impacts and risks. Climate change impacts and risks include impacts and risks to physical systems and biological systems. Impacts and risks to human and managed systems are impacts and risks to national interests.

Overarching national interest and response-related themes consisted of political will and policy, social change and behaviour, business and economic activity, technology and practices as well as other factors that influence decisions. International cooperation responses consisted of changes to the international regime as well as international cooperation on climate change. The diagram does not address exogenous factors or other scenarios that might influence climate or the global response to climate change.

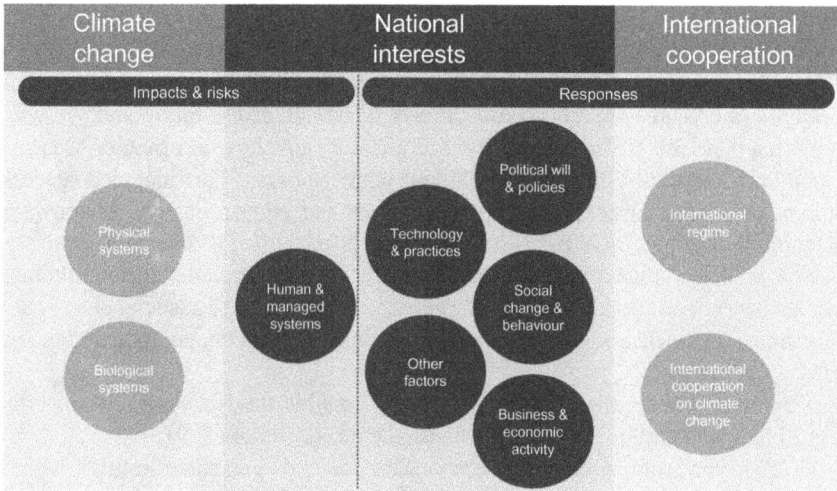

Figure 4.2 Overarching themes situated in relation to the CCNIIC Model.

It should be noted here that while the overarching themes in Figure 4.2 attempt to be comprehensive, they are also a reflection of sample of possible futures collected. The extent to which the sample of possible futures presented in Chapters 5–9 are representative of the range of possible futures, or are biased, is difficult to judge (Appendix D). Likewise, the extent to which the sample of possible future biases overarching themes identified or the limits the extent to which these overarching themes are comprehensive is also difficult to judge. However, these overarching themes seem consistent with the range of themes that could be drawn from comprehensive IPCC assessments.

Themes and scenarios related to these overarching themes are presented in Chapters 5–9. These overarching themes are analysed in Chapter 10.

4.5 Conclusions

The two previous chapters looked at the pieces of the climate change problem and global response system, but how this system might respond to climate change and other influences remained a mystery. We needed to find methods for looking at how the system, and actors within, might respond to climate change.

After reviewing foresight methods including scenario and stress test methods, it was decided to conduct semi-structured interviews using a backcasting and reverse stress testing approach to collect information on possible futures. In total 27 experts from around the world were interviewed. Theory of change and thematic analysis methods were then used to process, plot, and thematically analyse 175 scenarios, forming a searchable sample of possible futures. This searchable sample of possible futures was analysed using a new method called "thematic chain analysis", synthesising information from across multiple scenarios. Thematic chain analysis involved keyword searches (e.g., social unrest) and the tabulation of conditions leading to the keyword, and the conditions following the keyword.

As noted in the introduction, having a searchable sample possible future is akin to having a crystal ball, where you can ask questions of the future and see what might happen. The methods set out in this chapter show how information was collected and compiled filling the crystal ball with scenarios. This included success, failure, branching success and failure scenarios and other scenarios. Meanwhile, overarching themes included risks and impacts on physical and biological systems as well as human and managed systems. Responses included the overarching themes of technologies and practices, political will and policies, social change and behaviours, business and economic activities, international cooperation and the international regime.

Now it is time to dive into the sample of possible futures, look deep into the crystal ball, and explore these overarching themes (Chapters 5–9).

Note: if you want to know more about the methods presented in this Chapter, please see Appendix D. For more information on respondent sentiment see Appendix E and for a summary of scenario types and complexity, see Appendix F.

Note

1 The problem with searching for "extreme but plausible" scenarios is that plausibility is subjective and often judged according to historical experience (Borio et al. 2014), limiting the exploration of possible futures which may be probable or improbable (GOS 2016), fantastical (Bishop et al. 2007), or unprecedented (Borio et al. 2014).

Part II

Exploring 175 global response scenarios

Part II consists of five chapters, each of which takes a deep dive into the searchable sample of possible futures, exploring the themes identified.

Chapter 5 explores climate change impacts, risks, and responses.

Chapter 6 identifies actors, interests, and related themes.

Chapter 7 explores domestic response options including GHG removal options.

Chapter 8 explores international cooperation options and themes.

Chapter 9 explores other unusual scenarios including the influence these scenarios and themes might have on climate change or the global response.

Note: There are many section headings in Chapters 5–9 reflecting the diversity of themes and scenarios identified. Each section heading represents a theme for which there may be multiple related scenarios. Sections have not been joined (i.e., conflated) where there are distinct themes, even if a section is very short.

Many of the quotes are taken from much longer sentences collected from interviews. Instead of burdening the document with "…" at the beginning and end of quotes, the following convention is used. If a quote starts with a capital letter, that statement was taken from the beginning of a respondent's sentence. If a quote starts with a small letter, this is a statement taken mid sentence. If a quote finishes with a full stop inside the quotation marks, then the respondent had finished their sentence. If a quote finishes with the full stop on the outside of the quotation marks, then the quote is taken from a sentence that was not completed.

DOI: 10.4324/9781003465911-6

5 Climate impact, risk, and response scenarios

5.1 Introduction

Chapter 4 talked about a "searchable sample of possible futures" being similar to a crystal ball where you can ask questions and see how the future might unfold. Instead of having a crystal ball, the sample of possible futures consists of 175 short stories, shared verbally by 27 different respondents. In this chapter, we look into the crystal ball (i.e., sample of possible futures) and explore climate change impacts, risks, and responses, asking the question: What influence might climate change have on actors and the global response to climate change?

In the sections below, we dissect the scenarios provided by respondents, looking for climate response drivers, often quoting the respondent's own words. By contrasting what respondents said, a wide range of climate change responses are revealed.

So, how might we respond to climate change?

Section 5.2 looks at the range of possible climate change impact and risk scenarios, unpacking the issues of climate sensitivity (Section 5.2.1), impacts and risks (Section 5.2.2), responses where human and managed systems are threatened (Section 5.2.3), and threat distribution in relation to the power of actors (Section 5.2.4).

The importance of "responsiveness" is investigated in Section 5.3, including two preconditions for effective responses, consisting of timeliness (Section 5.3.1) and scale (Section 5.3.2). Response triggers and drivers are identified in Section 5.4, consisting of non-response and responses driven by other interests (Section 5.4.1) ranging from cynical through to hypocritical responses. Responses driven by climate change are addressed in Section 5.4.2, ranging from responses to impacts and risks, to enlightened responses and cooperation.

Response attitudes and inclinations are explored in Section 5.5, contrasting defensive versus cooperative inclinations (Section 5.5.1), technology versus practice-focused inclinations (Section 5.5.2), and competitive attitudes (Section 5.5.3).

Lastly, Section 5.6 concludes with a summary of themes from this chapter and indicates where these themes are analysed further in Chapter 10.

5.2 Climate change impact and risk scenarios

While analysing the sample of possible futures, a wide range of climate change hazards were identified. These are presented in Table 5.1. As these climate change

DOI: 10.4324/9781003465911-7

related risks and impacts neatly fit with the IPCC's categories of physical, biological, and human systems (see light grey subheadings in Table 5.1), the IPCC categories were adopted for this book, along with IPCC definitions (Table 5.2).

According to the IPCC (IPCC 2018b), human systems are any system in which human organisations and institutions play a major role. The IPCC gives examples of food production, livelihoods, health, and economics (see Section 2.2.2). The IPCC does not define physical or biological systems but does give examples. As such, the definitions for physical and biological systems were drafted by the author but with examples from IPCC Fifth and Sixth Assessment Reports (see Table 5.2). The theme of risk is also from the IPCC but the term impact is defined by the author.

Table 5.1 Examples of climate change hazard-related themes (in light grey rows) and related notes from the scenarios

Climate risk and impact-related themes from the scenarios collected

Climate change and related risks and impacts on physical systems

Bigger hurricanes	Increased flooding
Changing coastlines	Ocean acidification
Even more rapid melting of Antarctic and Greenland ice sheets	Only lose a few islands (as opposed to many islands)
Fast feedbacks	Permafrost melt
Flooding	Positive feedbacks, where we have a massive rapid warming
Flooding of China Coastal Plain	
Flooding of Nile Delta	Sea level rise
Greenland ice sheet (melt)	Tipping point
	Weather variability

Climate change and related risks and impacts on biological systems

Concerns on whether ecosystems can adapt	Mass die-off of species
Ecosystems not able to adapt	Widespread species loss
Loss of biodiversity	

Climate change and related risks and impacts on human systems

Bangladesh ceases to be a viable country	Famine
Collapse in agricultural systems	Food crisis e.g. double bread-basket failure type event
Death from heat	
Disrupted food system	Food security and insecurity
Entire nations disappear if the sea rises	Losses in yields of food in crops.
Failure of agricultural systems	Perpetual retreat from the coasts
	Water security and insecurity

Unspecified climate change and related risks and impacts on human systems

A crisis that is directly attributed to climate change	Crisis scenario
	Crisis years
Catastrophe	Environmental catastrophe
Catastrophic events	Not sure it will be internationally catastrophic
Catastrophic impacts	
Climate-driven catastrophes	Unforeseen crisis
Crisis	

Table 5.2 Climate change risks and impact-related themes

Theme	Definition
Risk	The potential for adverse consequences where something of value is at stake and where the occurrence and degree of an outcome is uncertain.[a]
Impact	Effects on natural and human systems.[b]
Physical systems	Any system in which physical processes play a major role.[b] This includes glaciers snow, ice, permafrost, rivers lakes, floods, drought, coastal erosion, and sea level effects.[c]
Biological systems	Any system in which organisms play a major role.[b] This includes terrestrial ecosystems, wildfire and marine ecosystems.[c]
Human systems	Any system in which human organisations and institutions play a major role.[a] This includes food production, livelihoods, health and economics.[c]

Source: Author, IPCC (2018b), IPCC (2014a), and IPCC (2022a).

[a]Definition from the IPCC 1.5 Degree Report Glossary (IPCC 2018b).

[b]Definition by the author.

[c]Examples from the Fifth and Sixth Assessment Reports.

5.2.1   Climate sensitivity

Climate sensitivity is very important as it regards the extent to which the accumulation of greenhouse gases (GHGs) in the atmosphere will result in climate changes. As such, climate sensitivity influences the scale of climate change and related hazards that people will need to deal with for a given concentration of GHGs. In a scenario where climate sensitivity is lower than expected, Respondent 12 stated climate "changes will not be as fast as expected" hence "from a political point of view, that's good, because we… have a better chance of organizing." Respondent 6 was less optimistic when it came to climate sensitivity. After providing a couple of success scenarios, Respondent 6 stated

> both [success scenarios] need to be qualified by the fact, they assume climate sensitivity is kind of what people think it is. Our bet would be climate sensitivity is far higher than people think it is… We think people are massively overestimating how much we understand about the climate system and the feedbacks.

5.2.2   Climate change impacts and risks

From the sample of possible futures, there's a broad range of climate change scenarios including scenarios with low climate sensitivity through to scenarios with "fast" climate feedbacks. There are scenarios that are benign and scenarios involving other disasters, crises or catastrophes that could influence climate or the global response to climate change.

Impacts and risks presented by respondents tended to be illustrative and less detailed than IPCC assessments of impacts and risks (see Sections 2.2.2 and 2.2.3). Respondents tended to mention specific impacts and risks for the purposes of illustration. For example, Respondent 9 stated, "One scenario would be that in the next few years we experience what we are already are experiencing, bigger hurricanes, sea level rise that effects people, et cetera..." The point is that something happens that could generate some sort of response.

The clearest exception was Respondent 9, who stated that the current level of global warming "has already set us on a track to many meters of sea level rise" Respondent 9 then focused on directly related risks, including for example "sea level rise happens at a rate that is slow enough for our political and social economic systems to adapt to" or "In the worst case scenario, the rise is going to happen too quickly to avoid large swathes of our population are losing their properties and their livelihoods."

In many scenarios the source of the impact or risk was unspecified, except to say climate change related impacts constitute a "disaster", "emergency" or "catastrophe". For example, Respondent 10 stated "...pressure and disasters influences [policy maker] decisions more than... anything." without giving further context on what such disasters might include. Respondent 13 gave scenarios where a "global emergency" related to climate change was a basis for an international response but the nature of the emergency was not specified. Respondent 14 stated "if there's some catastrophic, a series of certain catastrophic events" then there could be a response. Meanwhile, Respondent 8 mused "I wonder about if there is a major environmental catastrophe that somehow triggers people's consciousness to that we really need to change." Again, the nature of the catastrophe is not specified beyond being environmental. At a minimum, terms such as "disaster", "emergency", "crisis" or "catastrophe" indicate climate change is impacting human and managed systems, and hence the interests of actors.

From the sample of possible futures, some scenarios included "tipping points" and "fast feedback". For example, Respondent 20 noted "something could happen as a combination of positive feedbacks, where we have a massive rapid warming... fast feedbacks" and highlighted "one thing is a major concern is the release of a lot of CO_2 and methane from the melting permafrost in the Northern Hemisphere." Meanwhile, Respondent 2 noted with regards to climate change and related impacts "we don't understand whether these are straight line in effect [or] whether [there are] tipping points." In a failure scenario, Respondent 9 stated, "The questions are: will climate hit some tipping point so we can't recover from? Our ecosystems aren't able to adapt." Specific examples include GHG release due to permafrost melting (Respondents 21, 20, and 17), changing ocean currents (Respondent 15), ice sheets melting (Respondent 15), disruption to the monsoon (Respondent 6), methane hydrate release (Respondent 17). These concerns are consistent with issues raised by the IPCC in its Fifth Assessment Report which consisted of Atlantic Meridional Overturning Circulation (AMOC) collapse, ice sheet collapse, permafrost carbon release, clathrate methane release, tropical forests dieback, boreal forests dieback, disappearance of summer Arctic sea ice, long-term droughts, Monsoonal circulation (Section 2.2.3). The IPCC noted a great deal of uncertainty around these possibilities (Collins et al. 2013).

The extent to which future climate impacts might be a problem differed across scenarios. For example, Respondent 15 stated

> Climate change is real. It's gonna be a significant problem by the end of the century according to the UN. We're probably talking about an impact that's equivalent to about 2 to 4% of GDP. So, it's the equivalent of each person, on average, on the planet being 2 to 4% less well-off by the end of the century.

The respondent also noted, "Overall and in the long run, global warming will be predominated by bad things which is why it's a problem. But, actually, right now, there's probably about equal good and bad stuff happening with global warming." Interestingly, Respondent 15 also provided a scenario involving tipping points, acknowledging the possibility of changes that would presumably be outside the loss of economic activity estimates discussed.

Other respondents were less sanguine, for example Respondent 2 noted that a worst-case business-as-usual scenario "takes us into the world of four degrees plus against pre-industrial– Why is that our worst-case scenario? Because we don't actually know what will happen."

Several scenarios highlighted impacts and risks to specific areas of the world. For example, Respondent 14 noted

> There's certainly going to be places in the world where it won't be much fun to live and where food security is going to diminish significantly. The issue of food security and water security is going to flow on from these, the international issues around national borders and conflict between states over resources.

Likewise, Respondent 25 stated,

> If you're in Northern-Eastern US or Central Canada or... New Zealand, you're probably fine. Africa is a disaster... it gets worse and worse, mass famine, immigration, refugees. Bangladesh ceases to be a viable country. The whole Nile River delta is under water... China coastal plain. There are hundreds and millions of refugees wandering around.

Respondent 24 noted "Look at the United States. We are facing some pretty hefty impacts of climate change increasingly attributable to climate change" suggesting the USA is also exposed and vulnerable to climate change.

It is important to note that the IPCC has mapped impacts and risks around the world in broad terms (see Sections 2.2.2 and 2.2.3) and has addressed reasons for concern ever since its Third Assessment Report (see Section 2.2.3). The impacts and risks highlighted by respondents are consistent with IPCC assessments although the language used by respondents to describe scenarios was a lot less formal, given the nature of a semi-structured interview.

5.2.3 Responses to impacts and risks to human and managed systems

From the sample of possible futures, only 25 scenarios indicated physical climate change drivers for a response. All 25 of these response scenarios involved impacts and risks to human and managed systems. This strongly suggests impacts and risks to natural systems (i.e., physical and biological systems) will have very little influence on the global response to climate change, while impacts and risks to human and managed systems will have a much stronger influence on the global response to climate change. Even the respondents that lamented the lack of a global response to date given evidence of climate change, only gave success scenarios where the global response is driven by impacts or risks to human and managed systems. This may be due to the study presented in this book being focused on "what would happen" (i.e., scenarios) rather than "what should happen" (i.e., ethics).

5.2.4 Responses and the distribution of impacts, risks, and power

The distribution of impacts was highlighted in several scenarios. For example, when it comes to climate change impacts, Respondent 8 suggested it could take "lots of local ones around the world to collectively trigger a change", the premise being local impacts are more likely to influence people's perceptions of climate change, and hence, it would take lots of local impacts around the world to create an effective global response.

In some scenarios, the distribution of impacts and risks was presented in the context of geopolitical power. For example, Respondent 14 stated that for impacts to influence that global response they would have to impact "nation states bigger than Tuvalu." because "They've been pleading for action for a while, and have got very strong basis for the plea, but, you can, very easy to dismiss frankly, internationally." The situation for powerful countries might be very different. For example, Respondent 7 put forward a stringently enforced climate agreement scenario, where "some as yet an unforeseen crisis" related to climate change affected a powerful state. This resulted in a situation where the agreement is "enforced by a group" described by the respondent as an "elite over a minority" and "that elite could be a China over diminishing Western World." The respondent also noted "power is the exhibition of authority, and authority allows you to maintain what you've got and increase [what you've got]" i.e., defend and pursue one's interests. As such, the influence climate change and related impacts might have on the global response to climate change depends at least in part, on the power and influence of the actors being impacted.

In addition to the distribution of impacts across states, some respondents and scenarios included the distribution of impacts across wealthy and poor people. With regards to climate change impacts due to sea level rise, Respondent 20 stated "it will probably only have true political ramifications when it starts to impact wealthier parts of the community". The respondent also noted

we already know that people that are wealthy and better educated, especially those people on the coast are very forceful in advocating for public money to be spent on coastal defences for example, like sea walls and installing pump stations and those sorts of things. Whereas people who are less educated and poorer are not very good advocates for themselves as far as that is concerned.

5.3   Responsiveness and preconditions for effective responses

From the thematic analysis, the timeliness and scale of the global response were highlighted by respondents as being important for an effective global response to climate change. Together, these characteristics constitute responsiveness, i.e., the extent to which human actions, or inactions, are timely and of sufficient scale to fulfil the UNFCCC objective.

5.3.1   Timeliness

Timeliness refers to the extent to which the global response to climate change is happening quickly enough to achieve the UNFCCC objective without calamity or the need for extreme interventions. In this regards, Respondent 2 noted "The challenge is not to agree to do something, it's to do it and the pace at which you do it." In total, 17 scenarios from the sample of possible futures directly addressed the timeliness of the global response to climate change. This included three branching success or failure scenarios, nine success scenarios, and four failure scenarios.

The need to respond at scale between now and 2030, or even 2025, was highlighted in scenarios as being critical (Respondents 26, 13, and 9) when it came to global responses that appeared to rely on mitigation of GHG emissions. Respondent 6 noted, "2025 it's critical because, you know, the trajectories you set in 2025 towards 2035 and 2040 pretty much determine if you're gonna get... anywhere near two and a half [degrees] [or] at 1.5 or well below two [degrees]". Meanwhile, 2023 was identified by Respondent 13 as being important for the voluntary submission of revised NDCs. Respondent 9 suggested the same thing but by 2025. These scenarios follow a path similar to the IPCC (2018a) "stabilises at or below 1.5°C", although the level of warming could be higher. The period between 2020 and 2030 was also identified as being very important in the IPCC 1.5 Degree Report and Sixth Assessment Report (IPCC 2018a, 2023).

Respondent 18 put forward a scenario where GHG emissions are mitigated in the second half of the century and GHG removals are used around the end of the century to lower atmospheric concentrations of GHGs (see Figure D.3 in Section D.3.1). Respondent 18's scenario involved overshoot and drawdown, a much more extreme version of the IPCC's "temporarily exceeding 1.5°C and returning path". Overshoot and drawdown scenarios come with the risk that climate changes might not be reversible if feedbacks or tipping points are breached.

5.3.2   Scale

Scale refers to the size of the response with respect to the size of the problem and what needs to be done. Respondent 27 noted, "the scale of the problem seems to be substantial". Along the same lines, Respondent 18 noted, "We're starting, we're already experiencing impacts of climate change associated with roughly 1°C warming, and those impacts are growing in scale and in frequency and across sectors like agriculture, coastal areas, health." The scale of impacts might trigger or drive responses, for example Respondent 14 suggested "Particularly the ocean current [impact] which, because it's a very clear cause and effect there, that would galvanize action in Europe on a very large scale" referring to the possibility that the Atlantic Meridional Overturning Circulation current could slow or stop altogether, leading to rapid cooling in Europe (Collins et al. 2013).

The scale of GHG emissions mitigation and GHG removals are related. For example, Respondent 24 noted

50, 60 gigatons which is where we're heading now of carbon [removals] per annum is impossible. So, we will have to do the number one priority which we need to do anyway which is reduction of emissions, yes, to zero. Initially, to net zero and hopefully eventually to zero.

As such, mitigation (i.e., reduction of emissions) is related to the scale and viability of GHG removals. This is consistent with the IPCC findings (Section 2.3).

Given the scale of the problem including quantities of GHGs in the atmosphere, Respondent 27 noted "large scale carbon dioxide removal… is gonna have to be part of the mix" of actions in response to climate change. Respondent 14 noted

It's a few years since I stopped following greenhouse gas removal technology, but the ones that I was familiar with, all seemed to use an awful lot of energy. And I… know nothing to make me optimistic that this could be done on a grand scale.

Respondent 23 noted

I have been following this, maybe more than most co-participants. I think it is delusional, to think that it can be done at the scale that is needed. I mean, it's technologically possible, of course, but the amount of energy that needs to go into this, the amount of, where do you stuff the carbon once you've sucked it from the atmosphere and so on?

Highlighting the relationship between timing and scale, Respondent 24 noted "the longer we wait with deployment of CDR, the more challenges that might arise. Because they will take a long time…to bring these technologies to scale would require years and decades. So, let's get going as soon as possible". Respondent 18 suggested, that it may be "more towards the end of the century [that we] would really have the scale of investments in carbon removal that we need."

Respondent 18 also noted

one scenario, and the most plausible in my mind is that happens after 2100, because I think, in reality it'll probably take that long before we've gotten serious enough about the policy action and had all the investment and the technologies in manufacturing at scale so that it's more easily economic, and more easily, can direct the capital needed. But dramatic impacts up until that point, until we get stabilization.

As such, Respondent 18 identified preconditions for stabilisation of atmospheric concentrations of GHGs as being "dramatic impacts", "investment", "technologies" and "manufacturing at scale", and GHG removal technologies being "economic".

With regards to afforestation and GHG removals by trees, Respondent 3 noted "It's around tenure and it's around shifting in incentives. You know, let's do that in a scaled-up way." Taking into account land use for removals, Respondent 6 noted

So, you can only see [greenhouse gas removals] really working [at the] global scale under very high cooperation scenario because there's likely to be a need for quite significant transfers especially if you expect the value of farmland to go up.

Respondent 9 highlighted the need for scale when it comes to adaptation and mitigation, and stated

We could all but see the, you know, large companies, philanthropists, people who have access to real finance, view, or have climate change become more of their priority and they start financing climate adaptation, mitigation projects on a large scale.

Respondent 9 also addressed the scale of resilience and adaptation when it comes to food production, and stated,

food production wise, I think we have a lot of experience in terms of large scale food production in certain places that can be applied to others and we have drought resistant crops et cetera that are more able to cope with our changing climate that [if] shared more widely could mean food security for more people around the world and again in a best case scenario nations would work together to utilize these technologies to everyone's benefit.

Respondent 9 also addressed the scale of behavioural change, noting

We could see large scale change on the household level where people are eating less meat and having more carbon neutral lifestyle, fuels like prosperity of kind of a green economy that trickle down to the household level. [It] could be a bright new world [laugh].

Along the same lines, Respondent 6 highlighted the possibility of "EVs taking off [on] a massive scale" or "the technology around fake meat taking off at very significant [scale]".

5.4   Response triggers and drivers

The theme "triggers and drivers" was identified during the analysis of scenarios. Triggers and drivers refer to things that create or sustain reactions (Section 5.4.2), or non-reactions (Section 5.4.1), to climate change and its impacts. With the issues of timing and scale from Section 5.3 in mind, an important question is when might actors respond to climate change and related impacts? Triggers and drivers can help address this question.

5.4.1   *No trigger response – driven by other interests*

Cynical and non-responses are no trigger responses, where climate change impacts and risks are not considered in decisions. In such scenarios, other actors need to do more to limit climate change and its impacts, compensating for deliberate and incidental negative contributions to the global response from cynical actors and non-response actors respectively. Presumably if climate change and related impacts get sufficiently severe, then the triggers and drivers for cynical and non-response actors might change, for example discovering some ambition and taking action due to impacts or an emergency.

No trigger responses identified from the sample of possible futures (Table 5.3) consist of cynical responses, non-responses with negative incidental contributions to the global response to climate change, non-responses with positive incidental contributions, apathetic or hopeless responses, and hypocritical responses.

Table 5.3 No trigger responses and related definitions

Theme	Definition
Cynical response	Action based on self-interest, disregarding evidence on climate change and related issues, making negative contributions to the global response to climate change.
Non-response with negative incidental contributions	Human actions, and inaction, that affect the climate, where decisions to act are made for reasons other than climate change or related impacts. In this case, incidental contributions to the global response to climate change are unhelpful, for example reducing climate resilience or increasing atmospheric concentrations of greenhouse gases.
Non-response with positive incidental contributions	Human actions, and inaction, that affect the climate, where decisions to act are made for reasons other than climate change or related impacts. In this case, incidental contributions to the global response to climate change happen to be helpful, for example increasing climate resilience or limiting atmospheric concentrations of greenhouse gases.
Apathetic or hopeless response	A lack of interest, ambition or concern[1] for climate change and related issues or despair regarding the possibility of addressing climate change.
Hypocritical response	Human actions or inaction that affect the climate or the global response, by an actor that has expressed concern about the very same actions or inaction.

Cynical response

Epitomising the cynical response, Respondent 6 provided a scenario where "there's gonna be a fundamental push... promoting fossil fuels using coal, and the kind of you know incumbents climate denial type scenario." Furthermore, incumbents could "use that as a kind of organ of state craft." Such an attitude has no trigger for a response to climate change, but instead there are deliberate negative contributions to the global response to climate change.

Non-response with negative incidental contributions

The non-response theme was identified when Respondent 1 stated "it will be exogenous factors... that drive that" referring to the global response to climate change. As such, climate change is not a consideration in this scenario, yet these exogenous factors are driving the global response to climate change. Like cynical responses, this is a no-trigger response, but it is possible there could be incidental positive contributions to the global response to climate change, although given the current situation with regards to technologies and practices, it would seem more likely that there will be negative contributions to the global response to climate change overall.

The lack of a global response to climate change was highlighted in many responses as being the reason climate change is a problem and as such, non-response with negative contributions was identified as being an underlying theme. As noted by Respondent 5, "We had the science, there and then, to say that [is] the level of problem and... we need to take action. If not to take action, those are the consequences we'll be facing".

Non-response with positive incidental contributions

In the scenario provided by Respondent 1, non-response makes a positive contribution to the global response to climate change because "climate change now is just so embedded in the green economy, everyday policies and practices" and "actors don't necessarily care about climate change. I mean, it's just... This is what businesses do. This is... how we make money these days. This is the latest thing that we all want to do if we want to get funded." As such, actors are making a positive contribution to the global response to climate change but for reasons other than climate change.

Apathetic and hopeless response

Apathy and hopelessness are variants of non-responses, highlighting underlying attitudes. Apathy refers to a lack of interest, ambition or concern for climate change and related issues, meanwhile hopelessness refers to despair regarding the possibility of addressing climate change. Respondent 9 stated

further catastrophe would have probably the effect of fuelling more and inaction because then you reach a point, if you do reach a tipping point then it's even more difficult to fix the problem. So, you might even get more apathy to act where nations have said it's too late and the problems are too great.

Hence there is non-response due to an attitude of apathy. Meanwhile, Respondent 14 noted that some very large release of GHGs from the environment "sort of introduces a hopelessness at an individual or societal level which is not particularly helpful." Respondent 14 also noted that a tipping point might also end up with a non-response and attitude of "hopelessness".

Hypocritical response

The hypocritical response was not explicitly stated, but rather was derived from something Respondent 23 said while being interviewed at the climate negotiations in Katowice, specifically,

> the key actors is us who produce fossil... fuel emissions. You have a car, I have a car, I have a heating system, I travel, I fly we all do that. And, of course, we don't make those decisions totally irrespective of society. ...if everybody goes to the Bahamas for holidays then why should I not go. Right?

Given the context of the response, it highlighted the conflict between expectations, consumptive behaviours, and climate concerns. Another important reason for including hypocritical response as a category, was because cynical response was already in the list and it seemed hypocritical to omit hypocritical responses, given that most of the author's contributions to the global response are hypocritical.

5.4.2 Trigger responses – driven by climate change and other factors

Trigger responses involve decisions and actions based at least in part on climate change or related impacts. The scale and timeliness of responses depend in part on triggers that initiate a response, and drivers that sustain a response, including the urgency and ambition that come with particular responses.

Trigger responses identified from the sample of possible futures (Table 5.4) consist of impact responses, risk responses, emergency responses, enlightened responses, and cooperation responses.

Impact responses

20 impact response scenarios were identified from the sample of possible futures. Impact responses are actions, or inaction, where climate change and related effects on physical, biological or human systems are a factor in the decision to act or not. For example, Respondent 8 wondered if a major environmental catastrophe could trigger people's consciousness about the need for change (Section 5.2.2). Meanwhile Respondent 3 stated, "I guess what drives progressive action is social acceptability of climate action and what drives that, series of very high impact

Table 5.4 Trigger responses and related definitions

Theme	Definition
Impact response	Actions, or inaction, where climate change and related effects on physical, biological or human systems are a factor in the decision to act or not.
Risk response	Actions, or inaction, where the potential for climate change and related effects on physical, biological or human systems are a factor in the decision to act or not.
Emergency response	Emergency response is an urgent reaction to climate change and related impacts or risks.
Enlightened response	Enlightened responses come about from awareness that creates change. Enlightened responses can be driven by awareness of climate change and related impacts and risks as well as awareness of other things not directly related to climate change, for example the value of international cooperation.
Cooperation response	International cooperation on other issues leads to a realisation that international cooperation could help limit climate change and its impacts.

events". With regards to creating a global response, Respondent 21 stated, "I think it would probably end up having to be some extreme event happening to businesses." For example, "Some shock to the stock exchange that was clearly due to climate change. So, something happened and it wipes a bunch of, you know, companies profit zone."

Given the limited time available for effective mitigation responses without GHG removals, some fairly significant impacts would need to be experienced very soon if there is to be a timely mitigation response. Interestingly, Respondent 6 highlighted "A double bread-basket failure type event." and stated "We kinda expect to see one of those before 2025…" However, one of the problems of impact responses is that the hazards generating the impacts being experienced are likely to be locked in and further changes in the distribution of hazards experienced are also very likely to be locked in as well due to inertia (Steffen 2011). Adaptation will be required and rapid GHG removals may also be required.

Risk responses

Risk responses are actions, or inaction, where the potential for climate change and related effects on physical, biological or human systems are a factor in the decision to act or not. For example, Respondent 9 put forward a hyperbolic success scenario where there is a global response because "the scientific community says we really can't afford not to act. We are very sure that we'll reach a tipping point in 20 years and you have to… decrease emissions or we are all going die." Other risk response scenarios were less extreme, including for example a scenario that Respondent 6 referred to as a "risk-rules based scenario" where actors respond to risks.

Risk responses are based on expectations about the future including for example fears. In this regards, Respondent 14 noted

people in the society have to be scared. There has to be some fear to drive change. They've got to be scared of losing property, losing food. Not necessarily an immediate ... it's got to move the level of understanding, concern has to notch up quite a bit.

There is a question as to whether risk response scenarios relying on "fear" can drive attitudes that are cooperative rather than defensive (Section 5.5.1).

Respondent 5 put forward a scenario where there is a "massive campaign on informing people... on the climate issues to be sensitive because each individual... person has to act." The respondent gave an analogous example of advertising the risks of cigarettes, and noted that despite the wealth of the industry, people have been informed of the risks of smoking cigarettes and "many people have stopped smoking." Such a campaign would have to be initiated very soon if it was to lead to a timely risk response.

Risk responses react to potential impacts before they happen and as such risk responses, at face value, appear more timely than impact responses. However, this statement does not take into account costs or the effectiveness of responses, or the possibility of new technologies being developed that could help with the global response later. Another tricky issue raised by Respondent 15 is that "you can always, for any policy, make up very low probability scenarios and say, 'Give me all your money'" but at the same time, "we have no idea of how much our policies will... change that probability".

Emergency response

An emergency response is an urgent reaction to climate change and related impacts or risks. Respondent 13 provided a scenario where climate change and related impacts constitute a "global emergency" and noted the possibility of "emergency powers" or "emergency planning" being used to address climate change. Respondent 25 noted the possibility of responsibilities for climate change being "handed... to the emergency services". The differentiating characteristics of an emergency response are urgency and ambition. Emergency responses could be risk responses or perhaps more likely, impact responses.

Respondent 6 noted

there's no way it's a cost-benefit analysis issue that they [will] do anything. The numbers are not big enough in a cost-benefit world where everything is fungible. The only way you deal with it is if things go fundamental[ly] economic, or national security, and your ability to maintain stability internally or on your borders.

As such, Respondent 6 made a clear distinction between the effectiveness of responses based on a cost-benefit criteria, compared to responses where climate change is treated as either a fundamental economic or national security issue. In short, an impact or risk response relying on cost-benefit analyses is unlikely to be sufficient, but an emergency response might be able to generate responses of sufficient scale to limit climate change and its impacts.

Enlightened response

Respondent 26 stated, "basically you would need a flash of genius, awareness, enlightenment across all societies." and then "there's radical transformation and then, everybody just goes, 'Yeah, that's, we're all going to pull in the same direction.'"

Respondent 23 offered three enlightenment scenarios. For example, Respondent 23 noted, "It sounds like science fiction, but I really think that there needs to be reckoning. I think we need to get sufficiently close to the edge of our humanity" before we "recognize that we need to work together to help each other." In the second scenario, Respondent 23 stated that if the global response succeeds in fulfilling the UNFCCC objective, "it's likely that we get there not because of our own climate people doing, but because of a broader cultural change in planetary forces." The respondent also noted that "if we don't care about the planet, we're unlikely to care about people." And in a third scenario, Respondent 23 quoted Saleemul Huq who said "Slavery did not end because it became uneconomical or because, you know, certain new technology. It was also very centrally an embrace of a moral imperative." Respondent 23 went on to say,

> until we recognize that there needs to be a sense of justice, a sense of humanity as a driving force, if we continue to aspire to have economics or policy or regulations as the thing that takes care of our global atmosphere, I think we're going to continue to be too slow.

It should be noted that it is debated whether slavery was ended due to a moral imperative, for economic or other reasons, or a combination of reasons (Engerman 1986). However, a moral imperative was at a minimum an influence.

Respondent 8 focused on enlightened leadership, and suggested, "maybe we have a huge change in global leadership somehow and people's mentalities change, and their priorities." Related to the theme of enlightened response is the cooperation response.

Cooperation response (e.g., due to a pandemic)

Cooperation-driven responses refer to responses to climate change driven by a realisation that international cooperation is useful and can be used to limit climate change and its impacts. In the case of cooperation-driven responses, the decision to respond to climate change comes about because of some crisis, catastrophe or disaster, such as a global pandemic that gets people thinking about climate change as well. For example, Respondent 22 gave a scenario where the global response to a pandemic could "demonstrate the value of global cooperation" and this "may enhance global cooperation to deal with it." referring to climate change. With the benefit of hindsight following the COVID-19 pandemic, the assumption that a pandemic would lead to global cooperation was optimistic (see Chapter 13 for more on COVID-19).

5.5 Response attitudes and inclinations

Response attitudes and inclinations refer to ways of thinking about climate change and related issues, including predispositions influencing options considered and actions taken. Response attitudes and inclinations identified from the sample of possible futures consisted of cooperative responses, competitive responses, technological responses, practice-focused responses, and defensive responses.

5.5.1 Defensive versus cooperative

Defensive responses are reactions to climate change where actors attempt to preserve what they have and limit loss and damage. Meanwhile, cooperative responses are reactions where actors work together to address climate change and related impacts.

In two different scenarios, defensive and cooperative responses are shown as branching options. For example, Respondent 6 put forward a scenario where there is "a big fuck off event, climate-driven event" and noted, "There's a real question though whether that would drive defensive... or actually more cooperation". Likewise, in a scenario involving a spike in GHGs releases from permafrost, Respondent 21 noted this could lead to people having the attitude of "let's start doing something about it together" or the attitude of "look after ourselves". Respondent 26 suggested there could be "emergency level global cooperation where everybody is on the same page" but then noted, "It's not gonna happen. [laughter]".

With regards to cooperative attitudes and inclinations, Respondent 21 stated "I think cooperation is always going to... be much more beneficial". However, Respondent 1 noted that there has been a "top down" approach which focused too much on cooperation. Respondent 1 noted that governments have indicated

> Just wait for us, we're going to cut the grand political scheme that we're all going to agree on and once we've got that grand political scheme, we'll be able to tell you what you need to do in terms of emissions reduction.

The respondent noted "that's hugely inhibiting. I think it prevents innovation, encourages a kind of a slow gradual emissions decrease rather than thinking more creatively." The time required to organise and agree to cooperate is an issue when it comes to timeliness. Issues of ambition, non-participation (e.g., withdrawal), and non-compliance (i.e., cheating) also mean cooperative attitude alone, may have trouble creating a global response of sufficient scale to address climate change.

With regards to defensive attitudes and inclinations, Respondent 6 noted that a crisis or catastrophe can drive people into security mode, and "The general thing that happens when people go into a security mode is they get short-termist, distrusting and [make] defensive investment[s] and so it's never good for anything long term or cooperative."

Respondent 6 noted concerns around security can drive a defensive attitude and that a defensive response can lead to a situation that is "distrustful, populist, very much about we wanna keep our cake, we don't wanna share our cake with anybody else. Lots of walls. Lots of fences. Generally, quite unpleasant." In a scenario involving the fragmentation of international cooperation, Respondent 21 noted "The response to that is to become much more defensive and trying to sort things out by yourself and your own region."

Respondent 11 noted, "people with power, people who are in the defence industry, they want to present this as a security problem, not as a climate change problem." The reason being that, they "can build more aeroplanes and build bigger walls" and as such, "they make money out of that." Respondent 11 also noted that when things get bad, that's when people with the means might take a defensive attitude, for example, "when food is not available, or when temperatures become unbearable or when wealthy people are perceived to be protecting themselves behind high walls". This might also happen at the national level according to Respondent 3 who put forward a scenario where "countries that are doing better... are seeking to retain that by building more insular sort of, you know... you get islands of wealth and... expanding areas of governments falling apart". Defensive attitudes, at the cost of cooperation, could generate adaptive responses and building of resilience, but only for those individuals, groups or states that have resources to support this. A defensive attitude seems very unlikely to be able to generate an effective global response of any sort, but rather appears likely to generate a fragmented response, one that is unlikely to be timely or at sufficient scale.

5.5.2 *Technological versus practice focused (including nature-based solutions)*

From the sample of possible futures, a technological response attitude or inclination is where actors expect there are physical, chemical, or biological properties, related objects, knowledge and processes that can be used at will to solve the problem.

Respondent 24 noted, "there are some people who believe that tomorrow there will be a technological miracle that will solve our problems". Similarly, Respondent 17 noted some people think "There must be a technical fix for everything, right?" As such, some actors seek technology-based solutions to climate change.

With regards to timeliness, Respondent 15 noted the need to "dramatically increase... investment in research and development.", specifically in "the long shot things that'll actually work out over the next couple of decades." so that there are geoengineering and other options to address any unexpected climate changes such and feedbacks or tipping points.

With regards to the feasibility of technological options, Respondent 21 noted

it seems to me that it would be simpler to try to scale negative emissions technologies, than it would be to rely on us all changing who we are, 'cause we know that it's very difficult for people to change.

However, while Respondent 21 claimed "I'm with the sort of technology optimists" the respondent also noted "I think it can achieve a lot, but it's not gonna achieve the fundamental underlying changes that are required for us to stop worrying about you know, ecosystems and that sort of thing."

While Respondents 24 and 17 clearly addressed technological attitudes in their responses, practice focused responses appeared to be more of a respondent inclination, seemingly mentioned as a counterpoint to technological attitudes. Initially, practice-focused inclinations were not highlighted as part of the thematic analysis. But given that practice focused attitudes and inclinations were demonstrated by the respondents they have been included below.

A practice-focused response attitude or inclination is where actors expect there are human, ecological or environmental systems knowledge and processes that can be used to solve the problem. Practice-focused responses include behaviour-focused responses as well as nature-based (i.e., environmental) responses and often involve an mix of personal responsibility, policies or resources management.

Respondent 20 stated "Now, I'm quite optimistic that technologies will help us but also, and maybe changes in philosophy and behaviour." Respondent 1 went further and asked the question "…why not just focus on actions? Why not just focus on practices? Why not just focus on making environmental practices more appealing, more positive more incentivized, whatever we want to?"

Respondent 3 said "…we've had a technology that we've known about… it's the cheapest technology" and stated that the "Stern report achieved a level of consensus that forests [are] a good thing so we should have more of them". Likewise, Respondent 14 stated "Well, [the] best removal system is photosynthesis and green growth…" and suggested, "you've probably got some potential for co- opting the public into helping with that… [It] might be digging up your path and planting it for instance…"

With regards to challenges, Respondent 8 highlighted tree planting and soil management require government policy (Section 6.3.12) and Respondent 24 noted nature-based solutions require a lot of work (Section 7.3.3). Meanwhile, Respondent 15 expressed scepticism regarding practice-based responses and stated "the amount of impact from a personal choice is going to be fairly small…" due to the "rebound" effect discussed in economic literature (Section 6.2.5).

5.5.3 Competitive

Competitive response attitudes or inclinations are visible where actors have expectations of some reward, which drives activities and engagement in technologies or practices that make a positive contribution to the global response.

Several scenarios included competitive attitudes, while several others highlighted how markets might be incentivised, driving a competitive response to climate change with positive contributions to the global response (i.e., as a co-benefit).

In a scenario featuring competition, Respondent 27 highlighted that a competitive attitude could include "[the] perception of first mover advantage", "natural one upmanship or one up personship" and "looking at what your competitors doing". Importantly, "competition is stimulated" by seeing "who can invest the most in a bunch of projects, because they're gonna make the most money in the future". Hence competition could influence and generate a coherent global response to climate change. Importantly, competition creates "a race to the bottom of emissions". In such a scenario, where a "winning team" is formed around a technology, including business and government, it is possible that "there's a magic moment... or there's just a tipping point, you know, in terms of the proliferation of clean tech and the other things."

Respondent 6 contrasted competition with cooperation noting "There is an alternative scenario... it gets us quite close to [the UNFCCC objective], which is perhaps less cooperative so more a slightly competitive" scenario with a "much stronger role for cities and companies".

From the scenarios, competitive attitudes hold the potential to scale action that can make positive contributions to the global response, including for example technologies, but this depends on incentives structures. Conversely, a competitive attitude might even drive a cynical response and the scaling of GHG-emitting or climate-vulnerable technologies, business models or interventions.

5.6 Conclusions

The results presented in this chapter help address the question of what influence climate change might have on actors and the global response to climate change (Section 1.5). From Section 5.2 there are a range of possible impact and risk scenarios, and responses too. Factors influencing these scenarios included climate sensitivity and distribution of impacts and risks, as well as the power of actors responding. Section 5.3 highlighted the extent to which responses might be timely or at scale, noting these are essential characteristics of responsiveness. Meanwhile, Sections 5.4 and 5.5 showed there are a range of possible response triggers and drivers as well as response attitudes and inclinations, each of which influence the type of response and its likely effectiveness. These range from cynical responses and apathetic responses through to impact and risk responses, or even emergency responses. The influence climate change impacts and risks might have on the global response to climate change are analysed in more detail in Chapters 10 and 11.

The themes from this chapter form a basis for the analysis in Chapter 10 and the development of a climate change signal response model in Section 10.2. Furthermore, the themes from this chapter also form a basis for the discussion in Chapter 11 regarding climate change signals (Section 11.5), as well as the influence of climate change on responses (Section 11.6).

Note

1 Adapted from the Oxford Dictionary definition for "apathy" which is "Lack of interest, enthusiasm, or concern." In the context of the global response to climate change, the word ambition substitutes for enthusiasm.

6 Actors and interests in climate scenarios

6.1 Introduction

Like Chapter 4, this chapter looks into the crystal ball (i.e., sample of possible futures) but this time we explore climate change related actors and their interests.

You may recall from the Introduction in Chapter 1 that "We are all actors" in the global response to climate change. Chapter 3 noted the importance of state and non-state actors in the global response to climate change and acknowledged the influence of international institutions. From Section 3.3.4, much of the global response to climate change happens at the domestic level by actors in states and the global response to climate change is assessed as the sum of responses by these actors.

This chapter looks at state and non-state actors identified in scenarios by respondents, including social, political, and business actors. You may also recall from Chapter 4, that each scenario is like a short story on how the future might unfold. In many of these scenarios, the respondent explained the things actors cared about, and how these interests influenced their actions.

From the analysis of domestic response scenarios and national interests, social actors contributed to climate change through their behaviour (Section 6.2), political actors contributed to climate change according to their political will (Section 6.3) and businesses contributed to the global response through their economic activities (Section 6.4). In each of these sections, between 9 and 15 themes are identified. Related scenarios from different respondents are contrasted. These themes and scenarios are diverse including, overconsumption and social contracts, energy and land use, technology and insurance withdrawal. Section 6.5 concludes with a summary of the themes from this chapter and indicates where these themes are analysed, and brought together, in Chapters 10 and 11.

Given that the themes and scenarios presented below are much like a compendium of short stories, you may want to read a few examples and then move straight to Part III to see how lessons taken from these scenarios link up. Alternatively, you might take a deep dive looking into themes such as social change and behaviour, political will and policy or business and economic activity, and read all related scenarios.

DOI: 10.4324/9781003465911-8

6.2   Social change and behaviour

The overarching theme "social change and behaviour" addresses the interests of people, as individuals, households and communities, and the actions, or inactions, these people might individually or collectively take (Table 6.1). Illustrating the concept of social change, Respondent 23 suggested "there needs to be a sense of justice, a sense of humanity as a driving force" in society for there to be an effective global response to climate change. Respondent 6 noted the need for a "social contract". With regards to behaviour, Respondent 9 noted the impact of behaviours is small on an "individual scale, but large scale [if] everyone does them."

From the sample of possible futures, social change and behaviour can help put the global response to climate change on path towards fulfilling the UNFCCC objective, for example in the case of enlightenment. Alternatively, social change and behaviour can keep the global response to climate change on path towards failure in the case of overconsumption or social unrest.

Social change and behaviour-related themes consist of: triggers and drivers (Section 6.2.1); leadership and policy-driven social change (Section 6.2.2); enlightenment (Section 6.2.3); youth-led social movements (Section 6.2.4); behavioural responses (Section 6.2.5); overconsumption (Section 6.2.6); social disruption (Section 6.2.7); inequity, exposure and vulnerability (Section 6.2.8); and social contracts (Section 6.2.9).

6.2.1   *Triggers and drivers of social change and behaviour*

Based on the sample of possible futures, it is possible that social change and behaviour can be triggered or driven by impacts, risks, or enlightenment including cooperation responses. For example, Respondent 14 stated that if people are scared of climate risks, this will "get people to take individual action" and "get people to …drive full political action." Respondent 5 provided a risk response-driven social change and behaviour scenario involving a massive campaign informing people about climate change (Section 5.4.2). Respondent 9 noted, "I think there's more pain mainly for the lower classes before the upper classes, the power holders, are forced to change." Similarly, Respondent 20 noted "it [i.e., climate change] will probably only have true political ramifications when it starts to impact wealthier

Table 6.1 Social change and behaviour themes with descriptions

Theme	Definition
Social change	Social change refers to a shift in common values, norms and expectations that influence the behaviours of individuals, households, groups and communities, businesses and governments within a society. In short, it's a shift in the way individuals and groups live together.
Behaviour	Actions, and inaction, by individuals and households. This includes purchasing choices as a consumer. This can also be thought of as the options taken or not taken by individuals and households.

parts of the community", and "in a western essentially liberal society... it could still lead to political turmoil".

At the same time, some things could undermine a response, for example, Respondent 3 noted that

> there are all sorts of, you know, scandals that could emerge that could severely undermine the trust that everyone has within the system and the trust of science, the trust of whether actors are doing what they say they would do. A big scandal with, you know, an emission trading scheme.

Respondent 12 noted that "certain groups are claiming that it's gonna lose our islands" and

> If you really look at the literature about SLR[1], there's still a lot of uncertainty about islands disappearing and we need to be careful around that so what if even if we're so-called runaway climate change, more than two degrees, you'll only lose a few islands.

In such a scenario, "that's not gonna be good with respect to the credibility of the groups who are using that narrative to drum up support for climate policy and action."

6.2.2 Leadership and policy-driven social change

Only one of the success scenarios involving social change and behaviour was not triggered or driven by climate change impacts or risks. Rather, it was a response to leadership, political will, and policy. Respondent 21 noted it is possible that "someone comes in as a Democratic candidate that has more extreme views on a Green Deal" and once the policy is implemented and people see "the benefits of when this gets put into place" then, "there's sort of a groundswell of the population which are actually on board that no longer see this as sort of a costly burden on the US economy but see the benefits that sort of green revolution can have."

6.2.3 Enlightenment

In the previous chapter, Section 5.4.2 addressed enlightened responses to climate change. While an enlightened response may be part of a wider global shift, each state and set of domestic actors forming a society would have a slightly different enlightened response. For example, Respondent 26 noted that there could be a recognition that "there needs to be a change in how we live in society in terms of travel, how we eat", "all businesses shareholders recognize that you know, you can't get your pension or your investment returns from things that are contributing to climate change.", and "There is a recognition that there needs to be some redistribution of wealth". As a result, "there's radical transformation" because "everybody just goes, 'Yeah, we're all going to pull in the same direction.'"

6.2.4   *Youth-led movement (social change)*

The possibility of youth-led social change was featured, for example in the branching success or failure scenario from Respondent 23. In this scenario "The generation of our children is going to not give a damn what is expected of them" and they will "take what they want by taking it." This could result in overconsumption due to a competitive attitude of "I want to succeed and have, you know, the most expensive vehicle". Alternatively, this could lead to "socially-oriented youth, becoming a force of nature" and a "radical collective transformation for the wellbeing of people and planet."

Likewise, Respondent 18 put forward a scenario where "there's really widespread public mobilization throughout all countries to act in a much more urgent and ambitious way than is happening now". And Respondent 25 suggested that it is possible

> Enough people in the United States, perhaps UK, and Brazil, South Africa, India, young people, hopefully, of my kids' generation… just say, 'You guys have made a big fucking mess, get out of the way we're taking over and this is what we're gonna do.'

6.2.5   *Behavioural responses*

A range of possible behavioural responses were highlighted that could make a positive contribution to the global response to climate change. For example, Respondent 18 noted behaviours individuals and groups could take including, "personal responsibility for the actions that we take", "demand accountability from their leaders, both political leaders and business leaders" and "responsibility to speak through voting." Likewise, Respondent 8 had a scenario where there is "a shift from a lot of consumption to reducing general consumption and being more conscious about what people purchase…" From Section 6.2, Respondent 9 noted that large-scale changes can happen if everyone changes their behaviour. Respondent 9 also suggested behavioural changes "like eating less meat and [heat] saving or having more energy efficient products in our lives, having better, and using more sustainable transit options, or divesting their finance." From Section 5.3.2, Respondent 9 noted the possibility that philanthropists and large companies could start financing adaptation and mitigation at scale.

Respondent 8 noted for there to be changes in consumption there needs to be "a confidence [in]… technologies to replace fossil fuels and things like that". Respondent 8 also highlighted that "we need… consumption signalling… a change." to businesses. Respondent 18 included a range of roles for individuals and households, civil society, business in a successful global response scenario (Table 6.2).

Respondent 15 doubted it is possible to have a global response to climate change at scale coming from social change and behaviour. Respondent 15 noted "Look, there's a lot of small things that can have a tiny impact." including virtuous things

Table 6.2 Behavioural responses including actors, roles and actions

Actors	Responses (roles and actions)
Individuals and households	"personal responsibility for the actions that we take." "a critical role for educating people, for people to motivate and inspire each other" "speak through voting." "demand accountability from their leaders, both political leaders and business leaders."
Civil society	"educate themselves." "hold their politicians accountable" "hold the business community accountable through their investments, through liability and through their purchase decisions."
Business and economy	"a huge shift towards purchasing based on carbon impacts and carbon leadership associated with products and services."

Source: Author and Respondent 18.

such as switching "off the light" or having "short showers". The respondent noted "there is a tendency for people to also believe that 'Oh, now I've done something virtuous. Now, I'm also allowed to do something wicked'." which undoes the benefit of the virtuous act. The respondent also noted if money is saved by these virtuous behaviours "I'm gonna use that for something else which is also going to emit CO_2." Hence, the "rebound" effect discussed in economic literature. The respondent stated,

the fundamental point is that even if you manage to get people really concerned or really focused on this, the amount of impact from a personal choice is going to be fairly small... So, fundamentally this is not about changing behaviour... society shouldn't... have to worry about all kinds of things.

However, Respondents 6 and 26 provided a list of behaviour and related scenarios that might contribute to effective global responses to climate change (Table 6.3).

6.2.6 Overconsumption

Not all social changes or behaviours in the sample of possible futures were positive. For example, a couple of respondents (5 and 21) expressed concern on issues of consumption. Respondent 5 noted "...there's no discussion on over-consumption..." or "overproduction" and highlighted that "overconsumption has not been addressed or discussed over the UNFCCC" and has "not been given any prominent role in the SDGs". The respondent noted that overconsumption is considered a "good thing for society" as it drives economic activity, while the only "bad thing is poverty". Respondent 5 also expressed concern about China's "over-consumption" following the patterns of Western countries before them. Respondent 21 expressed concern about consumption in "Asian countries" noting that a successful scenario would

Table 6.3 Possible behaviour-related scenarios

Themes	Quotes
Behaviour, technology	"EVs taking off [on] a massive scale"[a]
Social change, behaviour	"big shifts in consumer preferences"[a]
Behaviour, technology	"the technology around fake meat taking off at very significant [scale]"[a]
Social change, behaviour	"so, a large youth based social movements around diet."[a]
Leadership, political will and policy	"I'll say, kind of, cities really taking a lead, for multiple reasons, to drive development and deployment of new technology"[a]
Social change, political will and policy	"Trump doesn't get elected and then [laughs]" and "America re-joins"[b]
Social change, political will and policy	"they listen to the young."[b]
Legal precedent	"There's been two or three litigation cases with children or young people holding to account their governments and businesses"[b]
Social change, political will and policy	"there's been a recognition that the climate justice movement is right"[b]
Social change, political will and policy	"there's been a recognition that... the rights of Mother Nature needs to be respected as well"[b]
Enlightenment	"we become much more aware of our interconnection with global ecosystems"[b]

Sources: Author, [a]Respondent 6 and [b]Respondent 26.

include these countries "skipping or jumping ahead..." and consumption trends taking "...a slightly different direction" from the West.

In Section 6.2.5, Respondent 8 noted the possibility of a behavioural response where consumption is reduced but also noted that society needs to be confident that substitute technologies will provide the same utility. Meanwhile, Respondent 11 noted

> it's hard to resolve the amount of consumption, and the consumption, kind of, is related to our economic utilization of resources, which is related to climate change. Can we change our thinking about that without a more profound sort of spiritual change?

6.2.7 Social disruption

In a situation where there is food insecurity and wealthy individuals and households are defending themselves and their property, Respondent 11 stated "then you see that turning into social disruption.", and "I think that that could happen in our cities. That could happen in different places". Likewise, Respondent 24 put forward a scenario where "the economic costs and therefore the hardship on people is substantial and as these extreme events will increase... at some point... worst case scenario, I think governments will break down." Respondent 11 also noted "the people who are feeling the pain are not the ones that have the power." Hence "there's more pain mainly for the lower classes before the upper classes, the power holders, are forced to change."

6.2.8   *Inequity, exposure and vulnerability*

In a sea level rise scenario where there is insurance and public service withdrawal from coastal areas, Respondent 20 noted "more wealthy people can just avoid the problem by moving out", meanwhile,

> people who are owner-occupiers of their coastal properties, maybe left behind and in fact, precisely because house prices, the value of coastal properties will decrease. People who don't have much money may actually go to these areas... because they are now affordable to them.

Furthermore, it is possible that "perversely we actually might increase the hazards on the coast because of the stratification" "because of market forces and people are going to where there is affordable housing even if the housing is in potentially a hazardous area". Likewise, Respondent 10 noted, there could be a scenario where, "Maybe some places they're better adapted than others."

6.2.9   *Social contracts*

Respondent 24 noted that "there are pathways that are feasible, but it requires a kind of social contract of the global level that we just simply don't have." This social contract needs to acknowledge that

> ... We do need economic development... we do need to figure out our problems, we do need to have jobs for people all of that is a reality. Now, I know that [at] the theoretical level it is totally possible to do that.

Linking social change and behaviour with political will and policy as well as business and economic activity, Respondent 6 stated,

> it's critical to have a clear social contract on the transition, which is... a clear sense of agreement between the population and government and affected groups both affected by climate change and affected by the transition - that there is a fair deal between those elements.

However, Respondent 6 also noted that currently "we don't have the social permissions to get to well below two" degrees Celsius of global warming. In short, without a social contract, the global response to climate change will lack scale and timeliness.

Respondent 17 noted, "the Trump voters, are more in the distrust camp... they don't want to help too much because they feel like somebody's going to be a freeloader." Meanwhile the respondent noted "I'm like, 'Yeah. There's going to be freeloaders. Who cares? That's life, right?'" As such, the question of freeloading is something a social contract will need to address.

6.3   Political will and policy

Political will and policy refers to the ambition level of government leaders, and others in government when it comes to positions on climate change, and the interventions they make (Table 6.4). Respondents expressed disappointment regarding political will and policy responses to climate change and frustration that political leaders had not done more to address climate change given the threat of climate change. Respondent 25 stated, "We have the technology to solve the problem... while you can argue 20 years ago that it was a technological issue, it's not anymore. It's purely a political issue". The respondent went on to state "I think the main thing that we're missing at the moment is leadership, political leadership." Likewise, Respondent 2 stated,

> The impact of a changing climate is likely to pose one of the greatest strategic threats or challenges, perhaps is a better word, to nations. The question is are nations, and in particular leaders, up to, and capable of, making the necessary decisions, some of which will be painful, and have impact on our lifestyle?

Respondent 24 was very succinct regarding political will and policy, "I think the leaders have failed to do that. It's as simple and as complicated as that."

The overarching theme "political will and policy" includes an analysis of political will and policy as a response to stress signals highlighted in IPCC assessments (Section 6.3.1). Other political will and policy themes identified from the sample of possible futures consist of: emergency responses (Section 6.3.2); social and business influences on political will and policy (Section 6.3.3); institutions and development (Section 6.3.4); policy directions (Section 6.3.5); technology-driven political will (Section 6.3.6); policies on technology and practices (Section 6.3.7); energy policy (Section 6.3.8); infrastructure policy (Section 6.3.9); public service withdrawal (Section 6.3.10); localisation (Section 6.3.11); agriculture, forestry, and land use (Section 6.3.12); planning (Section 6.3.13); failure of political will and policy (Section 6.3.14); and defence (Section 6.3.15).

Note: Political will and policy scenarios included central government and local government. Scenarios addressing international cooperation between states, or between local government bodies from different states, are addressed in Chapter 8.

Table 6.4 Political will and policy themes with descriptions

Theme	Definition
Political will	The ambition level of government leaders, and others in government, to act on climate change and related options.
Policy	Positions and interventions.

6.3.1   Political will and policy as a response to climate stress signals

From the sample of possible futures, scenarios involving political will and policy included emergency responses to impacts on human and managed systems (Section 6.3.2) as well as responses to social and business interests (Section 6.3.3). With regards to impacts on human and managed systems, Respondent 24 also noted

> the attribution science is getting better and the probability assigned to a particular event because of climate change is actually increasing. So, I guess the point I'm making is that what we are seeing today in many parts of the world is increasingly linked to climate change directly.

This is consistent with Table A.1 from Appendix A which summarises the changing state of knowledge and information on climate science, risks and impacts as published in successive IPCC assessment reports. Given the sample of possible futures has no scenarios where there is a response to impacts or risks to physical systems, information on climate science, in and of itself, is unlikely to generate substantial political will or policy response[2].

6.3.2   Emergency response

Respondent 2 indicated the possible preconditions for political will to emerge include impacts on human systems when a warning light "starts flashing amber or red [for] the politicians around the world". The respondent also highlighted social change where "a government might find itself... under huge pressure from its citizens because it's done nothing to ensure quality of health... to the point at which they know they won't be re-elected or something like that." As such the link to impacts on human systems is made along with social change and behaviour in the form of voting. Respondent 2 noted that political will for policy comes starts at "The point at which the risks associated with doing nothing are greater than those of doing something" and

> a recognition that... there is an issue in whatever form it takes, that unless you... take action to do something, the impact on your well-being and prosperity will be so great that it'll outweigh any potential constraint imposed on you by acting.

6.3.3   Social and business influences on political will and policy

Respondent 27 noted politicians respond "to the appetite that their constituents, in the form of businesses, investors, as well as citizens" and the things constituents have "mandated them to do" and as such politicians respond to social change. Respondent 27 also noted that

> In terms of sort of de-carbonizing very quickly... there are signs of political will to drive this. Political will that's based on, not just politicians out there

for a legacy, but politicians responding to the appetite that their constituents, in the form of businesses, investors, as well as citizens, and have kind of mandated them to... from very small towns to cities to even... national governments as well.

Contrasting behaviour with political will and business interests, Respondent 21 suggested: "Individual behaviour, like that doesn't really have much to do with it. That is about choosing what your policy response is". The respondent then stated "it comes down to probably government regulation and a handful of companies deciding that this is the approach that should be taken". As such, the respondent indicated that a precondition for an effective global response is having powerful and influential political and business actors driving the response rather than social change or behaviour as a precondition.

Respondent 5 noted "the policy maker would only act under public pressure, if you have a strong public opinion," hence social change can create political will and the voting behaviours in democracies could also create political will. However, Respondent 5 also noted, "public opinion can come only if they're informed, if they're aware." Respondent 18 noted that political will created through social change would make it possible to "put in place the policy regime and really the mandatory commitments needed, and we're holding the businesses accountable and businesses you know, are fully kind of invested in this solution".

Respondent 9 provided a scenario where impacts on human systems and a humanitarian outcry "pushes people to elect officials that think addressing climate change is a top priority in nations across the world", i.e., "Voting for people who think climate change is the priority." Respondent 9 went on to say

and then we see countries not only fulfil their commitments under Paris Agreement but after a review of those commitments when science tells us that we still have more to do, they put forward in 2025 a commitment that will bring us in line with the temperature reductions we need. Then they work to meet those.

Respondent 11 stated that if "the situation becomes desperate enough that... the public demands change of their leaders" then "leaders can enforce change even when certain companies... are gonna suffer from that." The respondent provided the analogy of

the breakup of AT& T, the breakup of Standard Oil, you know, it's not that the government can't address, large economic interests, they can. They have the power. But it has to be quite a clear issue and a lot of political support.

Respondent 21 noted that when there has been social change, society could state "what's acceptable and what's not acceptable to do and what needs to be done and put pressure on governments". Respondent 21 also suggested that if some sort of Green Deal is successfully implemented in the United States it could lead to a groundswell of support for a green revolution (Section 6.2.2). In this global response scenario there is a proactive US and European response with "support from industries and individuals", and with enough support, change could be "quite fast".

6.3.4 Institutions and development

In one scenario, development levels and domestic institutions were highlighted as possible preconditions to an effective domestic response. Specifically, Respondent 21 noted that for some countries responding to climate change is "going to be very difficult cause they're not quite at the same level of development that, or even have the same political systems that would allow that to happen." and

> because they're still going through the levels of development and require high levels of growth, in India, that sort of thing. Then... how does that growth happen? ...that probably comes down to a number of things with finance being a huge thing, people being willing to invest and also some levels of regulation.

In a scenario where impacts on human and managed systems have resulted in internal displacement and migration, Respondent 6 stated "elected officials have to respond" including supporting "adaptation to the extent they can" and this could include "solar radiation management". In this scenario, ultimately there is further "hardship on people" and in a "worst-case scenario" there would be social unrest and institutional failure as "governments will break down."

6.3.5 Policy directions

Many scenarios included policy options or actions, but few described the characteristics of what constitutes policy. However, back in the first paragraph of Section 6.3, Respondent 2 noted that policy addressing strategic issues such as climate change is usually decided by political leaders and can have impacts on every part of life, including individual lifestyles. Meanwhile, Respondent 25 mentioned "direction from governments" in the context of markets, highlighting an important aspect of policy, which is direction. Policies identified in different scenarios included a "carbon price", a "carbon tax", "subsidies", "NDCs", the funding of "research and development", "massive tree planting", as well as positions related to international cooperation.

6.3.6 Technology-driven political will

From Section 5.5.3, Respondent 27 provided a scenario where technology drives competition and "a race to the bottom of emissions". Respondent 27 also noted there could be "competition between nation states" as well as competition between economies, businesses, and investors.

6.3.7 Policies on technology and practices

Respondent 11 stated "I think that there is a place for government to create the right incentives and investment and promote investment in those kinds of technologies but leave space for the private sector to actually accelerate that and to produce that." Respondent 11 also noted "the free market could play a very positive role... but I don't think that the free market will unilaterally create the technologies or invest in those technologies, to bring about that level of change with... negative [emissions technologies]".

In Section 5.5.2, Respondent 15 noted there need for governments to invest heavily in research and development of "long shot things" rather than "close to market technology". Respondent 15 noted "If we do that, we have a much greater chance" of achieving the UNFCCC objective.

Respondent 11 put forward a scenario where "some of those countries that have a history of innovation... get them to be thinking about this, the way that universities and departments of engineering are incentivized by government grants." The respondent stated, "it should be investing and giving the signals to the private sector in those countries or globally that this is the direction we're going and we are going to be investing, we are going to be procuring." Maybe, "change will happen much faster than we think away from a carbon-based economy". And as a result, there is a "tipping point" for example with "renewable energies" or "food production systems and the reaction against, sort of, what I call industrialized food".

6.3.8 Energy policy

Respondent 6 stated with regards to achieving the UNFCCC objective, "the only way you get into the position [is] to take on and retire existing assets." Furthermore, it is possible that incremental energy demand "is met by renewables and efficiency.", but "What is necessary is we get rid of coal power stations, which are 15 years old." As such, some technologies may need to be excluded from use.

In another scenario, where a defensive policy position is adopted, according to Respondent 6, governments "don't do the aggressive" climate change related policies required, for example they don't unravel "existing fossil actors because of their role as national champions". Then with time "the price on the economy" makes action "much harder", hence "you might be deploying clean tech[nology] a lot but you are probably not getting rid of dirty [technology]." Respondent 1 highlighted that actors may

> Retreat to what you can provide internally or with people you trust.... renewable energy technology is not produced everywhere... for many countries around the world, fossil fuels remain most easily accessible national resource of energy... people would maintain their own economic self-interest more.

Then, "All of those would speak to a decline rapid decline in any ability to achieve emissions targets."

In Section 5.4.1 Respondent 6 put forward a cynical response scenario where states promote fossil fuels despite climate change. Respondent 6 noted this scenario could include "US, Australia, Russia, Saudi (after the Saudi leadership change)... India on a bad day, Japan". As such, it is possible states could privilege special interests in the energy sector ahead of addressing climate change.

6.3.9 Infrastructure policy

Respondent 27 noted that sub-national governments could be important investors in low greenhouse gas emissions infrastructure, along with businesses and financiers. The timing of investment was noted as being very important. Specifically, the respondent noted, "If that's not done in the next decade... we're locked into higher emissions ...for another 30, 50 years" and noted, "some might argue that it's game over in some ways...". However, this assumes that the infrastructure is actually used for its design life and does not become "stranded".

6.3.10 Public service withdrawal

In a scenario where there is insurance withdrawal in advance of climate change impacts, Respondent 20 noted cascading effects might include, "city authorities themselves may start to withdraw maintenance of their water and sanitation and certain other services and road maintenance and things like this". Such public service withdrawal could be abrupt especially as climate change related impacts start manifesting themselves unambiguously. In situations where sea level rises, and other impacts, are anticipated and predictable, public service withdrawal may be scheduled years in advance.

6.3.11 Localisation

Localisation is where non-state actors are empowered to develop and respond to climate change and other issues at the local level. For example, Respondent 17 noted

> if you just ask... how people would want to design their communities if they could... what you find is that even though they might be focusing on something like safety in the case of cycling, or just a clean environment, which doesn't necessarily have to be, pro-climate... the outcome of a lot of this will be in fact, very sustainable. And so, the cumulative effect of allowing local people to decide is that suddenly you end up with a system where nobody wants to have anything dumped on them, right? And so suddenly, we have to rethink our industrialized world, right? People will want to have healthy communities, they want to have good jobs.

Respondent 17 also noted that "if you really focus on this local decision-making, you would have greater satisfaction overall, and you would have a discussion about the values across these rifts that we have today." The respondent also stated "From what I understand, from democracy theory or whatever, there's a bit of a consensus that democracy works best in spoonfuls. At small levels. And the more you go to these gigantic state things like the European Union and whatever, the more cumbersome it becomes. It's more distant" to people.

With these things in mind, the respondent suggested that when it comes to "International trade laws… the last word should be at the local level." However, the respondent also acknowledged that there are certain things such as "standards for industry" that may be best addressed globally. The respondent noted that "there's a role for corporations." For example, "community solar projects or wind farms or whatever are installing turbines from big corporations.", and as a practical matter, "local citizenry… might come up with a solar roof for the school, but they're not going to come up with an electric vehicle." As a result of localisation, Respondent 17 thought local businesses "would thrive pretty well." The scenario put forward by Respondent 17 focused very much on social and community empowerment.

From Section 5.5.3, Respondent 6 provided a success scenario involving a level of localisation and competition. According to Respondent 6, in this scenario, businesses and local government "would need to be given the ability to raise money and spend money and make choices, to exercise their so-called purchasing power." and there would be "distribution of power away from centralized decision-making allowing people to adopt technologies much faster." Furthermore, "at the moment, it's seen as a complement for government but it needs to get a lot more radical to be a big replacement for central state power." Then, "large corporates and cities using that purchasing power [on] both infrastructure and energy and food to drive change into the markets as they are the major drivers of economic growth and wealth." This could lead to "deploy, digitize resilient joined up infrastructure" and coupled with a "bottom-up disruption and for that has to [be a] much stronger pushed from consumers to EVs and resilience in agriculture.", then it "doesn't perhaps get you all the way to well below two, but it gets you kind of near it nearing between the two and the two point five."

With regards to motivations, the respondent stated, "For cities it's about being liveable and attractive [for] talent and investment. For companies, it's about having supply chains for investors it is about judging their portfolios." Respondent 6 also noted

we can't do a lot about geopolitics, but [we] can do quite a lot about decisions made at local level and company level as non-government activists and social movements so it's a complement of plan A at the moment. But… if it was really deployed as a plan B, it could make a significant difference.

The respondent did note that "if we're in a really bad geopolitical situation [with] the rise of autocracy… it's very hard to see that one working."

In contrast with Respondent 17's social and community empowerment, Respondent 6's scenario is much more focused on empowering existing local government and business actors to address issues locally. In either case, sovereign states would need to empower these actors, and decentralise decision making – hence the theme "localisation".

6.3.12   Agriculture forestry and land use

With regards to land use related responses and actors, Respondent 8 noted that "ultimately it's the national governments... if you are gonna do massive tree planting... you've got all these issues around land tenure and compensation... I think it's quite complicated". Only by addressing these things is it possible to have "A lot of tree planting, I guess a lot of soil management techniques." are required for an effective global response to climate change.

6.3.13   Planning

After highlighting the likelihood of failing to adapt to climate change, Respondent 6 went on to say, "Coherent resilience plans that stress tests to three [or] four degrees and look to build in national and international resilience in those things" are required, and "of course, that has the other benefit of making the impacts of carbon policy failure much clearer at the national level." because risks of policy failure are identified, for example in reports related to stress tests. Respondent 6 stated, "success would look like for us is a kind of coherent [resilience plans], like we have, coherent low carbon plans in countries."

6.3.14   Failure of political will and policy

Failure scenarios included a lack of social change and political will. For example, Respondent 21 noted there could be a lack of social change because people are "not feeling the pressure on them to make the changes" and voters have "other priorities". There is a lack of political will and policy "not anywhere near as sufficient or fast to achieve stabilization of CO_2." In such a scenario we could "see a path where we would end up looking at a sort of three and a half degree [Celsius] world by the end of the century."

Respondent 20 highlighted the issue of sea level rise, rising groundwater and flooding in the short term as well as changing coastlines which "our decedents will have to deal with" Respondent 20 noted that in "the best-case scenario... sea level rise happens at a rate that is slow enough for our political and social economic systems to adapt to." However, "the worst case would be political stagnation and trying to deal with this at a policy level" indicating political will may be lacking and policy may not be effective in addressing rapid sea level rise.

6.3.15   Defence

With regards to defensive attitudes and policies, Respondent 25 noted "democratically or whatever elected governments... have a few years to sort this out because if we don't, we assume the whole issue gets handed, first to the emergency services" and "then ultimately, to the military... I mean, it becomes one where militaries from the rich countries are employed for a crowd control." resulting in "fortress Europe, fortress North America, fortress Japan, fortress North China". In such a scenario

> Business is fucked... The only thing that will matter is that they, how they can cozy up to whoever is in power for their own national markets. I mean, the global marketplace under which most modern businesses seeks to thrive or whatever, ceases to exist in any real way.

Meanwhile, when it comes to to civil society, "They mostly get lined up and shot; if they speak out of turn." The respondent cautioned that an emergency response could be defensive, fragmented, and very difficult for many actors, including the loss of current institutions and freedoms.

6.4   Business and economic activity

Business and economic activity refers to actions, or inaction, by individuals or groups undertaking productive activities, in many cases driven by a profit motive (Table 6.5). Respondent 6 noted that "the business community is a bit tricky because the business is to make profit. And it is not easy... to have any business sector going... against [its] interests.", suggesting that responding to climate change might be against some business interests.

Business and economic activity themes and scenarios consist of: triggers and drivers of a business response (Section 6.4.1); market-driven responses (Section 6.4.2); research and development (Section 6.4.3); technology (Section 6.4.4); re-industrial revolution and green economy (Section 6.4.5); infrastructure investment (Section 6.4.6); localisation (Section 6.4.7); agriculture and food production (Section 6.4.8); insurance withdrawal (Section 6.4.9); and, defence (Section 6.4.10).

6.4.1   *Triggers and drivers of a business response*

In a scenario involving social change and behaviour, Respondent 8 noted that it is possible that these changes could feed "into the company's ethos." involving "a change in, I guess, mentality among everybody and shareholders and about what's important and what's valued." However, the respondent also noted that there needs

Table 6.5 Business and economic activity themes and descriptions

Theme	Definition
Business	Actions, or inaction, by individuals or groups undertaking productive activities, in many cases driven by a profit motive. This includes finance and investment and decisions by business leaders and shareholders.
Economic activity	Production and consumption of goods and services by individuals and groups.

to be confidence in "technologies to replace fossil fuels" and "confidence that actually the whole economy won't fall down if they try and phase out or replace fossil fuels."

Respondent 18 thought that "There'll be more conflict over resources and more scarcity of key resources that they need as part of their manufacturing processes. There'll be reduced economic output, so which reduces demand for their services". As such, "most businesses will see that they're getting hurt economically." and "there'll be an impact on free trade agreements and greater security issues that'll make it harder to do business." Then the "majority of businesses will clamour for more political leadership, more aggressive government action". According to Respondent 21, this could include "all these big companies come together and say we want to set a global carbon price." as well as "Trying to influence, trying to get some more rules WTO related to import tariffs. Carbon related import tariffs, something like that." Respondent 18 noted that "there still will be a few recalcitrant industries, but I think that'll become much smaller." indicating cynical responses will have less influence on the response to climate change over time.

From Section 5.4.2, Respondent 21 noted the possibility of an extreme event attributable to climate change impacting businesses and stock exchanges.

According to Respondent 8, it is possible that business could be part of an enlightened response to climate change. However, the scenarios provided by Respondents 21 and 18, it is impacts on profitability that are assumed to be a driver for business responses.

6.4.2 Market-driven response

Respondent 22 stated, "I think industry is actually, the major leader, because, they have a pure motive to the meet market demand." Furthermore, "there's new technologies being developed and have been developed and industry now sees a lot of, potential in green technologies, green products because the people want it, you know." The respondent noted the need for demand from developed countries to drive this while at the same time making life easier for people i.e., "get an economic break".

Respondent 1 noted

if we think about the green economy and we think about climate change finance, and we think about all the things that businesses are doing, that's where the action is happening... it's not necessarily all being done in the name of climate change. It's not all necessarily directly linked to climate change. It will be done for all kinds of other reasons but climate change in many cases.

Respondent 25 stated, "I think most of what we need, or at least the preconditions for most of what we need is now in the process of being delivered by the marketplace. But without direction from governments it will not happen quickly enough." hence, "Climate change is one of the great market failures of all time."

Respondent 18 put forward a scenario where low carbon energy technologies get "near zero costs energy." then there could be "completely market-driven solutions we'll never even get to mandatory requirements mandatory emission or actions." In a related scenario, Respondent 18 noted

> what would be most effective is that markets are driving the transition... renewables have become so cheap that we're producing not only all electricity and all heat but all chemicals and manufacturing processes are driven by renewable energy and it may not only be renewable, we may have some progress in nuclear fusion and other zero-carbon sources. And ultimately, we won't actually need policies because it's that's the economic preferred low cost and alternative.

Respondent 18 also cautioned, "in the land use sector, I think we still will need some continued regulations to better manage carbon, even though there [a]re lots of co-benefits" because land use is "such a distributed sector." In this scenario, energy prices change so much that prices alone drive actions. As such, a non-response based on energy prices includes positive contributions to the global response to climate change but land use remains an issue.

Respondent 25 noted that in a defensive policy environment "the global marketplace under which most modern businesses seeks to thrive... ceases to exist in any real way." As such, "The only thing that will matter" for businesses is "how they can cosy up to whoever is in power for their own national markets."

6.4.3 *Research and development*

Respondent 15 noted that political leaders need to greatly increase investment in research and development (Section 5.5.2) to have a greater chance of fulfilling the UNFCCC objective (Section 6.3.7). However, Respondent 15 cautioned "Businesses, obviously, want you to invest in stuff that's close to market but if it was close to market, they would already be investing in it" Furthermore, "If we spend lots of the research and development money on companies, we're gonna get close to market technology perfected which is not really what this challenge asks for." Instead, scenarios with research and development in long-shot things are more likely to help fulfil the UNFCCC objective (Section 6.3.7).

With regards to technologies for greenhouse gas removals, Respondent 20 put forward a scenario where "government" is "mandating it, and subsidizing" these technologies. In this scenario, there's "a mix of government, government-supported research and also research and development invested by the private sector." making it possible that there could be "a new technological revolution around, direct removal."

6.4.4 *Technology*

From Section 5.5.3, Respondent 27 provided a scenario where competition between actors around a new technology helps limit climate change and its impacts. Using renewable energy as an example, Respondent 27 stated that "Once there's enough, you know, the 'Elon Musks,' those investors, that have divested from fossil fuels and try to get into renewables in some shape or form."[3] becoming a "winning team" including the "big investment houses and big political actors" and "consequently finance, will drain away from these losing, old school." technologies.

6.4.5 Re-industrial revolution and green economy

Respondent 18 noted in an enlightenment scenario described in Section 6.2.3 that it is possible there could be a realisation by shareholders and others about the need for divestment from businesses that make negative contributions to the global response. Respondent 9 noted it is possible that "large companies, philanthropists, people who have access to real finance, view, or have climate change become more of their priority and they start financing climate adaptation, mitigation projects on a large scale." Then "we could have a re-industrial revolution [laughs] of clean technologies and fuels renewable energies, and clean transport and greater efficiency et cetera." In this scenario,

> We could see the economic benefits that renewable energy technologies and other low emissions technologies are already reaping kind of hit the market in the United States and other large developing and developed economies and the real benefits of those could kind of fuel [a] re-industrial revolution.

Furthermore, it is possible that there could be "prosperity... a green economy that trickle down to the household level."

6.4.6 Infrastructure investment

Respondent 22 noted that businesses "could be putting money into infrastructure investments that help us address and reduce emissions or not." and highlighted the importance of timing, specifically "If that's not done in the next decade now, that window's half-closing. Then we're locked into higher emissions" for a long time to come and its essentially "game over" (Section 6.3.9) with regards to the "viability to cut emissions" assuming infrastructure is used for its design life. However, it is possible that fossil fuel-reliant infrastructure investments could become stranded, in which case investments may be abandoned before the end of design life, hence limiting greenhouse gas emissions from the investment. As such, fossil fuel-reliant infrastructure investments might not lock in as much greenhouse gas emissions as anticipated.

6.4.7 Localisation

A very different scenario involves localisation where "the last word should be at the local level." for example when it comes to international trade laws. Respondent 20 went on to say, "I mean, local businesses would, I would expect they would thrive pretty well.", while at the same time, "there's a role for corporations" for example manufacturing technologies like electric vehicles or related to renewable energy. As such, businesses would operate under a new global trade and domestic governance regime.

6.4.8 Agriculture and food production

With regards to behavioural change and food, Respondent 8 noted there "may be a shift from... animal products". Furthermore, the respondent highlighted that "because we, humans depend on it", then "I'm relatively positive... that we will ensure that and continue producing enough food" at the "global aggregate level". However, the respondent noted that "there are examples of food production failing at the moment". From Section 5.3.2, Respondent 9 noted that there are many large-scale food production technologies that could be shared and applied around the world to increase food security. Respondent 7 noted that "Getting cheap bulk foods to masses might be straightforward but getting niche products might become much more difficult." As such "getting middle-class products to middle-class people is going to become a lot more difficult." Respondent 26 stated "if we can't adapt quickly enough to ensuring supply of and distribution of food" then "large areas of the world will go into extreme hunger".

6.4.9 Insurance withdrawal

Respondent 20 stated it is possible that "insurance industry may withdraw insurance to coastal communities well before the actual physical environment starts to be a problem", and then there could be "cascading and flow-on effects for people being able to afford to live in those places". Insurance withdrawal, like government service withdrawal, includes issues of equity.

6.4.10 Defence

From Section 5.5.1, Respondent 11 indicated that people with power in the defence industry have an incentive to present climate change as security problem because they have a role, can sell equipment and make money.

6.5 Conclusions

After looking deep into the crystal ball (i.e., searchable sample of possible futures) at domestic response scenarios, three overarching sets of actors, interests and responses were identified consisting of social change and behaviour, political will and policy as well as business and economic activity. There are many possible response scenarios involving social, political, and business actors that could contribute towards fulfilling the UNFCCC objective (i.e., success). Scenario combinations are mapped out and analysed in Section 10.3 including socially driven responses (Section 10.3.1), politically driven responses (Section 10.3.2) and business-driven responses (Section 10.3.3). Importantly, effective responses involve multiple actors and interests, so combinations of actors and interests are analysed further in Section 10.3.4. Meanwhile, Section 10.3.5 analyses actor interests and incentives to mitigate and adapt to climate change. The discussion in Chapter 11 also addresses actors and interests (Section 11.7) along with the need for coalitions (Section 11.8) supporting effective global responses to climate change.

Notes

1 SLR = Sea Level Rise.
2 It is important to note here that while climate science by itself does not appear to be a compelling signal to political actors, climate science is often an important input to impact attribution studies and studies of climate change related risks to people, their property, and livelihoods.
3 Elon Musk being a prominent entrepreneur behind the company Tesla famous for its electric vehicle, solar PV, and electricity storage technologies.

7 Climate change response options and scenarios

7.1 Introduction

The previous chapter addressed broad categories of actors and interests. This chapter addresses the types of options actors might take, contributing to the global response to climate change.

From Chapter 2, the options available to actors will have an influence on the global response to climate change and some options are controversial. Options available and the actions taken depend on technologies and practices (Section 7.2) including greenhouse gas (GHG) removals from the atmosphere (Section 7.3). Importantly, there are a range of other factors (Section 7.4) that can influence incentives for actors and the actions they take. In total, 20 types of options and related scenarios are addressed including, for example, technologies, overproduction, GHG removals, leadership, and conflict.

Like in Chapters 5 and 6, for each theme scenarios from different respondents are contrasted. And as noted in Chapter 5, the themes and scenarios presented below are akin to reading a series of short stories. You might want to read all of them, or just a few, and then skip to the conclusions in Section 7.5 which summarises themes from this chapter and indicates where lessons from these scenarios are analysed further and linked together in Chapter 10.

International cooperation options are addressed in Chapter 8.

7.2 Technology and practices

Technology is an important theme in the context of options and possible actions. For example, when giving an effective global response scenario, Respondent 21 noted "Technology can achieve a good deal of this" (Section 7.2.4). In the context of over-production and people's perceptions, Respondent 5 noted "It looked like technology will solve the problem".

Practices are another important theme when it comes to options and possible actions. Respondent 1 asked whether it is possible to focus on environmental practices and make them more appealing (Section 5.5.2). Meanwhile, Respondent 18 noted the possibility of "knowledge sharing and policy, good practices". So, in addition to technologies, we also need to consider practices.

DOI: 10.4324/9781003465911-9

With regards to definitions, technology, and practices are "methods" used when attempting to achieve something (Table 7.1), including objects, activities, and rules.

Technology and practice themes identified from scenarios include: technology breaks the box (Section 7.2.1) and fundamentally shifts systems; technology can create coalitions and movements (Section 7.2.2); utility and substitution (Section 7.2.3); overproduction and economic growth (Section 7.2.4); and technology and climate risk (Section 7.2.5).

Technologies and practices also include GHG removals and solar radiation management options. These are two very different approaches to addressing climate change and global warming (Table 7.2). Solar radiation management is highlighted below in the context of technology and climate risk (Section 7.2.5). GHG removals are addressed later in the chapter (Section 7.3).

7.2.1   Technology breaks the box

Respondent 24 noted "You know sometimes [it is] technology that break[s] the box." indicating that technology can fundamentally change systems. Respondent 25 put forward a technology scenario where

Wind and solar are the cheapest way to add new power generation to the grid. Electric vehicles reduce primary energy consumption in the transport sector by two thirds, while... solving pollution problems. You know, smart grids happening everywhere. The evolution of technology in eating, in all

Table 7.1 Technology and practices definitions

Theme	Definition
Technology	Methods used (i.e., techniques) when attempting to achieve something, that rely on physical, chemical, or biological properties. This includes objects used for a particular purpose, the processes used, and information related to these things.
Practices	Methods used when attempting to achieve something, that rely on properties of human, ecological or environmental systems.

Table 7.2 Greenhouse gas removals and solar radiation management definitions

Theme	Definition
Greenhouse gas removals	The extraction of chemicals that drive global warming and climate change, from the atmosphere.
Solar radiation management	Solar radiation management refers to the intentional modification of the Earth's shortwave radiative budget with the aim of reducing warming. Artificial injection of stratospheric aerosols, marine cloud brightening and land surface albedo modification are examples of proposed SRM methods.[a]

[a]IPCC (2018b).

sorts of other areas. The ability to actually monitor the planet in terms of what's happening in ways which would allow us to control deforestation, illegal deforestation if anybody wanted to. Lots of exciting new techniques in agriculture which dramatically reduce the requirement for nitrogen input and hence nitrogen oxide emissions, et cetera, et cetera. I mean, it's technology. We have the technology. That's, and I think that's been, has been created through a combination of government incentives and innovation within business to provide more efficient, better products.

As such, Respondent 25 indicated technology can have multiple functions, can generate new response options, and as such, can fundamentally shift systems.

7.2.2 Coalitions and movements

After addressing political will and policy, Respondent 27 put forward a scenario where a coalition of actors, i.e. a "winning team", forms around an amazing new technology that helps the global response to climate change rapidly scale up (Sections 5.5.3 and 6.4.4). Respondent 24 noted, "you need a coalition of the key forces that govern society to do this better and of course the private sector, and the governments, and to some extent the various civil society movements." And observed,

So, what we've seen in the solar energy revolution, I think it's remarkable. That's some of the more positive developments, you know. There is not only negative, right. I think that's a good example where technology can actually create movements.

As such, technology can influence options available, and the things people rally around. This includes influencing social change and political will.

7.2.3 Utility and substitution

Respondent 16 noted it is possible that there could be "new technologies that can easily be used, particularly if they are appropriate for the developing world" and these would be quickly adopted. Then, "that would make things completely obsolete, you know, the oil economy quite obsolete for instance." Such technology could even "take the world by storm".

Respondent 14 stated "There needs to be a bit more of technological breakthrough" resulting in the use of "electricity cars... using electricity generated by the non-fossil fuel sources" and this becomes the "norm everywhere." The respondent noted that for the required global response and related behavioural and political changes to happen, technology "needs to be available... it needs to be in prospect as well." At some level there needs to be faith in the technology and its ability to meet people's expectations and aspirations.

7.2.4 Overproduction and economic growth

With regards to overconsumption and technology, Respondent 5 noted "we produce and we consume far more I think than we need for the wellbeing of the people.", but because people think technology will solve climate change and other problems (first paragraph of Section 7.2) this creates the impression that "we can forever continue and increasing our consumption production patterns." Likewise, Respondent 21 noted that even though technology can do a lot "if you have increasing demand for goods and services, I don't see how that's compatible with the real best-case scenario." Respondent 21 also noted that other fundamental changes are needed to improve the outlook for ecosystems for example (Section 5.5.2). Respondent 17 noted concerns regarding carbon capture and storage, in particular "the storage part because of that kind of creates this waste product that now we also have to monitor it and whatever, if there's a leak and stuff like that." Importantly, technology can have unintended consequences including the creation of wastes and generation of demand for further inputs for the production and operation of technologies.

7.2.5 Technology and climate risk

Respondent 2 noted

> If two degrees is what the scientists judge as the point at which we can manage the risks 2.1 degrees is the least worst if you follow me. Now there's going to be a margin of error in that, clearly. And a lot of it is driven by what we believe technology can offer us by way of mitigation or adaptation.

As such, faith in technology influences the perceived risks that come with climate change.

Respondent 20 noted the possibility of using "solar radiation management" in the context of addressing fast feedbacks. Likewise, Respondent 24 provided a scenario where "the only option is to use solar radiation modification to relatively quickly bring the temperature down". As such, solar radiation management is a possible option. Note: See Section 9.5 for more on solar radiation management.

7.3 Greenhouse gas removals from the atmosphere

GHG removal scenarios address the extraction of GHGs from the air by any means including the use of technology or by natural means (e.g., resource management practices encouraging the growth of trees, biological activity in soils or the sea). Some respondents noted, without prompting, that GHG removals are required to limit global warming. For example, Respondent 1 noted that a scenario where the UNFCCC objective is achieved "puts you in negative emissions... territory." which is consistent with the literature (Section 2.2.4). Other respondents were asked a specific follow-up question, asking for a scenario where GHG removals are used to achieve the UNFCCC objective.

7.3.1   GHG removals as a supplement to mitigation

Respondent 21 noted "there's two options", consisting of, "either you reduce demand" of things generating GHG emissions "or, you use negative emissions technologies." The respondent also included a third option, "I guess the other option… is that neither of those are required." essentially suggesting that if climate sensitivity is lower than expected, then mitigation or GHGs may not be required.

Respondent 24 noted "I don't see carbon dioxide removal as… the factor to stabilize the emissions. It's a supplementary activity that we, unfortunately, now have to do because we have gone so far." due to the volumes of GHG emissions (Section 5.3.2). Hence the need to reduce emissions to zero. Respondent 24 went on to state "And until that time comes we have to make use of carbon removal first to offset, to balance, and then to remove whatever level of carbon we think needs to be removed." with the aim of stabilising atmospheric GHGs at safe levels. As such this is an overshoot and remove scenario, consistent with research summarised in IPCC assessments (Sections 2.2.4 and 2.3.1).

Respondent 24 also noted,

> the longer we wait with the reduction of emissions the more CDR we have to use… And it gets pretty big… the longer we wait. The second is that… the longer we wait with deployment of CDR, the more challenges that might arise. Because they will take a long time…to bring these technologies to scale would require years and decades. So, let's get going as soon as possible, because anyway, the carbon needs to be removed so, it doesn't matter if you remove it too early. But it matters if you remove it too late.

Therefore, working on GHG removal technology and practices is also a risk management strategy, one that helps ensure technologies, practices and capacities to implement these things are ready in case of need.

7.3.2   GHG removals as a contingency option

Respondent 6 suggested GHG removals be kept as a backup option in case climate sensitivity is higher than expected. The respondent stated, "we have BECCS in our scenarios [as a] way of dealing with the high climate sensitivity rather than a way of dealing with the fact we didn't insulate houses quick[ly] enough." As such, the respondent noted

> I would like to leave negative emissions as a way of dealing with surprises on climate science rather than failures of climate abatement. Because I worry a lot more about surprises in the climate system, and if we use up all the BECCS to deal with our climate policy, then we got no… other leavers to pull that'll keep us on the safe side.

In short, GHG removals could be kept as a contingency option.

7.3.3 GHG removal practices

A couple of respondents put forward GHG removal scenarios that involved biological GHG removals without technological GHG removals. For example, Respondent 14 and Respondent 21 both highlighted the capacity of photosynthesis and forests to remove GHGs from the atmosphere (Section 5.5.2).

However, Respondent 24 noted

> Even with the some of the technologies that are referred to as nature based you actually have to do some pretty hard work to get them done. Afforestation we know is challenging. It doesn't happen by itself. And maintaining high levels of carbon in the soil doesn't happen by itself.

By contrast, Respondent 3 noted "we know how to shift the incentives for communities around forests. We have done it in many countries now." Respondent 3 went on to state "We know how to shift the incentives for farmers around managing their soils and woodlands and their forests better." From Section 5.3.2, this should include land tenure if scale is to be achieved, and as such, Respondent 3 even suggested the "UNFCCC should have some statements around tenure." Respondent 14 suggested it is possible to engage the public into "digging up your path and planting it" (Section 5.5.2), hence generating more "carbon dioxide absorbing greenery around the place, and that potentially could be done in a large scale, subject of course to what the climates doing by way of delivering rainfall."

7.3.4 Practical limitations

Respondent 6 noted

> I know there's a scenario where you have a small amount of mechanical removal but a large amount of natural replenishment particularly around soils. And restoring the degraded... forest plan, which gives us a bit more wiggle room but I think it's a small proportion of the game.

By way of contrast, in Section 5.3.2 Respondent 14 expressed pessimism regarding the scale at which GHGs might be removed due to challenges such as the energy required and where to put removed GHGs. Likewise, Respondent 26 noted that GHG removals "needs big time investment in large scale physical technologies" and

> There are lots of questions about carbon capture and storage in terms of its effectiveness, in terms of the physical investment, the energy requirements to even build the number of plants that you would need. So, getting to the point of stabilization, it's you know, it's a big question whether that's possible.

Respondent 23 noted that they think it is "delusional" to think GHG removals can be done at the scale required to limit climate change even if it is technologically possible. (Section 5.3.2). However, the respondent also noted

> If we find a way to create energy very affordably, like some truly genuine revolution in photovoltaics or in going deep into the guts of the earth or something, maybe energy becomes so cheap that [GHG removals] can be done.

With regards to the cost of energy, in Section 6.4.2 Respondent 18 suggested that it is possible renewable energy becomes so cheap that it addresses all types of energy needs, including energy needed for heat, chemical processes, and manufacturing. However, Respondent 23 noted that with cheap energy "it would also be easier to adapt to the consequences because you can desalinate water and you can do all sorts of things with energy." Furthermore, "I think that the amount of land and energy needed for removing enough carbon from the atmosphere is, if that is profitable or affordable, then many other things involving adaptation also will be."

7.3.5 Trade-offs and unintended consequences

Respondent 27 noted, "there will be trade-offs." that need to be addressed if these GHG removal methods are to be applied. Other respondents specified what some of these trade-offs might include. For example, Respondent 26 noted with regards to "carbon capture and storage" that "it's a very industrial strategy that will take away from more localized smaller scale [actions]. It would... either write off completely or undermine efficiency strategies." Respondent 26 noted a possible scenario is that "people have been pushed off the land, for example, [for] bioenergy, carbon capture and storage strategies." Respondent 3 stated,

> when we have looked to carbon capture and storage in real ways, there are very few sites have been identified where it's a realistic prospect. Where it is a realistic prospect in terms of the geology often it is not in terms of water intensity. So, if you add six years to a coal station basically it consumes more water. In a lot of water scarce locations you know, like Saudi Arabia which has the best geology for CCS, what are you gonna do? You know. This is not a win-win by any means, it's a trade-off.

With regards to "bio-energy carbon capture and storage", Respondent 26 noted this could result in "appropriation of land and land use in certain ways that will definitely displace people, either physically or culturally from their activities". It was noted that GHG removal is "one of those big solutions that will you know, from a sort of a climate justice perspective, I think will be really quite problematic, for the poorer and more vulnerable." From Section 7.3.4, Respondent 23 highlighted the possibility that the land and cheap energy required for GHG removals might be used for other purposes including adaptation highlighting that decisions are made based on options available and that options change as technologies, practices and prices change.

7.3.6 *Political will and GHG removal policy*

Respondent 20 noted "CO_2 scrubbing from the atmosphere, that's expensive technology", and as such, removals scenarios require "government mandating it, and subsidizing", as well as "a mix of government, government-supported research and also research and development invested by the private sector." perhaps leading to "a new technological revolution around, direct removal." From Section 6.3.7, Respondent 11 noted there is a role for the government to create incentives for the private sector and markets to develop technologies that help address climate change. The respondent also noted the private sector won't develop negative emissions technologies of their own accord. Hence government support for research and development is required.

With regards to the conditions required for actors to actually conduct atmospheric GHG removals, in Section 5.3.2 Respondent 6 noted GHG removals require high levels of cooperation including large financial transfers to compensate actors for opportunity costs including land uses in the case of BECCS. However, from Section 5.5.2 Respondent 21 stated that it is simpler to scale negative emissions technologies than change people's behaviours. This is a path of least resistance approach.

Respondent 1 asked, "unless we have bioenergy carbon capture and storage, where are the negative emissions coming from to deliver those climate policy targets?" The respondent also noted "All the targets you see today are about reducing emissions. The national targets [INDCs]. They're not about going to negative territory."

7.3.7 *Timing of GHG removals*

With regards to timing, Respondent 18 put forward a scenario where

> sometime after 2050, maybe right after 2050, we'll really start to see dramatic declines in emissions and then more towards the end of the century would really have the scale of investments in carbon removal that we need. I'd hope that it would be quicker than that, but I [am] trying to be realistic.

According to the scenario, "geoengineering and carbon removal from the atmosphere" is one of a mix of actions, that also included "widespread adoption of renewable energy and energy efficiency", "low carbon agriculture and forestry", and "methane captured from landfills and industry" for example.

Respondent 18 stated,

> one scenario, and the most plausible in my mind is [greenhouse gas removals] happens after 2100, because I think, in reality it'll probably take that

long before we've gotten serious enough about the policy action and had all the investment and the technologies in manufacturing at scale so that it's more easily economic, and... can direct the capital needed. But dramatic impacts up until that point, until we get stabilization...

This last scenario assumes no tipping point is passed or crisis is caused by climate change and climate change is reversible.

Respondent 17 provided a scenario that started with "Let's assume that this negative emission stuff, whatever it is, is going to work." then "we might get back down to 1.5 after touching 1.8 or whatever". However, the respondent also noted "So, in 2100, we're good, but we're good in a potentially different world" that might be missing "coral reefs" lost due to "ocean acidification" or something else (Section 4.4).

7.4   Other factors influencing actions

Other factors are other things that can influence the global response to climate change. From the sample of possible futures, other factors include things such as: leadership (Section 7.4.1); prices and the costs of technology (Section 7.4.2); carbon prices (Section 7.4.3); finance and investment (Section 7.4.4); law, contracts and legal precedent (Section 7.4.5); migration (Section 7.4.6); and conflict (Section 7.4.7).

7.4.1   Leadership

From the thematic analysis and for the purposes of this book, leadership is the ability of individuals from any walk of life to make things happen and have others follow. Importantly this includes people from government, business, civil society and communities for example. From the sample of possible futures, leadership themes and issues related to the global response to climate change included: the need for leadership; urgency and ambition; cultural and historical connection; and, the idea that "leadership breaks the box" and can shift systems. It should be noted that leadership also featured at the international level, but this is addressed in the next chapter in Section 8.3.2.

The need for leadership

Respondent 24 said

I think that we need to have the opposite of what's happening today in the world. In terms of what's happening in the geopolitical sphere... But it is this incredible shift of focusing on yourself. When we, none of us exist in isolation.

Therefore,

one needs to do almost the opposite. But for that, you need leadership who can sell this to the public... There are different ways societies are set up.

China is very different than the United States or Switzerland or Sweden. They all have their particularities but in each, there are different ways that ultimately the public has to be engaged.

Respondent 14 noted that for the UNFCCC objective to be fulfilled, "There needs to be a bit more of technological breakthrough... there has to be leadership from countries... people in society need to be scared".

Urgency and ambition

From Section 5.3.1, Respondent 2 highlighted the need for a timely response and in the same scenario noted climate change is a strategic threat that needs to be addressed by political leaders, including making tough decisions and convincing the populace (Section 6.3).

Respondent 23 had a scenario with "some very charismatic and informed and good-willing and motivating person or group of persons, whether it's a president or a new head of COP, or youth movements storming in demanding a change that actually is desired". With regards to social change, this could "infuse a level of energy that is more analogous to religious movements, or Gandhi, or civil rights and anti-slavery." meanwhile the global response to climate change could "get closer to the 1.5" while at the same time getting "closer to the stated aspirations of common but differentiated responsibilities and so on, and development and everything".

In Section 6.4.2, Respondent 25 noted, that to fulfil the UNFCCC objective most of what is needed for a global response is being delivered through the marketplace, but this response is too slow. Hence, Respondent 25 went on to state the market is "Sending, sort of, signals in the right direction... but... it "needs a Jacinda equivalent in most countries to just say, 'Fuck you guys, get on with it. I don't wanna hear it, your excuses and your bullshit.'" In short, urgent government direction is needed for a timely global response. Jacinda in this context refers to the New Zealand Prime Minister at the time of the interview.

Cultural and historical connection

Respondent 14 stated, "there has to be some leadership from countries that people look up to and that leadership has got to be dramatic, visible and seen to be transferable". The respondent also suggested that "ideally you'd have regional leaders, because the different countries, different cultures, different religions look to different leaders for guidance." The respondent suggested leadership could come from China, an African country but it is difficult to know which country, a European country perhaps Italy or Greece rather than a Nordic country, and the Democratic Party in the United States but only "if they can get their shit together and decide that they are truly a supporter of democracy and can unhook themselves from their traditional connections with big business and big lobbyists." The respondent noted that leadership would need to include governance "that accelerates the technology development from concept to delivery."

Leadership breaks the box

Respondent 24 also noted that an effective global response to climate change

> requires an understanding by the global community that we're in deep shit
> and only the radical action can take us on a path which is both from a carbon
> point of view, matching the UNFCCC objectives and does not destroy sus-
> tainable development.

Furthermore, "this needs to be sold to societies." For this to happen you need

> political leadership... you need people who are charismatic, strong, who are
> able to get out there and basically... sell the idea that we have a problem.
> And we will fix it. We can fix it. It is not the end of the world but we need to
> do this together... I don't just mean government leaders mind you. We need
> leaders in NGOs, in private sector, in religion, all those different groups be-
> cause it is the leaders who break the box in which we are located.

With regards to breaking the box, this seems similar to the concept of "growing the
pie" or "expanding the pie" in negotiations where more and more issues are linked
and addressed together, so trade-offs can be made and decisions agreed upon by
all parties (Wetlaufer 1996, Basadur et al. 2000). Kingdon (1995) also noted that
leaders can link and reframe issues, redefine problems and agendas so that some
solution can be adopted (Kingdon 1995). As such, leadership can fundamentally
shift systems and the options considered by society, government or business.

7.4.2   *Prices and costs of technology*

Respondent 15 stated,

> fundamentally, the only way I can imagine that we're gonna achieve some
> sort of stabilization on a fairly low temperature is if we manage to get tech-
> nologies that are cheap and produce about the same benefits as fossil fuels
> do now.

As such, "The fundamental point here is much more about getting green energy
to be so cheap that it's close to taking over." and "we need technology in order to
solve this if you're gonna, if it's both gonna be politically realistic and if it's gonna
be realistic economically to stabilize at a low level." "But if we can make technology that's gonna be cheaper... that are green then we can get this takeover. If we
don't, it won't happen." These are essentially conditions for a non-response with
positive incidental contributions to the global response to climate change.

Respondent 18 put forward a scenario where "we get quickly to the zero carbon
near zero costs energy.", then, "most of the problem can be dissolved." but "We
still have land use and few other issues to deal with." Alternatively, "If it's the case

that it's not really zero cost energy, but it's, you know, a cost that developed countries can afford, developing countries can't or the infrastructure and technical skills needed are significant", then that would "be good news for the global climate, bad news in terms of regional disparity." In either scenario, the respondent thought "it'll move us towards completely market-driven solutions we'll never even get to mandatory requirements mandatory emission or actions." The respondent also stated that "I don't think we have the political will for it to be policy-driven. I think it'll be technology-driven supported a little bit with policy." This indicates that cost is not the only factor influencing the options considered and actions taken, but other issues such as the capacity to act may also have an influence.

7.4.3 Carbon price

Respondent 17 noted "There is a carbon price out there, different regions have it, it's very low, it doesn't cover enough space.", so "let's have a carbon price that requires actually a global government, that's always like the step that they leave out is you need for the carbon price to work." Then, "we would actually draw a line and say, 'Here's the budget for this year and next year and all that.'" and, "say, 'What would it require for us to get this carbon price?' It would require actually 190 or whatever how many countries we have sitting down and saying, 'Yeah, that's right.'" Then, "if we move from carbon price to global government, and then we say, 'Okay, we acknowledge that there's gonna be winners and losers.'", and "Then you have to start saying, 'Okay, now we can talk about' We've sort of described the problem, what would be the remedies for that situation so we can actually work this out." Lastly, "you would have to probably come up with something like transfer payments or border taxes or... something along those lines".

Respondent 17 noted "Because under a carbon price, there're gonna be winners and losers... we would need to talk about, "Okay, what kind of compensation mechanisms, transfer payments?" You know, the whole thing becomes super complex. But saying, 'Let's have a carbon price,' is a simplification."

7.4.4 Finance and investment

Finance is the source of money, and this could be for any activity or investment. Finance was mentioned, along with terms such as investment, usually with the source (i.e., actors) indicated for example business or government. Finance can be an important part of change, for example in Section 6.4.4, Respondent 27 noted that it is possible finance could rapidly move away from fossil fuel technologies into "amazing" new technologies such as those used for renewable energy generation. Finance, in this scenario, is important for sustaining and scaling up technologies and business models. Finance and investment might also support research and development, infrastructure or energy, meanwhile finance could support international cooperation, or other things such as transfer payments, all of which were mentioned in the sample of possible futures.

7.4.5   *Law, contracts and legal precedent*

Respondent 26 noted that for effective contributions to the global response from the law "you have to have a good legal system that can respond as well" and, "you need a scattering of cases in different jurisdictions" in particular when it comes to "private sector law around, investment and, say foreign direct investment". With these preconditions it is possible that legal systems may help with "ensuring that all investments, don't contribute to climate change"; "flagging up this fiduciary responsibility of shareholders and investors"; "they can start to influence the standards and requirements, within private sector law."; rule on "domestic private legal systems, in relationship to, investment, company responsibility et cetera."

Respondent 26 also noted "I think there's some really big issues at the international level and at the national level about ensuring that all investments don't contribute to climate change so, I mean, we're seeing this in the sort of flagging up this fiduciary responsibility of shareholders and investors, so sort of, to de-carbonize their investments, both for, for the real fact that they'll, the investment won't be realized if, if you don't de-carbonize, because there won't be a future." The respondent noted "that that's big area, which is very different from this litigation, which is much more trying to hold to account, government." Respondent 26 also noted that legal "systems learn from other systems".

In addition to private sector law, Respondent 26 addressed "public interest litigation to call to account that the government is failing to do enough in terms of either mitigation or adaptation for climate change." This could be "a means in to using the legal institutional structures to try and change the policy framework in a country."

7.4.6   *Migration*

Respondent 24 noted "when the immediate changes in climate which have impact on agricultural production, things like that, will produce mass migration." Then there will be "initially, internally displaced, some of which will then result in externally displaced. And that is where things are really going to hit the fan, so to speak." Respondent 21 made the link between climate impacts, domestic institutions and migration, noting

> if there isn't enough of a, sort of credible institutions within their own country, these people may leave their own country… People are going to start going elsewhere and then you're going to see sort of responses to migration and immigration that are potentially going to become more and more strained.

7.4.7   *Conflict*

Two failure scenarios featured the theme of conflict. For example, Respondent 26 noted there could be "more tension and conflict in urban spaces" and "there's gonna be real fights over natural resources." Respondent 26 also noted social change that could result from conflict, stating "conflict brings violence and… extreme politics, and sort of complete social breakdown."

According to Respondent 14, possible preconditions leading to conflict include "The issue of food security and water security", "international issues around national borders" and "conflict between states over resources." Respondent 26 also noted conflict could result from conditions such as "if we can't adapt quickly enough to ensuring supply of and distribution of food" or due to demand for "Primary natural resources even to create a green eco, you know, to create a low carbon world." Furthermore, it is possible "people have been pushed off the land, for example, [for] bioenergy, carbon capture and storage strategies." hence, "there's gonna be this movement of people into urban centres." creating conflict.

7.5 Conclusions

When it comes to response options and related scenarios, technology is an important theme as are practices. The availability of technologies and practices is an important influence on the options that might be taken by actors. Given that the GHG budget is likely to be breached, the availability of GHG removal technologies and practices can either be part of efforts to offset, stabilise or reduce atmospheric concentrations of GHGs or serve as contingency options. Other factors can influence the uptake of technologies and practices and the effectiveness of responses to climate change such as leadership, prices or finance. Other influences include legal and contractual considerations, migration, and the possibility of conflict. Each of these factors can have a bearing on actor interests and their incentives to act on climate change.

For further analysis of the scenarios and themes in this chapter, the influence of technology, prices, and the commercial viability of options are analysed in Section 10.3.5. The scenarios and conditions under which actors might undertake GHG removals are mapped out in Section 10.4 and discussed in Section 11.9.

8 International climate cooperation scenarios

8.1 Introduction

Even though much of the global response to climate change happens at the subnational level, international cooperation is an important influence on actors and their responses, especially states (Chapter 3). This chapter looks into the crystal ball (i.e., sample of possible futures) asking the question: What are the preconditions for effective international cooperation on climate change?

So, what themes and scenarios were found? Themes included the international regime (Section 8.2) and international cooperation (Section 8.3). Scenarios included globalism, nationalism, issues of power and influence, and more.

Given the importance of stringency and compliance from Section 3.4.3, the semi-structured survey included a follow-up question asking respondents to give a scenario where there is a stringent enforced international agreement on climate change. These scenarios and themes are presented in Section 8.3.4.

In many of the sections below, contrasting scenarios from different respondents are presented. Like in Chapters 5–7, the scenarios in this chapter are akin to a set of short stories. You may want to read about some of the scenarios and themes. Alternatively, you may want to go straight to Section 8.4 which concludes with themes from this chapter and indicates where these themes are analysed further and linked up in Chapter 10.

8.2 International regime

As noted in Section 3.3.4, climate change and related responses are part of an international regime complex. The response to climate change is among many issues on the international agenda. Agreements on trade and other issues can influence the global response to climate change (Stavins et al. 2014). When looking into the crystal ball (i.e., sample of possible futures) themes and scenarios include globalism versus nationalism (Section 8.2.1); climate change emerging as a premier issue (Section 8.2.2); a climate-focused international regime and what this might look like (Section 8.2.3); the UNFCCC as a coordination body (Section 8.2.4); and the possibility of a business dominated international regime (Section 8.2.5). It is also possible that there could be fragmentation or collapse of the international regime.

DOI: 10.4324/9781003465911-10

8.2.1 Globalism versus nationalism

Respondent 13 noted

> you've got to, gotta step back from climate and sustainability issues for the moment and see what's happening in the world politically and you've got massive movements going on. One is more of an intellectual realization that we have to cohere together as a species and however we structure ourselves politically for decision making. So, I just call that the globalist approach. And then you've got the nationalist-populist backlash against the migratory flows especially, and so you have it in the United States and in Venezuela and quite possibly Brazil, …you do have spots of populist backlash and for that matter inside Europe and the UK. So, it begs the question as to how those competing, and they are conflicting forces, play out to make decision making on anything whether it's military but also on sustainability and climate.

Wider movements, and tensions, between globalism and nationalism will influence international cooperation on climate change.

8.2.2 Climate change as the premier global issue

Respondent 14 noted that in a scenario where there is "Food insecurity, locally and increasing nationalization, leading to increasing conflict.", then the UNFCCC "may need to become the premiere… leading United Nations organization." In this scenario, "the UN might start to be centralized around that body because, if the negative scenarios play out, [and] climate change starts to dominate everything." As such, the scale of climate change impacts and risks could make climate change a central focus for international cooperation generally.

8.2.3 A climate-focused international regime

Respondent 13 stated "in a global emergency, you'd have to say, 'Right, the UN is taking a lead, and it's the Security Council, and it's working with the WTO and the World Bank.' And UNFCCC is simply negotiating machinery feeding into it." Because it is an emergency, "You can trigger chapter seven enforcement powers in the Security Council". For institutions to work together, there would "have to be an agreement, it would have to be economic sanctions through coordinated between UNFCCC, WTO, and Security Council. With leadership from the Security Council, with a strong Secretary-General, and a sobered collective resolve inside the Security Council." The respondent noted that "if the [Security] Council won't do it, then the secretary general could set up subsidiary bodies… and report into the Security Council… put proposals to the Security Council…" Importantly, in this scenario the international regime addressed aviation and shipping while also allowing for other institutions to be part of the regime. See Figure D.4 in Section D.3.1.

8.2.4   UNFCCC as a coordination body rather than a decision body

Respondent 6 put forward a scenario where "The UNFCCC …is not the body that decides, but organizes how we address it.", furthermore, "it organizes other organizations to address climate change in their work and… the response to climate change, but it's not… the thing that does everything…" The respondent suggested a scenario where the UNFCCC is "evolving into a much stronger global risk management organization. Much stronger focus on monitoring the earth system impacts, global system impacts with… it as an advisor in parallel to Security Council and the GA [and the UNSG's office]". The UNFCCC "looks at impacts overall and how the global system is responding…" The respondent went on to state "it needs to turn into a more of a[n] IAEA nuclear weapons convention style body not a WTO type body."

Respondent 6 noted

> the core piece that the regime adds to build agreement is an independent measure of what countries are doing, so what we don't want is to be arguing over the CIA's version of Chinese emissions verses the Chinese version of Chinese emissions. We've seen that in nuclear proliferation, [that] it really helps to have a kind of IAEA style [institution, where] we all agree what everybody is doing.

Respondent 18 didn't mention the IAEA, but did state "I don't think there'll be oversight of national policies… there will be sort of voluntary teams who go and check what countries are doing, but I don't think they'll be real enforceable policy reviews."

With regards to international institutions other than the UNFCCC, UNGA or UNSC, Respondent 6 suggested a scenario where there is growth "in the climate orientation of all the other bits of regime" including the IMF, multilateral development banks and FAO for example, with "integration of climate into their everyday business". In this scenario, the UNFCCC is "the delivery arm" rather than the driver of these things. Hence "there's a kind of… global climate agency model".

8.2.5   Business dominated regime

Respondent 7 provided a scenario where corporations dominate affairs including the global response to climate change. The respondent stated, "So we no longer have a Football World Cup [of] Nations, but a Football World Cup of Corporates.", and "that could happen… if society becomes so individualistic that governments… are hollowed out and lose their controlling power" and power shifts "almost entirely to the corporates." In such a scenario, corporations and financial institutions say "Look, UNFCCC you have not moved fast enough Governments you have not moved it. We're in 'schtook'. We are running out of places to move our businesses to." Then with the involvement of the G7 and BRICS, the global response could be "through financial instruments. Financial mechanisms. So, whether it's green taxing or shifting to a carbon, a genuinely carbon economy in other words carbon becomes the currency, not dollars." The respondent also noted that "the leverage for dissent has disappeared." meaning the opportunities for social change would be limited.

8.3 International cooperation on climate change

From the CCNIIC Model from Section 3.4.2, international cooperation on climate change is an important part of the global response. International cooperation on climate change refers to actors from different states acting in a coordinated way to address climate change and related issues. From the sample of possible futures, important themes and issues related to international cooperation on climate change include triggers and drivers for such cooperation (Section 8.3.1); global political leadership (Section 8.3.2); global political power and influence (Section 8.3.3); possible stringent enforced international agreements (Section 8.3.4); and international cooperation options (Section 8.3.5).

8.3.1 Triggers and drivers

An important part of understanding preconditions for an effective global response to climate change is understanding triggers and drivers for international cooperation on climate change. From the sample of possible futures, triggers and drivers might include: maintaining a rules-based regime; national interests; an emergency global response; business interests; and enlightenment.

Maintaining a rules-based regime

Respondent 6 provided a scenario where "international cooperation is essentially driven by major powers doing state-to-state diplomacy and realizing that they both have an interest in keeping temperatures somewhere around two degrees." Furthermore, "they believe that maintaining the climate regime is a critical part of maintaining the overall rules-based regime... I mean the linkages, with trade and investments and other pieces. So that's the kind of top-down state to state geopolitical scenario." In this scenario, there is "a rebirth of multipolar cooperation" with "lots of lovely leadership, people get together, they coordinate, they keep markets open, central decisions in China and India."

In another scenario, involving a stringent enforced climate agreement, Respondent 6 put forward two preconditions consisting of "they want it because of the climate risk." And "...they believe in international rules, so that they will not renege on it because they want to show that international rules work and that's not just for climate but for other things."

National interests

With regards to underlying assumptions, Respondent 6 stated "we think there is a core realists national interest scenario where none of them can maintain the security and prosperity of current levels in a kind of two and a half degree and above world." Furthermore, "it's in a broad concept of security both hard and economic". As such, "it's a risk and rules based scenario... And both of those motivations are what drives action, it's not just climate risk driven, it is driven by geopolitical drivers". The respondent also stated, "it's kind of what people, you know, hope happens and that would give us a decent chance with a kind of tailwinds around the global economy of shifting investment, so that's the kind of international cooperation scenario." The respondent also noted that "We think climate change is a fundamental geopolitical issue now and that wouldn't have been true five years ago.", and as such climate change "is part of those things major powers need to cooperate on to maintain a kind of a bundle of cooperation which, again those are never done as single issues. They are done in bundles".

In the "risk-rules based scenario" Respondent 6 identified risks affecting the national interests of geopolitically powerful states,

> for Europe its migration and the stability of North Africa primarily, for China and Japan its primar[ily] food stability and internal stability for China. For Brazil there isn't really one because they're too complacent. For the US, they are one of the most exposed developed countries for internal damage, but let's see if... they kind of believe it again. India, most vulnerable country, monsoon rainfed agriculture, enormous disruption.
>
> (Table 8.1)

Despite these risks, these states have not yet responded and time is running out for timely mitigation-based risk responses.

Emergency global response

Respondent 13 noted

> I would anticipate that at some stage and there's going to be emergency powers kicking in at international level and on the question is do they kick in

Table 8.1 National interests that could drive risk responses for geopolitically powerful states or regions in the case of Europe

Region or state[a]	National interests driving risk response
Europe	Migration
China	Internal stability and food security
Japan	Food security
Brazil	None i.e., "too complacent"
US	Exposure to climate change loss and damage
India	Vulnerability including monsoon rainfed agriculture

Source: Respondent 6.

[a] States or regions in the case of Europe, that is big enough to upset a treaty if both sides don't get what they want. Russia and Australia were noted as being minor players by comparison.

somehow the global level Security Council whatever. Because of, humanity will suddenly start baying and saying, 'Enough [is] enough we got to change some things very fast.'

The respondent also noted, "Does that occur before 2030 or after 2030? By definition, if it is after 2030 it is too late."

Respondent 7 stated, "I think it can only come out of some as yet an unforeseen crisis. I can't see us working towards it incrementally. Incrementalism seems to have failed. So, it looks like the stepwise change would probably have to come through someone's yet unforeseen" crisis. Respondent 14 noted a stringent enforced agreement could happen, "When... more nation-states start to plead for [a stringent international agreement that is enforced], plead for action."

Respondent 9 noted that the trigger or driver for a stringent enforced international agreement could be

> the scientific community says we really can't afford not to act. We are very sure that we'll reach a tipping point in 20 years and you have to increase emissions or we're all gonna die. Decrease emissions or we are all going die. Or we have some kind of like catastrophe that can spur a similar response.

Respondent 27 stated, "I think it will be driven more by other reasons... if things start going really hectic and really disruptive and it might be driven by fear and by strong security imperatives and climate changes coming up more and more." Then

> Politicians just follow what their constituents and what their... ethnical grouping tell them so if that stuff shifts then the global agreement will suddenly become super strong and everyone will be super committed because that's doable and it's advantageous for other reason.

Business interests

Respondent 21 suggested business interests could drive a stringent enforced international agreement. From Section 5.4.2 Respondent 21 noted that an extreme event could affect businesses and stock exchanges. In the same scenario, Respondent 21 went on to state that "the response the next day is... all these big companies come together and say we want to set a global carbon price." The respondent noted, "people can respond and make changes very quickly if they are threatened."

Enlightenment

Respondent 20 noted that an enlightenment scenario could lead to "an atmosphere of intense global cooperation and goodwill". In this atmosphere,

> the key players globally, that have the highest carbon intensity emissions, are agreeing on a global technology transfer and cooperation and mitigation,

and also, facilitating the transfer of those technologies, and also agricultural practices, to what we mostly call the developing world, other developing countries.

Furthermore, "externalization of fossil fuel impacts, combustion, is brought into the fold through carbon taxes, and... subsidizing green technologies. So, basically, reining in and, neutering the independence of these major international corporations that are behind many of these emissions".

8.3.2   Global political leadership

In a scenario focused on international cooperation and climate negotiations, Respondent 25 suggested it would help if there was "a Jacinda equivalent in most countries", and these leaders would "give their instructions to their negotiators to fix it. And they're not going to take no for an answer and they're not going to take excuses and they're not going to take blaming it on somebody else." Respondent 25 noted it is possible that

> the Chinese say, 'Okay, well, all right, we're gonna do this. We're gonna lead this. ...this is how we're gonna define the new rise of China.' And they could, easily. They have the money, they have the technology, they have the influence.

Respondent 25 also noted political leadership "will depend upon the outcome of the 2020 election." And

> We're due for a renewal in generational terms in political leadership in an awful lot of places. Donald Trump is an old man. Shinzo Abe is old. Bolsonaro is old... I think that's what's required, and I think that's the only thing that's going to make a difference.

From Section 7.4.1, Respondent 14 provided a scenario where there is multipolar leadership driving the global response to climate change, including regional leadership and leadership from countries with cultural connections and influence. Similarly, Respondent 10 suggested, "The other scenario is... big developing countries like China and India, are taking the lead." In another scenario, Respondent 12 suggested "[China] have to change how they consume materials and how they build their cities.", then they push similar policies through "Their network that Belt and Road... seaways initiative." In this scenario, China say to ASEAN "you have to make sure that those laws will be sustainable."

8.3.3   Geopolitical power and influence

Several respondents highlighted the role of geopolitically influential actors in their success scenario. With regards to China, India, Brazil, and South Africa,

Respondent 9 noted "most important factor is not only to their economic size and the rate they produce emissions [but] also their geopolitical power and influence." Furthermore, "where they lead the world follows. Especially with regional powers like Brazil, South Africa, China, et cetera. They set the tone of Africa's response, South America's response, et cetera." This is a multipolar leadership scenario.

In Section 5.2.4 Respondent 14 noted that a stringent enforced agreement would require pleas from states bigger than Tuvalu. The respondent went on to state

> the pleas have to come from countries that are potentially dramatically affected and whose culture or population, people, are of value by other nations states, otherwise they'll say, 'Sorry, we're gonna look after our own self.' So, who might that be? I don't know, I'll think about it.

Respondent 7 also noted

> Well, so where's, who's got the power? Who's got the authority? And what would cause the G7 to screw it up enough to give it away? What would cause say the BRICS to get enough clout, what would be their counter move to side-track with G7? I think it's difficult to see a consensus... the power dynamics would change... it's going to have to be quite a different move.

As highlighted in Section 5.2.4, "power is the exhibition of authority" allowing those with power to "maintain" and "increase" what they have.

With regards to the sample of possible futures, two different types of coalitions were identified, consisting of a coalition of powerful states and coalitions of less powerful states.

Coalition of powerful states

Respondent 21 noted "globally at the moment, it's really a handful of large players that are causing the majority of emissions and really", and "you need them around the table agreeing on a way forward to solve that". As such, "it's getting China and the US and the EU and India and Brazil or not, maybe a handful of others to agree on a way forward for this." In short, it's about getting an agreement "That's as beneficial across the, you know, across them on this as possible." Then, "everyone else has to fall in line anyway because they want to trade with the big trading partners".

Respondent 6 noted that for a stringent enforced agreement to work it would need "a club of 6 to 10 countries who are the core of the agreement and don't renege". With regards to other Parties, "anybody else just comes along for the ride. And if you do get a rogue nation, like Australia, they get slapped down with trade sanctions again, but, again that's minor stuff... that's not a major part maintaining stability." The respondent noted, "It's strategic stability done by people sitting in rooms talking to each other."

Respondent 7 stated that a stringent enforced agreement could be "enforced by a group." The respondent suggested that this would involve "an elite over a minority"

(Section 5.2.4) and with regards to a stringent enforced climate agreement, "it's going to be very uncomfortable" particularly for less powerful states and actors.

Respondent 16 put forward a scenario where "the big emitting countries want to be part of it, but... they feel the others are not contributing enough... because it shouldn't be only on them. So, they can actually say, 'No, we'll only do it if these countries also come on board.'" thus forcing smaller less powerful states to participate.

Coalitions of less powerful states

Respondent 16 put forward another scenario where "big countries that are emitting don't want to be part of" a stringent enforced climate agreement, then "75% of the countries can go there and say, 'You know, no this is how it will happen.' Although in terms of warding off greenhouse gas emissions, they might be 30%, 40%." As such, a large coalition of small emitters could conceivably be formed leading to a stringent enforced climate agreement.

Respondent 12 noted

> If ASEAN is able to... have more power over its natural resources that's good leverage against consumers like China and the US. Or for the bigger countries. Maybe in our terms, 'Okay, you want our minerals? You want our fish? You want our gold? Want some of our forests, and the rubber? And the palm oil for the chips? You have to agree to these.'

Then at the regional level there can be "more stringent application or formulation of the climate agreement."

8.3.4   Stringent enforced international agreement

The semi-structured interview included a follow-up question asking for scenarios that would result in a stringent enforced agreement. In a few cases, respondents volunteered stringent enforced agreement scenarios without being asked the question. Important themes and scenarios identified regarded: national interests; business interests; voluntary stringency; climate change being elevated to the United Nations Security Council; transparency, reporting and verification; enforcement and compliance; and, non-compliance with agreements.

National interests

Respondent 6 noted, "A stringent international agreement rests on the foundation of national interest, so if the major powers want it to be stringent it'll be stringent." The respondent stated

> I don't believe trade sanctions, or so-called international law is what keeps order, agreements together. I think people stay in agreements because they are in their interest at international level, what the agreement does is make

it easier to stay together and to kind of fossilize the politics, and ratchets, in, and make defections costly in terms of broader international politics which is important, is not, everything else.

Furthermore, "trade sanctions and enforced environmental agreements only work if small countries defect." as opposed to large countries defecting.

Business interest driven scenario

From Section 6.4.1, Respondent 21 suggested a coalition of business interests could drive a stringent enforced international agreement for example including a carbon tax or climate change related trade tariffs.

Voluntary stringency

Respondent 26 provided a scenario where there is voluntary stringency. Preconditions included, "nationally determined contributions are well-planned", "the political will is there, then it could be, it could be achieved.", and "the Paris Agreement implementation framework is effective, with the report back, the stock-taking, the transparency framework". The respondent noted, "all of those things are very difficult. [laughs]". According to the respondent, this would mean eliminating the ambition gap by 2019 so "these national determined contributions get us to the target by 2050...."

Climate change elevated to the United Nations Security Council

Respondent 13 put forward a stringent enforced international agreement without prompting. The scenario starts with a situation where "the US comes to the party and even though the[y] are only, what is it 15% global emissions or something. Politically, it makes a huge difference if they are cooperative" and this "returns normalcy to the UNFCCC entire machinery". Then there also has to be "a new resolve based on the 1.5 report and other reports" and "we all agree that we have to stay on a 2.6 [RCP]" followed by "an extraordinary array of revised INDCs voluntarily submitted prior to 2023." At this point there is "another counting", the scientists note that there has been progress but not enough, and States increase their ambition, coupled with "a realization by Europe and North America and Japan, and North Korea and Australasia, that we do have a historical obligation." Respondent 13 then noted "The second step is do you succeed each of those... revised INDCs in that same year is getting close to 2.6 RCP and therefore, on track to two degrees. Then you have to make sure that somehow that's enforced." From Section 8.2.3, Respondent 13 noted sanctions would involve the coordination between the UNFCCC, WTO and Security Council. The respondent also noted that in case of a climate change driven global emergency, leadership would need to be taken by the Security Council with support from the World Bank, WTO, and the UNFCCC serving as a negotiating forum.

Respondent 25 also put forward a scenario where the United Nations Security Council is "revamped" and its members "put climate as a central issue there, which maybe doesn't get determined from the Security Council. But it's directed or led, shall we say." Then

> In terms of a particular punitive mechanism… if you get the Russians the Americans and the Chinese and the Indians, the Brazilians and a few others… a representative from Africa… [they] Wouldn't have to. They all said, do it. A list of consequences from trade embargoes to, you know. They would be able to make them a series of offers, which they would not be in a position to refuse.

Transparency, reporting, and verification

Respondent 6 also stated,

> the other key bit of the regime which is incredibly important is transparency… So, we [are] arguing about real, as opposed to arguing about the facts, which is the best way to show distrust and to fracture the agreement…. The core end of the regime is providing that transparency… The other bit is it gives a platform for the most vulnerable countries to put pressure on the big countries which was quite important in the first stages of the regime. It moves the needle a bit. Make defection more costly for China and India if loads of poor countries in Africa are saying, 'You are fucking us over.'

As noted in Section 8.3.4 in such a regime it is essential to have independent and credible verification of emissions from countries, including the possibility of "IAEA style" country inspections.

Respondent 14 stated,

> The best way to do [it] is from international scrutiny of performance and that requires, signatories to collect and publish accurate information on what they are doing and you will know well as I do that's a pretty patchy sort of system…if my crisis scenario is right, then the incentives for good behaviour on the countries or regions that are being most severely affected are fairly high for good behaviour. But I'm not quite sure how you would enforce it.

Enforcement and compliance

Respondent 23 stated

> I think that any form of enforcement is unlikely to be motivated by the document of the text, and by the doc, by the climate global aspirations. I think it's more likely to be about geopolitics and, you know, keeping someone in check with the excuse of a symbolic platform of the climate.

Furthermore, the respondent stated, "I think that as long as nation states, want to retain the concept of national sovereignty above global well-being, chances are that there will be at least one entity that will say, "Screw the collective. I want to be better." And if that entails, you know, violating the written text of the agreement, they're gonna try to get away with doing it."

Respondent 17 stated "At this point, I think because it's kind of five minutes past midnight", for a stringent enforced agreement "you probably would have to have some policy where you just started taking things away from people."

Respondent 14 noted "I'm not sure about enforcement. I've never been really confident that any of the UN agreements have been particularly well enforced." With regards to "What might it look like? It might involve sharing technologies, so quick transfer of technologies, where technology is part of the fix. It might involve relaxation on international barriers to trade, to migration, to financial systems, aid."

Respondent 9 highlighted possible mechanisms, for example for Parties to be

able to utilize the perks of the agreement such as market or you can also tie it to finance so if a nation didn't meet its obligations they would not be able to receive finance for climate change adaptation et cetera.

Furthermore,

you could go broader and link the Paris Agreement to other multilateral agreements and so if nations weren't compliant with them to reduce emissions, they could be subject to trade sanctions or economic penalties... or the loss of rights under other conventions like biodiversity etcetera, the loss of status on world heritage and stuff.

Respondent 18 suggested: "one enforcement mechanism could be trade. The access to markets could be contingent on proper action to reduce emissions and to support others in reducing emissions."; "other enforcement mechanisms could have to do with requirements in terms of payments to compensate others."; "Although, still a little bit of a challenge but dues responsible for international institutions could be higher for those that aren't taking, aren't meeting their mandatory commitments."; and "if you're a developing country, you're less able to receive aid, that could go in either direction based on your progress."

Respondent 3 noted

If you're holding governments to account to achieving Paris, first of all, there needs to be a way of portioning targets to those governments and that politically proved impossible as we saw in Copenhagen. So, we went with the bottom up, you know, everyone say what you can do.

Respondent 10 suggested, "Montreal protocol which has been implemented and the strong commitment with which it was implemented, you know? If I see a successful case, successful experience and similar thing can deal with greenhouse gas emissions."

Respondent 10 noted that for a

> stringent agreement you put also the tools for its implementation. It's laws, regulations and, different enforcement mechanism. I think there is, they can use experience from… Kyoto protocol because there are compliance and enforcement mechanism, yes, to make sure that this is implemented. So, I think there is enough experience if the world is committed they don't lack, mechanism. There is laws, there are everything. What is needed is the commitment to implement enforce, the political commitment.

Non-compliance with agreements

Respondent 6 addressed incentives and national interests, stating

> The incentives on countries that defect from a stringent environmental agreement are, if they have lots of cheap fossil [fuel] and they want to export lots of stuff. Now that solar is cheaper than fossil in most of those countries, it is really unclear whether we['re] gonna see that kind of small energy dense Saudi, Australia, Russia defection because actually quite a lot of them have lots of desert and lots of sun and so they can find low carbon alternatives which are as lucrative for their energy intensive industries.

However, this assumes that energy is substituting energy. In energy exporting states, it is export revenue that needs to be substituted rather than domestic energy consumption which is what renewable energy addresses in many cases. It may be possible that some countries can substitute fossil fuel-related export revenue with revenues from the export of minerals required to support an energy transformation (IRENA 2019).

8.3.5   *International cooperation options*

There are many possible international cooperation options. Respondent 18 alone provided a set of 13 possible international cooperation options related to climate change, addressing themes such as information, sanctions, penalties and compensation, finance, oversight and regulation, subnational government participation and private sector participation. These themes and others are discussed below, including the possibility of focusing on research and development; focusing on what works; carbon budgets and prices; international support for localisation; subnational participation in the Paris Agreement; climate finance; and the implementation of conditional and unconditional NDCs.

Focus on research and development

From Section 5.5.2, Respondent 15 indicated that dramatically increasing spending on research and development is important for an effective global response. As such, Respondent 15 suggested the

> UNFCCC should focus a lot more on getting nations to spend money on research and development. It's much cheaper, it's much easier, it's also much

easier to validate 'cause you can do it. You know, you can, say, just at the end of this year, you can actually see in the budget for the next year how much money you're gonna spend.

Focus on what works

With regards to focusing on what works, Respondent 3 noted the need for an international regime that consists "less of policing and more a celebration of progress that's real, sort of real economy shifts, rather than words." A key assumption is that

> if our attention is around positive deviance and what are the most effective interventions within a sector or within an economy, or how to incentivize significant and real behavioural change or change in decision rules... we begin to create a race to the top, rather than policing at the bottom.

In practical terms the respondent suggested that "we need the UNFCCC process, and instruments like the Talanoa Dialogue, the Global Stock Take and the compliance mechanism's to be doing is highlighting success, rather than policing failure." With regards to information and signals that might help, "the doing business rating is a good analogy for this. Ease of doing business rating highlights the countries that are doing best." It could be possible to "set benchmark, efficiency, carbon intensity of segments of the economy."

Carbon budget and price

From Section 7.4.3, Respondent 17 noted that an effective carbon price would require global governance to set a carbon budget each year. The respondent highlighted that this would require agreement between 190 states. Given that there would be winners and losers, a carbon tax would likely involve transfer payments, border taxes, and other mechanism all of which become very complex given the number of states and diverse interests involved.

Localisation

Respondent 6 noted that to initiate the localisation, "this requires horizontal diplomacy... this sort of city and business and investor grouping." and it is essential to have "a load stone in the international agreement, an idea of we're going somewhere, we can see everybody moving, and a clear sense of transparency." Respondent 6 also noted "you still need enough state-to-state glue to keep that going, then I think you can build out those coalitions quite strongly" and as such localisation is a matter of political will and policy regarding centralisation and decentralisation of power between state actors and non-state actors.

Subnational participation in the Paris Agreement

The sample of possible futures included the possibility of non-state actors participating in the Paris Agreement. For example, Respondent 26 noted it is possible that even with the United States withdrawing from the Paris Agreement, "you've just got states coming forward and saying, "Well, we're gonna stick with these targets and these goals then we're gonna find ways to working and feeding in what we're doing, our actions." Then and obviously cities". As such it is possible that there could be the establishment of "a framework within which that they can contribute and play a big part of" Respondent 26 also noted

> And that's why I think Paris is very interesting. It's like the architecture could exist without states. You have to have some stock taking and transparency with that reporting and engagement across different actors who are the people that are contributing to the reduction in emissions and the ones that are contributing to the adaptation.

With regards to adaptation, Respondent 18 stated "I think you'll bring in the sub-national governments, I think probably outside the UNFCCC, but there'll be processes for industry and for commitments and collaboratives."

Climate finance

Respondent 16 noted that developed countries needed to provide funds because "A lot of private actors, state actors changing their ways of doing business… catalyzed by the GCF or any other, scenario in both the business they do in the developing world or in their individual countries." The respondent also noted that developing countries could have a policy of "screening the right type of businesses to come to their country" for example including businesses with "resource efficient technology" or business "investment in the renewable sector".

Respondent 9 highlighted the need for large developed countries as well as large developing countries with high emissions, to engage in mitigation. Furthermore, the respondent noted the need for "developed countries to own up to their responsibilities reducing emissions". International cooperation could come in the form of "funding, climate adaptation and mitigation projects in developing states to share technology, share environmental technologies to build capacity of other nations so that more people can live in a safer world." The respondent also stated, "I think for finance historically developed countries, so the West, the US, Canada, the EU, Japan, Russia, historically the West but it's developed."

Conditional and unconditional NDCs

Respondent 3 noted, "because there was so much hot air, or you know, the country NDCs were actually quite low ambition, we're going to see a number of countries go further than what they said they'd do."

Respondent 10 noted conditional NDCs would require "developed countries provide… what they committed in terms of finance, in terms of the basic evolving technology". Furthermore, developed countries would see "developing countries also committing, at the highest level to contribute through the domestic tasks, using domestic resources and also efficiently use resources from, developed countries to implement, yes, measures that can contribute to green growth". Importantly, in an impact response scenario, there could be "a common goal.", i.e., common national interests.

8.4 Conclusions

International cooperation scenarios highlighted the tension between globalism and nationalism. Nationalism was seen as a threat to a rules-based international regime, limiting cooperation on climate change. In other scenarios, it is possible that climate change becomes the premier global issue, due to its impacts, and a focus of international cooperation. Scenarios included the UNFCCC becoming a secondary body coordinating climate action with the United Nations Security Council leading the global response. Other scenarios included the possibility that business interests dominate international cooperation and the global response.

In addition to these scenarios, possible triggers and drivers for international cooperation were identified, ranging from a desire to maintain a rules-based international order through to enlightenment. Scenarios included themes of leadership, geopolitical power, stringency, and international cooperation. These scenarios and themes are mapped out in Section 10.5 showing combinations that could contribute towards fulfilling the UNFCCC objective. Meanwhile, coalitions and effective international cooperation are discussed further in Section 11.8.

9 Other unusual climate-related scenarios

9.1 Introduction

Respondent 23 noted "we, climate people, put climate problems too much at the centre of the future." while the outlook is "very bad" there are "many other things that can go much worse including nuclear [war]." The message I took from this is we need to consider other scenarios including how they could affect climate or the global response to climate change. There will be many events and other changes that affect us, the climate and our global response, given the time it will take to resolve climate change. Some of these events will be historic and not widely anticipated.

When looking into the crystal ball (i.e., sample of possible futures), other scenarios and themes include cyborgization including socioeconomic implications (Section 9.2); enlightenment (Section 9.3); volcanic eruptions (Section 9.4); solar radiation management (Section 9.5); population decline (Section 9.6); conflict (Section 9.7); catastrophic cooling events (Section 9.8); responses to other events or crises (Section 9.9); and environmental feedbacks (Section 9.10).

Like in Chapters 5–8, this chapter is much like reading a set of short stories, some of which are related to each other, many of which are not. You may want to read all these scenarios or you may want to go straight to the conclusions in Section 9.11 which summarises the themes from this chapter and indicates where these themes are analysed and brought together in Chapter 11.

9.2 Cyborgization makes homo deus

After noting they had spent some time in San Diego, Respondent 11 said, "the capacity of biotechnology to change the way people look, live, their abilities to integrate technology into the human body to do genetic modification", then "we start to see sort of the elite become more separate from the rest of humanity. That's, I think, a possibility." Meanwhile, "the rest of humanity sort of suffers along with a more and more degraded environment." Respondent 23 put forward essentially the same scenario, and stated, "I think the developments in biotech, genetics cyborgization" make it "inevitable that within a generation, within 50 years, the possibilities of augmenting individual powers, like the premise of the book, the Homo Deus, man god" Then there is a situation where there is, "the bifurcation of a human species

DOI: 10.4324/9781003465911-11

between the haves and the have-nots". The respondent also noted that the split between the haves and the have-nots is already happening.

9.3 Enlightenment

Respondent 23 stated "I really think that there needs to be reckoning. I think we need to get sufficiently close to the edge of our humanity… [to] …recognize that we need to work together to help each other." The respondent suggested that "Breakthroughs in artificial intelligence" could "make people recognize that we have to become a more, solidarious species." Respondent 8 stated that "a war, a massive world war or something" is "probably going to take away the focus from the environment, and other things." but might ultimately result in "a huge change in global leadership somehow and people's mentalities change, and their priorities."

Respondent 22 noted that it is possible that there could be some other crisis such as global pandemic or bug causing population decline, which "would [affect] markets, it would affect politics". Furthermore, it could "demonstrate the value of global cooperation", and "it may enhance global cooperation" and "it might bolster the strength of the UN".

Respondent 25, when asked about other scenarios, noted "Well, my favorite, of course, is contact with another sentient species.", "Which… focuses the mind."

9.4 Volcanic eruptions

Respondent 14 said "Massive volcanic eruptions… could really skew the concentrations big time." increasing atmospheric concentrations of greenhouse gases (GHGs) through a natural process. The respondent went on to say, "The volcanic eruption is a worst-case scenario for me, because that introduces a feeling of hopelessness, you know, you've got no control."

Respondent 18 noted there could be "a period of lots of volcanic eruptions.", then "Dust and aerosols in the atmosphere… slow down the warming" for a period of time, but "then we'll have a big jump up in the warming." Respondent 23 also put forward a volcanic eruption scenario and made the link with solar radiation management-related responses (Section 9.5).

Respondent 23 noted it is possible that there could be another "Tambora-like volcanic eruption". This

> blocks so much sunlight that the planet goes through a phase of severe cooling where the science would say irrefutably that it's like a couple of years and then we go back to normal, so the climate change problem, the global warming problem will not go away.

The respondent also noted, "it may buy us a little bit of time." Based on the experience of this event, "it may be, 'Holy smokes' blocking sun that like can go so wrong that, you know, we enact a law that says it cannot be done." Alternatively, "It may end up… naturalizing geo-engineering".

9.5   Solar radiation management and geoengineering

Respondent 15 noted that to avert catastrophic risks such as the Gulf Stream switching off or the Western Antarctic Ice Sheet "tipping into the ocean" then

> ...the only way that we can actually do anything about this would be through geo-engineering. That's the only way that we can do something in, you know, say a couple of years or maybe even a couple of days... compared to any climate policy that would really take 20, 30, 40, 50 years to manifest itself in the climate system.

The respondent also suggested, "if you're worried about these black swan events you should really be focusing a lot more on at least tracking our ability to do geo-engineering."

Respondent 20 put forward a branching scenario where there is "even more rapid melting of Antarctic and Greenland ice sheets and those sorts of things." or fast feedbacks (Section 9.10.3). In response to this, there is "solar radiation management", specifically it is possible that "an individual country or even an individual wealthy person, a billionaire, could just start doing it". Respondent 20 explored possible geopolitical responses to such unilateral solar radiation management.

Respondent 20 noted that "If it was a smaller state, I think it would become an issue at the United Nations level and possibly at the Security Council." This could result in one of three possible situations: where the unilateral action could "force the issue, around reducing greenhouse gas emissions"; there could be a "fracturing of the United Nations body, of those countries that support them and those that don't."; and, it is possible that to "stop them" there is "military action", and as such unilateral solar radiation management "could lead to war."

Alternatively, Respondent 20 noted that "if it was the United States or a China, I think people would act vociferously, or vocally, but there would be nothing left they could do about it." Respondent 20 also noted that unilateral solar radiation management could lead "others to start doing the same".

9.6   Population decline

Respondent 11 noted it is possible that there could be "a radical collapse in the numbers of humans on the planet" for example due to "a health issue rather than say a nuclear war or rather than a slow progression of climate change leading to social disruption leading to you know collapse of the global economic order or food production systems". The respondent noted that something like this could be "the earth's solution".

9.7   Conflict

Respondent 23 noted that the possibility of war for example due to macho politics, and that this could include: "scrimmages"; "some really stupid decisions will lead to war at the regional level"; and, "some really stupid decisions will lead to...

devastation". Respondent 23 noted that dealing "with global climate change, requires global institutions." But cooperation would be undermined if other states and their leaders don't play by the rules i.e. "if someone is going to kick the board." The respondent also noted that conflict "will have the result of dissolving the trust in multilateral systems that keep us safe from each other."

Respondent 6 also highlighted the possibility of "an outbreak of hot war in various places". The respondent noted that "the distraction caused by hot war has already been a real problem for climate politics." as "bandwidth of prime ministers and security operators... It's completely been taken by hot wars." Such a scenario would "lower the willingness to take on [climate] fights."

9.8   Catastrophic cooling events

Respondent 25 noted there are multiple possible scenarios that could result in a catastrophic cooling event. These include: "A series of volcanic eruptions"; "a limited nuclear war, which would hopefully sober people up a bit."; and, "massive disruption or a meteor strike or something like that." These events could "precipitate rapid global cooling, knocking global economy back, or population back, through loss of agricultural production for a decade or so." This could also "take the population of the planet back from the eight or nine to three or four billion" constituting "an existential threat in the short term".

Respondent 25 noted that after the event, it is possible that there will be "some semblance of human societies" and perhaps "a chance to start over in a way.", because the event has happened

> without destruction of the whole fabric of civilisation as we know at present. Some of the institutions would survive, maybe the UN survives, maybe, you know, the telephone networks survives, the internet survives, or loses chunks of it for a while.

Respondent 25 also noted an alternative scenario, where the event would "blow our civilisation down to the level of, sort of, medieval warlords and shit like that." It is even possible that

> human civilisation is wiped out all but for a few leftover pockets of humanity going back to, neolithic hunter gatherer kind of things. If that happens, well, I mean, you would've fulfilled Article Two, but nobody would know what the fuck that was.

9.9   Responses to other events or crises

Respondent 22 provided a scenario where there is some other global event and asked if it is climate change related. If so, it may help generate a global response to climate change, but if not, it may distract from the global response to climate change.

9.10 Environmental feedbacks

From the sample of possible futures, several environmental feedback themes and issues were identified, including methane-related feedbacks (Section 9.10.1); ocean heat release (Section 9.10.2); and feedback-related ecosystem adaptation (Section 9.10.3).

9.10.1 Methane feedback

Respondent 17 expressed concern around permafrost melt-related GHG emissions and methane hydrate stability, noting it is possible that "you just end up with this you know, 10-year very dramatic spike because it doesn't have such a long atmospheric life". The respondent suggested that such an even "would be very hard to detect in a way because we're not emitting it. And the frozen tundras not a lot of people there." The respondent also asked, "And how would you stop it?"

Respondent 21 expressed similar concerns around permafrost-related GHG emissions and noted that "it could be responded to in one of two ways." including "look after ourselves." or "let's start doing something about it together". Panicking is also a possibility. The respondent noted "if you're in rich country ... or as an individual will be fine and hence that's why you don't do it". From Section 5.5.1 the respondent indicated cooperation is much more beneficial.

9.10.2 Ocean heat release

Respondent 18 noted "oceans have been making things better right now.", absorbing heat and taking it to the depths, but when this heated water comes to the surface again, atmospheric "warming is going to accelerate." The respondent noted, "Impacts will be even more severe [and] hopefully that means our actions and response will be quicker." It should be noted that there is a question as to how long such heat release would take to manifest.

9.10.3 Feedbacks and ecosystem adaptation

Respondent 20 noted, "something could happen as a combination of positive feedbacks, where we have a massive rapid warming... fast feedbacks" for example due to "the release of a lot of CO_2 and methane from the melting permafrost in the Northern Hemisphere." Then this, "begs the question whether ecosystems have the ability to adapt to changes and a fast feedback for greenhouse gas emissions rapidly increasing atmospheric concentrations it would be analogous to a super volcanic eruption or a meteorite impact for example, that we've seen in the geological past "and "we could see a collapse in agricultural systems and a mass die off of species." Respondent 20 also noted,

We know that the planet, is extremely sensitive to even minor environment changes in temperature, especially increases in temperature so we know the climate, it takes several thousand years for the planet to cool into an ice age, but over a few centuries, the temperature can rise significantly.

Furthermore, in previous warming events over a short geological period of time, "ecosystems have adapted because... plant and animal communities" were "still able to migrate." However, "with changes that are happening now potentially so quickly that they won't have time to" adapt, and, "humans have altered the planet so much we've got cities and roads and farm areas... which present barriers to migration and natural eco-systems, so it's a double whammy."

9.11 Conclusion

Given the long periods of time being considered when addressing climate change, it is important to be aware of other scenarios, many of which are unusual or not widely anticipated, might influence climate or the global response to climate change. From the sample of possible futures, a mix of wild and human events or processes were identified that could influence climate or the global response to climate change. Possible wild (i.e., unmanaged) events and processes include volcanic eruptions, environmental feedbacks and catastrophic cooling events. Possible human-driven events and processes include cyborgization of humanity exacerbating inequities, enlightenment, population decline, geoengineering as a backup option to feedbacks, conflict including nuclear war, as well as the possibility of solar radiation management technologies being deployed. These other scenarios could influence climate or the global response and are discussed further in Section 11.10.

Part III

Lessons learnt

Part III Lessons learnt consists of three chapters:

Chapter 10 analyses themes and scenarios from the searchable sample of possible futures, mapping possible climate change signals, responses, and pathways.

Chapter 11 discusses the key questions from the introduction with a focus on preconditions for effective global responses to climate change.

Chapter 12 summarises the preconditions for effective global responses to climate change, highlighting lessons learnt, new knowledge and further research needed. The chapter finishes with a question and challenge: What will we do? Hopefully our "better angels" prevail.

DOI: 10.4324/9781003465911-12

10 Analysing climate signals, actors, and responses

10.1 Introduction

Having looked into the crystal ball in Chapters 5–9, it is time to step back and consider what we have found across the "multiverse" of 175 possible scenarios. This chapter synthesises themes and pathways from across scenarios – identifying the system "conditions" needed for climate action, including relevant signals, responses, actors and interests. These system conditions help us identify "preconditions" for effective global response in Chapter 11.

Impacts and risks from Chapter 5 and response options from Chapter 7, are brought together in Section 10.2, addressing the question of how climate change impacts and risks might influence the global response to climate change? This leads to the development of a climate change signal response model and a conceptual framework for understanding climate response decision making and related actions.

Actors, interests, and actions from Chapter 6 are analysed in Section 10.3, identifying the relationships between society, government, and business (as a system), and mapping possible pathways involving social change and behaviour, political will and policy, as well as business and economic activity.

Based on greenhouse gas (GHG) removal scenarios from Chapter 7, Section 10.4 analyses preconditions for actors to remove atmospheric GHGs at a scale required to limit climate change to safe levels. International cooperation scenarios from Chapter 8 are analysed in Section 10.5 identifying preconditions for effective international cooperation on climate change.

The signals, responses, actors, and pathways from this chapter are used in Chapter 11 to help identify preconditions for effective global responses to climate change.

10.2 The global response system

The CCNIIC Model of the global response system was developed in Section 3.4.2 addressing interactions between climate change, national interests and international cooperation, but did not address the question of how actors might respond to climate change impacts and risks. Table 10.1 follows the CCNIIC Model, provides definitions for overarching themes from Figure 4.2 (Section 4.4), and makes a distinction "impacts and risks" and "responses".

DOI: 10.4324/9781003465911-13

Table 10.1 Summary of overarching themes and definitions organised in relation to the CCNIIC model

Overarching theme	Definition
Climate change	
Physical systems	Any system in which physical processes play a major role. This includes glaciers snow, ice, permafrost, rivers lakes, floods, drought, coastal erosion, and sea level effects.
Biological systems	Any system in which organisms play a major role. This includes terrestrial ecosystems, wildfire and marine ecosystems.
National interests	
Human and managed systems[a]	Any system in which human organisations and institutions play a major role.[a] This includes food production, livelihoods, health and economics.
Responsiveness	The extent to which human actions, or inactions, are timely and sufficiently scaled to fulfil the UNFCCC objective.
Social change and behaviour	The interests of people, as individuals, households and communities, and the actions, or inactions, these people might individually or collectively take.
Political will and policy	The ambition level of government leaders, and others in government when it comes to positions on climate change, and the interventions they make that influence fulfilment of the UNFCCC objective.
Business and economic activity	Actions, or inaction, by individuals or groups undertaking productive activities, in many cases driven by a profit motive.
Technology and practices	Methods including objects, activities, rules and knowledge used when attempting to achieve something.
Other factors	Other things that can influence the global response to climate change.
International cooperation	
International regime	The rules and norms that guide interactions between actors from different states. This includes state actors and non-state actors.
International cooperation on climate change	Actors from different states acting together to address climate change. This includes state actors and non-state actors.

The left margin labels: **Impacts and risks** (spanning Climate change and National interests sections) and **Responses** (spanning the lower sections).

[a]*IPCC (2018b).*

The distinction between climate change "impacts and risks" and "responses" by actors is important as it provides a basis for understanding how actors might respond to climate change. Based on this distinction, the results of the thematic analysis in Chapter 5 and concepts from Chapters 2 and 3, a "signal response model" for the global response system is developed, with Section 10.2.1 analysing the climate change signal including signal strength and Section 10.2.2 analysing possible responses.

10.2.1 Climate change signals

A signal is information about a situation, system conditions, or phenomena that influence decisions. While signals generally regard current situations or system conditions, they can be used to monitor phenomena and assess future risks. From Section 2.2.3, the TCFD identified two broad types of risk related to climate change, consisting of physical risk and transition risk (TCFD 2017). From the sample of possible futures, both physical and transition risks are identified along with impacts. The sections below analyse physical signals (i.e., impacts and risks); physical signal strength; and, transition risks including the influence these risks might have on actors and their ambition levels.

Physical signals

Physical signals regard climate and related physical hazards from the CCNIIC Model in Section 3.4.2 and the interactions these hazards have with domestic state and non-state actors. Conceptually, physical signals could include an actor's own experiences, news reports, research or risk analyses. From the literature and sample of possible futures, there is a range of physical signals identified, including climate science (Section 2.2.2); impacts on human and managed systems (Section 5.4.2); risks to human and managed systems (Section 5.4.2); impacts and risks to natural systems (Section 5.2.3); and the IPCC's five reasons for concern (Section 2.2.3).

Importantly, from the sample of possible futures, impacts and risks to human and managed systems are the only identified drivers of effective global response scenarios (Section 5.2.3). Following the IPCC's conceptual model for understanding climate change related risks (Oppenheimer et al. 2014 in Section 2.2.3) impacts and risks to human and managed systems consist of climate change and related hazards (i.e., climate stress) coupled with the exposure and vulnerability of actors and their interests (Figure 10.1).

It is important to note here that the concept of "climate stress" builds on work by Warner and van der Geest who identified and defined "climate change stressors" as the "Manifestations of climate variability and climate change in specific ecosystems (for example, rainfall variability, droughts, floods, cyclones and tropical storms, glacial melt, sea-level rise, etc.). This could involve extreme weather-related events and more gradual changes." (Warner and van der Geest 2013, p. 369). Given that "people are integral parts of ecosystems" (Millennium Ecosystem Assessment 2005, p. V), human and managed systems can be stressed by these manifestations of climate change.

Even though climate change signals generated by "climate science" or "impacts and risks to natural systems" don't appear sufficient to generate effective global responses to climate change based on the sample of possible futures (Section 5.2.3), that doesn't mean they don't create a signal. Climate change signals from "climate science" or "impacts or risks to natural systems" are related to actor interests by "concern" rather than "exposure" or "vulnerability" of actors, their property or livelihoods. Hence the dashed lines linking "climate science" and "climate change and related hazards to natural systems" to "actor interests" in Figure 10.1.

The IPCC's "reasons for concern" include the risk of impacts to a mix of natural, managed, and human systems (Section 2.2.2) and represent forms of climate stress. With regards to generating a climate change signal, the "distribution of impacts"

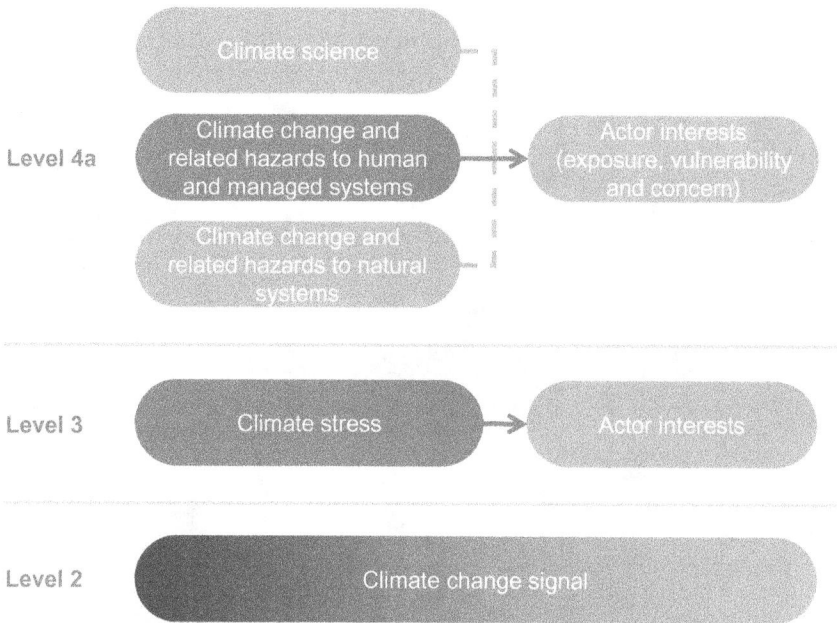

Figure 10.1 Climate change signal as a function of climate stress and the risk to actor interests.

(i.e., RFC 3) is particularly important as it is the distribution of impacts that dictates which actors and interests are affected and the extent to which they are affected by various "reasons for concern" (Figure 10.2).

Note: Levels in Figure 10.1, Figure 10.2, and subsequent figures refer to levels of disaggregation, with concepts broken down further into constituent parts with each additional level.

Physical signal strength

To understand the strength of the climate change signal, a qualitative analysis was made of IPCC assessment report summaries for policy makers, as well as the IP-CC's projections regarding reasons for concern and the risk of impacts. From Section 2.2.2 and Appendix A, it's possible to get a qualitative sense of the strength of the climate change signal, by focusing on impacts on human and managed systems. For example, the IPCC's first assessment did not identify any impacts, and the Second Assessment Report stated, "Unambiguous detection of climate-induced changes in most ecological and social systems will prove extremely difficult in the coming decades." (IPCC 1995, p. 6). It was not until the Fifth Assessment Report published in 2014, that the IPCC stated, "Some impacts on human systems have also been attributed to climate change, with a major or minor contribution of climate change distinguishable from other influences." (IPCC 2014a, p. 6).

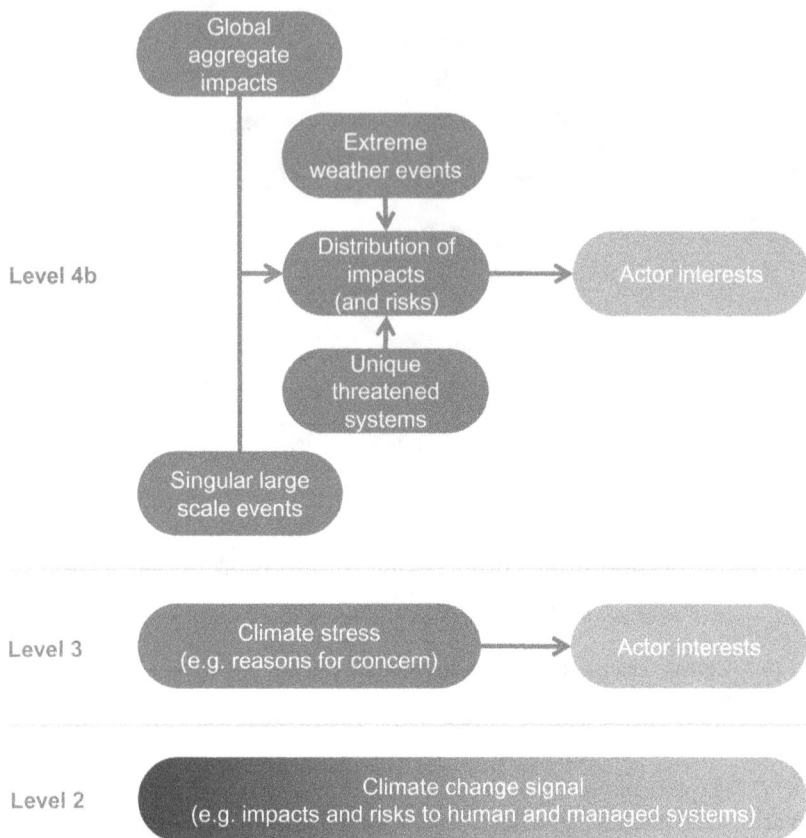

Figure 10.2 Climate stress based on the IPCC's reasons for concern and the distribution of impacts and risks to actor interests.

According to the IPCC's reasons for concern, with current levels of global warming at 1°C, the risk of impacts is undetectable to moderate. It is possible to get a sense of how the climate change signal might strengthen because the IPCC has the risk of impacts against global warming levels for the five reasons for concern and other concerns regarding natural, managed, and human systems (Section 2.2.3). While the IPCC (2018a) did not state which of these other concerns can be considered part of human or managed systems, small-scale low-latitude fisheries, coastal flooding, fluvial flooding, crop yields, tourism, and heat-related morbidity and mortality are taken to be human and managed systems that collectively contribute to climate stress signals. As such, Table 10.2 is based on the IPCCs assessment and shows physical stress levels at 1°C, 1.5°C, and 2°C.

The results from Table 10.2 are presented together with climate stress levels (see first paragraph of this section) to create a qualitative time series of climate stress levels in Table 10.3, including past, present, and projected estimates. Importantly, from the analysis, a definitive climate stress signal only emerged in 2014. There

Table 10.2 Summary of impact and risk levels on human and managed systems at 1.1°C, 1.5°C and 2°C of global warming, including generalised climate stress levels

Reasons for concern:	Current risk of impacts (~1.1°C of global warming)	Risk level at 1.5°C	Risk level at 2°C
Food production from crops, fisheries and livestock in Africa	Moderate	Moderate to high	High
Mortality and morbidity from heat and infectious disease in Africa	Moderate	Moderate to high	High
Delayed impacts of sea level rise in the Mediterranean	Moderate	High	Very high
Water quality and availability in the Mediterranean	Moderate	Moderate	High
Health and wellbeing in the Mediterranean	Moderate	Moderate	High
Water scarcity to people in southeastern Europe	Moderate	Moderate to high	High
Coastal flooding to people and infrastructures in Europe	Moderate	Moderate	High
Heat stress, mortality and morbidity to people in Europe	Moderate	Moderate	Moderate to high
Cascading impacts on cities and settlements in Australasia	Moderate to high	High	High to very high
Reduced viability of tourism-related activities in North America	Moderate	Moderate	Moderate to high
Costs and damages related to maintenance and reconstruction of transportation infrastructure in North America	Moderate	Moderate	Moderate to high
Lyme disease in North America under incomplete adaptation scenario	Moderate	Moderate	High
Changes in fisheries catch for Pollock and Pacific Cod in the Arctic	Undetectable (or barely detectable)	Moderate	Moderate to high
Costs and losses for key infrastructure in the Arctic	Moderate	Moderate	Moderate
Changes in krill fisheries in the Antarctic	Moderate	Moderate	High
Physical stress	**Undetectable to high (mostly moderate)**	**Moderate to high**	**Moderate to very high**

Source: Author compiled from IPCC (2022a).

has been evidence since the IPCC started making assessments that the climate is changing and this constitutes a risk, but the signal most important to actors, in the form of impacts on human and managed systems, only emerged the year before the Paris Agreement was negotiated. Now the signal is rapidly strengthening, and the signal is expected to continue strengthening. This marks a fundamental and unprecedented shift in system conditions, one where the climate change phenomenon increasingly and directly affects actors and their interests. The implications of a strengthening climate change signal, from Table 10.3, are discussed further in Section 11.6.

Transition risks

Transition risks and opportunities regard response-oriented options from the CC-NIIC Model available to state and non-state actors (Section 3.4.2). Transition risks include changes in policy, consumer behaviour and economic activities. The influence transition risks have on climate ambition and related actions depends on the situation and perspective of the actor. If a transition is perceived to be a cost, then a transition is a risk to the actor and the actor's climate ambition levels may be suppressed or be negative (e.g., cynical responses in Section 10.2.2). If a transition is perceived to be a benefit, then a transition is an opportunity and the actor's

Table 10.3 Qualitative assessment of climate change signal strength showing signal strength is expected to increase in the future

Year	Signal strength	Impacts on human and managed systems
1992	No signal	No impacts identified
1995	No signal	"Unambiguous detection of climate-induced changes in most ecological and social systems will prove extremely difficult in the coming decades." (IPCC 1995, p. 6).
2001	Faint signal	"there are preliminary indications that social and economic systems have been affected." (IPCC 2001, p. 6)
2007	Emerging signal	"There is medium confidence that other effects of regional climate change on natural and human environments are emerging, although many are difficult to discern due to adaptation and non-climatic drivers." (IPCC 2007, p. 3)
2014	A signal	"Some impacts on human systems have also been attributed to climate change, with a major or minor contribution of climate change distinguishable from other influences." (IPCC 2014, p. 6).
2023	A clear signal (mostly moderate)	"…widespread adverse impacts and related losses and damages to nature and people (high confidence)." (IPCC 2023)
2030–2035[a]	Moderate to high	Risk of impacts on human and managed systems moderate to high at 1.5°C of global warming (Table 10.2)
Before 2100	Moderate to very high	Risk of impacts on human and managed systems moderate to very high at 2°C of global warming (Table 10.2)

Sources: Author based on a qualitative analysis of IPCC assessment reports 1–6, summaries for policy makers or equivalent documents.

[a] The IPCC's Sixth Assessment Report expected the 20-year running average temperatures will show 1.5C of global warming sometime between 2030 and 2035 (IPCC 2023a).

climate ambition levels may be enhanced (e.g., "winning team" in Section 6.4.4). As such, transition risks are contentious and can generate a mix of support for, and resistance to, climate action and consequently may enhance or suppress collective ambition levels (Figure 10.3).

According to Rogelj et al. (2018), a transformation is required if adaptation and mitigation are to stabilise atmospheric concentrations of GHGs at safe levels with minimal GHG removals or risk of breaching tipping points. As such, there is a case for including "transformation risk" as a new category of risk, which is similar to transition risk, but much more disruptive to businesses reliant on GHG emitting technologies or practices; and consumers, individuals, and households (including smallholder family farms) reliant on technologies and practices that emit GHGs or reduce resilience. The possibility of unintended consequences of rapid climate action further amplifies risks associated with response options.

10.2.2 Responses

Understanding what makes an effective global response is a central part of the research presented in this book. As such, it is important to understand the range of possible responses and which of these responses are more likely to be effective. Figure 10.4 highlights the global response system (Level 1), the "climate change signal" and the "response" (Level 2). Level 3 shows the response has three parts consisting of "actor decision making", "actions" and "contributions to the global response". Level 4 is the most detailed level of the model and includes several variations depending on which elements of the system are being highlighted.

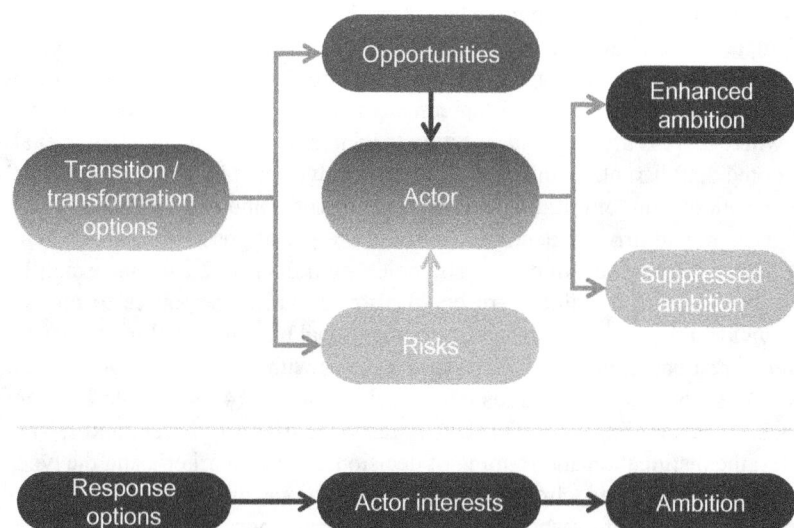

Figure 10.3 Ambition levels are influenced by the options available.

In Figure 10.4, at Level 4, actor decision making is made up of "triggers and drivers", "attitudes" and "options considered" as well as other factors, specifically "actor capacity" to implement and develop "technologies and practices" and "actor power and influence" to change "institutions" and hence the options available for consideration. From Section 7.4, actor capacity and power are important elements of the global response. The capacity to develop and apply technologies and practices influences the technologies and practices available to an actor as well as other actors in the future. Meanwhile the power to influence other actors and institutional arrangements can influence which options are encouraged, regulated or even prohibited and as such also influences options available to other actors.

Figure 10.5 unpacks "actor decision making" including "triggers and drivers" (Section 5.4), response "attitudes" (Section 5.5), and "options considered". Response triggers and drivers include "cost benefit responses", "enlightened responses" or "emergency responses". Meanwhile, no-trigger responses include "non-responses" for example due to hopelessness, apathy, or hypocrisy, or "cynical responses" where special interests are privileged ahead of climate change. Response attitudes include "defensive", "competitive", "technological", "practice focused", and "cooperative" attitudes. Importantly, triggers and drivers are related to levels of ambition, with cynical responses consisting of negative ambition and emergency responses being high ambition responses.

It is worth noting that triggers and drivers identified in Section 5.4 are related to Grubb et al.'s (2014) climate change risk conceptions and strategies (Section 2.2.4). These risk conceptions and strategies consist of: "indifferent or disempowered" which is a "non-response"; "tangible and attributed costs" where the "costs and benefits" of acting are weighed up; and, "disruption and securitisation" where there is a belief that there is a personal or collective "security" risk and climate change is seen as a threat multiplier (Grubb et al. 2014). Importantly, "Tangible and attributed costs" of climate change involve technocratic valuation and responses happen when "impacts rise above the noise" (Grubb et al. 2014, p. 48) i.e., when there's a discernible climate change signal. Meanwhile, Grubb et al.'s (2014) "disruption and securitisation" is equivalent to "emergency responses" identified from the scenarios, where there is urgency and ambition to address climate change. Grubb et al. (2014) noted that disruption and securitisation involve "containment and defence" strategies with the aim of mitigating as much as possible and adapting to impacts.

The triggers and drivers identified from the sample of possible futures extend climate change risk conceptions and strategies from Section 2.2.4. For example, from Section 5.4 it's possible there could also be cynical responses to climate change, incidental positive contributions to the global response to climate change, enlightened responses, or security-driven responses with the aim of "containment and fix". As such, Table 10.4 builds upon Grubb et al. (2014) and Table 2.1 from Section 2.2.4, addressing: ambition levels; response triggers, drivers and related strategies; the justification and framing of decisions; decision criteria and the types of options considered; and, climate change related actions. Importantly, the distinction between cost-benefit responses and emergency responses aligns with a discussion in the IPCC 1.5 Degree Report regarding decision criteria in the form of

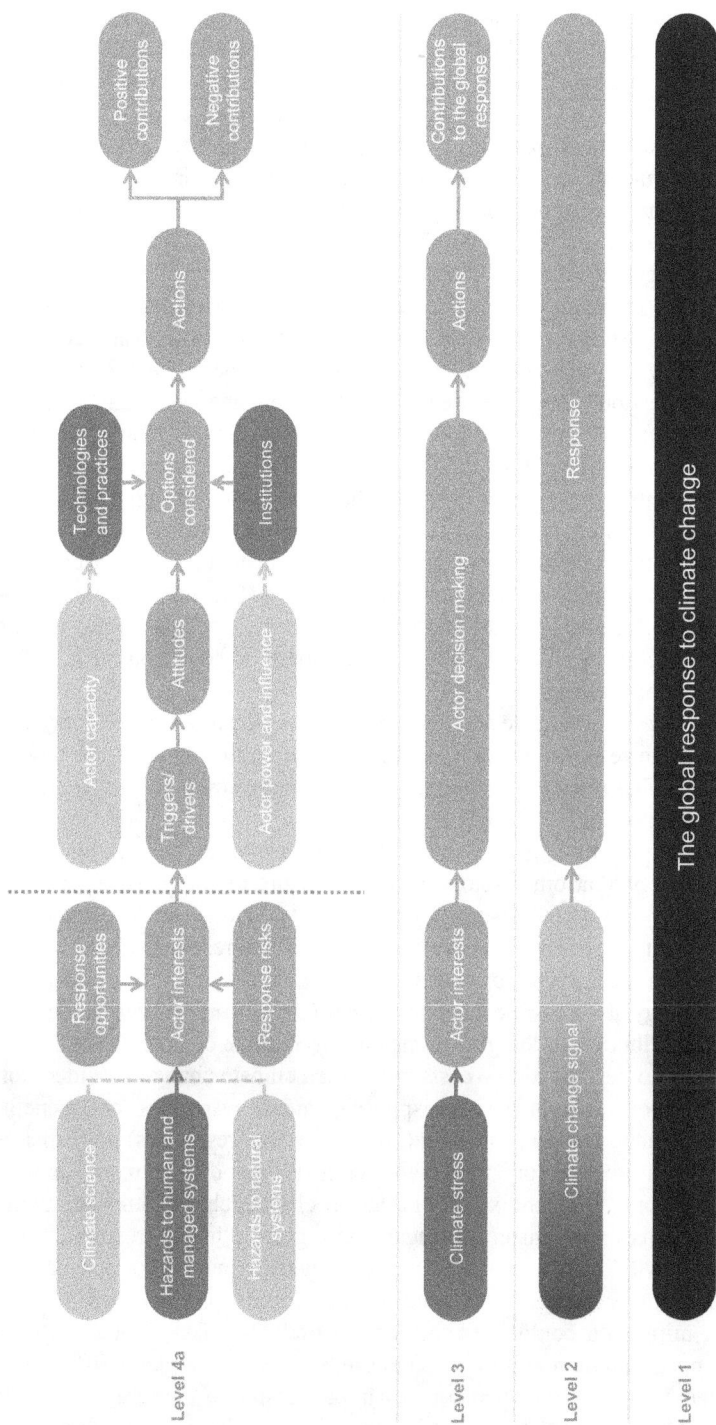

Figure 10.4 The global response to climate change depends on signals and responses.

cost-benefit analysis versus cost-effectiveness analysis (Table 10.5) as well as work by Grubb et al. (2014). Grubb et al. (2014) highlighted that when actors conceive the climate change problem in terms of tangible and attributed costs, then these actors "act at costs up to the 'social cost of carbon'", reflecting a cost-benefit analysis approach (i.e., economic response) to the problem of climate change. However, according to Grubb et al. (2014), when actors conceive of climate change in terms of "disruption and securitisation" then they follow a "containment and defence" strategy (see Table 2.1 from Section 2.2.4), which along with containment and fix constitute emergency responses (Table 10.4).

Importantly, emergency responses align with a cost-effectiveness analyses, where the goal is set and any economic analysis is around how to achieve the goal for least cost, given acceptable levels of risk, rather than whether to act or not which is the question being addressed with cost-benefit analyses. In such scenarios, "...the willingness to pay for imposing the goal..." can be interpreted as "...a political constraint." (Rogelj et al. 2018, p. 150). From the sample of possible futures, respondents indicated that it may not be until a warning light "starts flashing amber or red" that there is an effective global response (Section 6.3.2). It was also noted that responses to climate change will be inadequate if treated as a cost-benefit analysis issue, but might be effective if climate change becomes a national security issue (Section 5.4.2). As such, an emergency (i.e., a crisis or catastrophe rather than a rhetorical emergency) may be a possible important precondition for an effective response to climate change.

From Section 5.3, the scale and timeliness of the global response is very important. Each of the response types in Table 10.4 has implications on the timeliness and scale of global response to climate change, due to strategies involved and related decision criteria.

With regards to "no trigger" cynical and non-responses, the key issue for the global response is that other actors have to do more to compensate for negative contributions to the global response (Appendix G). However, it is important to note that incidental positive contributions to the global response are also possible due to changing incentives. For example, when it comes to the decreasing cost of renewable energy, it is possible that non-responses could make a very large positive incidental contribution to the global response to climate change, if actors engage in these technologies due to low costs rather than climate change considerations.

Trigger responses include risk responses, impact responses, cost-benefit responses, enlightened responses (including cooperation responses) and emergency responses. If all other factors are equal, then a risk response is more timely than an impact response (Appendix G). The lag in climate change impacts manifesting themselves is important in this regard, as a global "impact response" relying on mitigation will likely lock in the range of hazards being experienced as well as the risk of other changes in conditions and hazards. Global "risk responses" relying on mitigation could limit the lock-in of future risks, as the response is ahead of impacts being manifested. Likewise, a global "emergency risk response" is much better than a global "emergency impact response", where the conditions generating the emergency may already represent the new normal, unless there is

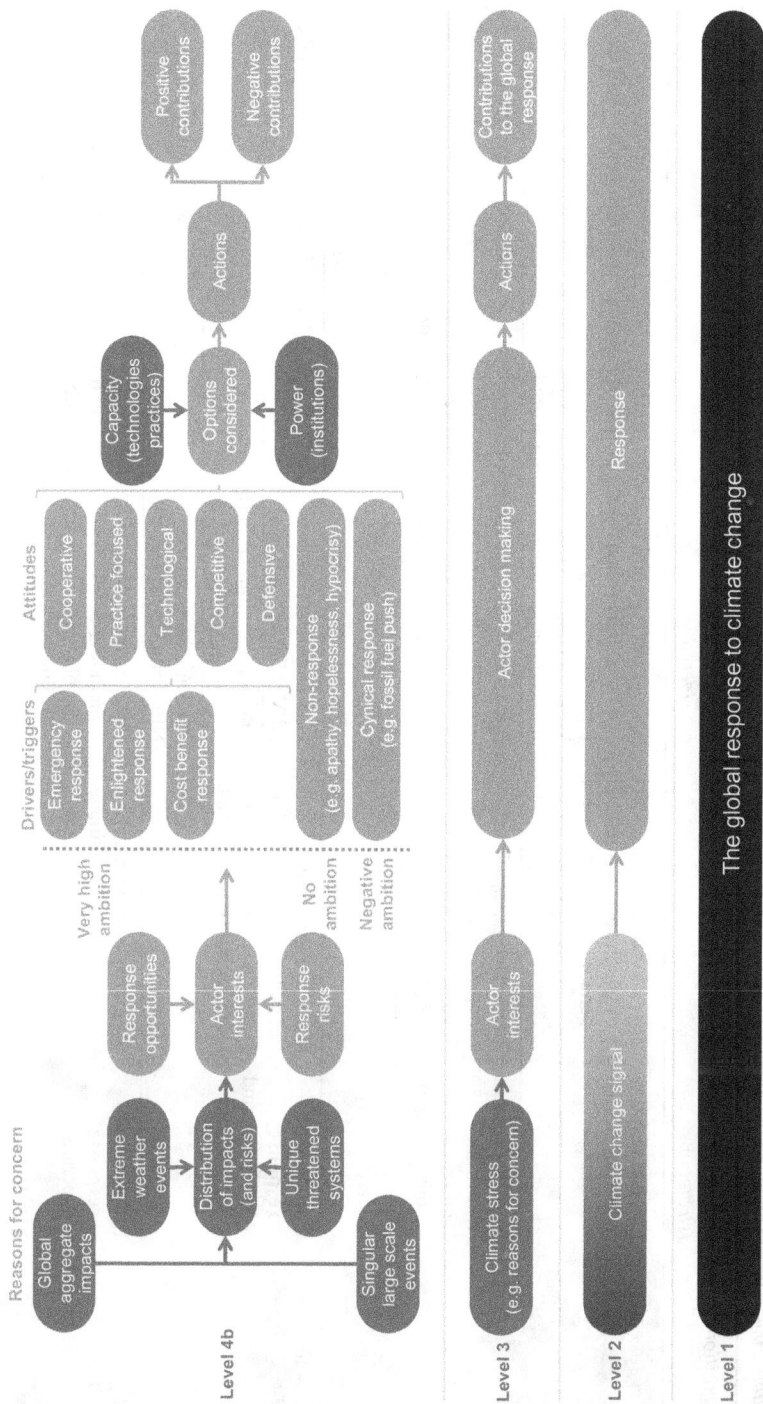

Figure 10.5 The global response to climate change depends on decision making including triggers and drivers, and response attitudes.

Table 10.4 Climate change related actions depend on triggers and drivers, strategies, justifications, and decision criteria

Ambition	Trigger/driver (strategy)	Justification and framing	Decision criteria and types of options considered	Actions
Negative	**Cynical response** (non-cooperation)	**Special interests:** the extent to which special interests benefit	Special interests and the extent to which these can be enhanced regardless of climate change. Mitigation, removals and low carbon development options not considered	Deliberate negative contributions, "drill baby drill"
None	**Non-response** (free riding and no cost contributions)	**None:** Denial,[a] apathy or hopelessness leading to inaction or incidental helpful contributions to the global response to climate change.	Mitigation, removals and low carbon development options not considered. Adaptation to impacts are considered and only acted upon for reasons of self-preservation.	Incidental negative contributions to the global response, "ignorance is bliss"[a] Incidental positive contributions to the global response
Low to high	**Cost-benefit response** (competition and cooperation)	**Cost-benefit analysis:** Decisions to act on climate change are based on costs and benefits of the options being considered	Low ambition criteria NPV (climate action) > 0 Moderate ambition criteria NPV (climate action) ~ 0 High ambition criteria NPV (climate action) < 0	"Act at costs up to social cost of carbon"[a]
High to very high	**Enlightened response** (cooperation)	**Evidence** of impacts, risks and need to limit climate change and its impacts	The extent to which collective action, including behavioural changes, will fulfil the objective of limiting climate change and its impacts.	Collective action on adaptation and mitigation

| Very high | Security response (cooperation and coercion) | Security: Climate change is a clear and present danger. It has been decided that climate change needs to be acted upon and now it is a question of what can be achieved with the resources available | A decision is made on a defensive outcome or approach (e.g. adaptation) taking into account willingness to pay for (and afford) the options available. Once an outcome or approach is decided, the most cost-effective options for achieving this within an acceptable level of risk are assessed and selected. | "Containment and defence"[a] |
| | | | A decision is made on an outcome or approach including containment (e.g. adaptation) and a fix (e.g. mitigation, removals or other geo-engineering interventions) taking into account willingness to pay (and afford) the options available. Once an outcome or approach is decided, the most cost-effective options for achieving this within an acceptable level of risk are assessed and selected. | Containment and fix |

Source: Author building on the work of Grubb et al. (2014).

[a] Grubb et al. (2014).

Table 10.5 Cost-benefit analysis, cost-effectiveness analysis and social cost of carbon from the IPCC 1.5 Degree Report

Concept	Purpose
Cost benefit analysis	Identify the optimal emissions trajectory minimising the discounted flows of abatement expenditures and monetised climate change damages (Boardman et al., 2006; Stern 2007)
Social cost of carbon	The total net damages of an extra metric ton of CO_2 emissions due to the associated climate change (Nordhaus 2014; Pizer et al. 2014; Rose et al. 2017)
Cost effectiveness analysis	Identifying emissions pathways minimising the total mitigation costs of achieving a given warming or GHG limit (Clarke et al. 2014)

Source: Rogelj et al. (2018).

substantial adaptation and measures to improve resilience, coupled with mitigation and GHG removals. Cost-benefit responses weigh impacts and the immediate cost of responding higher than the risks or the benefits of limiting future climate change.

Response attitudes can also influence the timeliness and scale of the global response as they have a strong bearing on the options likely to be considered, as well as the coherence of a global response (Appendix G). From the sample of possible futures, defensive response attitudes seemed likely to result in fragmented and ineffective responses to climate change (Section 5.5.1). Cooperative responses have the potential to generate coherent effective responses but only if there is participation, compliance and stringency (Bodansky and Diringer 2010, Bodansky 2012, Stavins et al. 2014, Wilson 2015) (Section 3.4.3). Interestingly, competitive responses could generate coherent effective responses to climate change, due to a common focus for example on a technology or business model (Section 5.5.3). However, whether competition makes positive or negative contributions depends upon the technologies and practices being incentivised and the extent to which these things make positive or negative contributions to the global response. Technological response attitudes tend to see every problem as being something technology can solve, but concerns were raised around the extent to which such attitudes simply drive greater consumption (Section 7.2.4) and wastes related to production. Conversely, practice-focused attitudes and inclinations see widespread behavioural change, nature-based solutions or policy responses as being required to limit climate change and related impacts. Practice-based options require considerable effort as they need to be grounded in local conditions and context, and hence may not be as scale-able as technological options. Practice-based options are best delivered as bespoke options.

It should be noted here that Table 10.4 presents the widest possible range of strategies, decision criteria, and actions. These strategies, decision criteria and actions are framed as "responses to climate change" and the language used reflects this i.e., is biased by the framing.

10.3   Actors, interests, and actions

One of the key questions from Section 1.5 regarded the conditions under which actors would act on these options. Actors, their interests and actions are central to answering the important question. From the thematic analysis of scenarios in Chapter 6, three broad categories of actors with coupled options and actions were identified, consisting of social change and behaviour, political will and policy, as well as business and economic activity. By having a limited number of actors and interests identified, rather than multiple sectoral breakdowns for example related to energy, agriculture, industry, various populations or demographics, analysis of the global response system is kept manageable and the main features of the global response systems can be identified.

Figure 10.6 presents actor–interest themes (i.e., social change, political will and business) as centres of decision making with related option–action themes in the form of behaviour, policy, and economic activity. From Chapter 7, technologies and practices featured in many scenarios as they provide actors with options. In addition to the availability of options, there is a range of other factors that influence decision making, including prices, the costs of different options, the availability of finance, or even contractual and legal arrangements. In addition to these things, there is the issue of leadership, for example from civil society, government or business, that can reframe and link issues, and influence which options are considered.

Drawing on themes and scenarios from the sample of possible futures, a thematic chain analysis was conducted (Section D.4) with possible preconditions mapped, along with possible responses and follow-on conditions, related to social change and behaviour (Section 10.3.1), political will and policy (Section 10.3.2) as well as business and economic activity (Section 10.3.3). These maps show possibilities (i.e. pathways) which could happen in various combinations. These maps generally don't attempt to show the level at which these possibilities might happen, for example at national or subnational levels. Furthermore, these maps don't attempt to contextualise or anticipate which combinations might happen in particular jurisdictions. Interactions between actors and interests are also analysed (Section 10.3.4) along with incentives, ambition, and related responses (Section 10.3.5).

10.3.1 Socially driven responses

Social change and behaviour regard the interests of people, as individuals, households, and communities, and the actions, or inactions, these people might

Table 10.6 The domestic response to climate change is made up of state and non-state actors, their interests, and actions taken

Domestic actors in a 2-level game	Actors mentioned in scenarios	Actor–action themes	
		Actor–interests	Options/actions
State	Central government	Political will	Policy
Domestic non-state actors	Local government		
	Business, private sector, shareholders, investors, finance	Business	Economic activity
	Society, people, individuals, households, groups, civil society	Social change	Behaviour

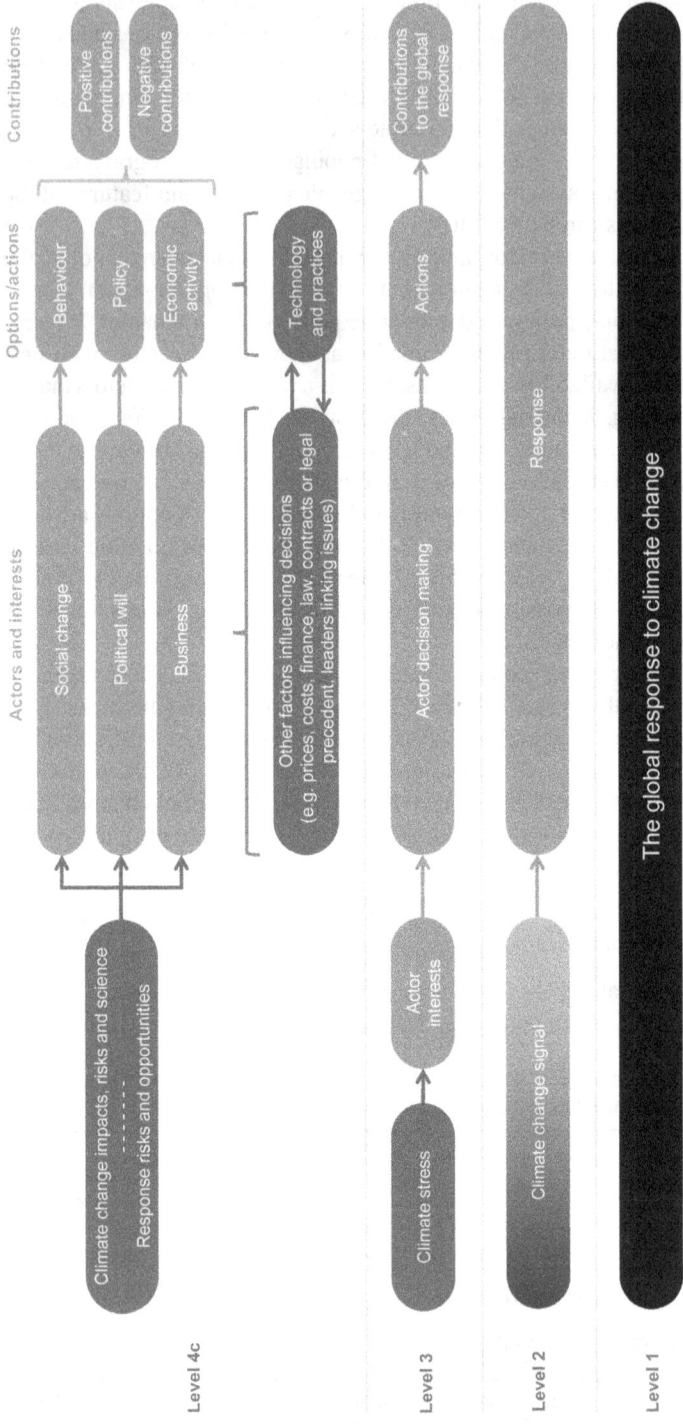

Figure 10.6 Other factors, technologies and practices and the influences these things can have on actors and interests as well as options and actions.

individually or collectively take. Figure 10.7 maps social change and behaviour-related themes and scenarios from Section 6.2, including climate change signals and preconditions for social change and behaviour, as well as follow-on conditions influenced by social change and behavioural responses.

Social change and behaviour scenarios include impacts and risks to human systems as triggers and drivers (Section 6.2.1). The distribution of impacts could be an important influence with regards to social change and behaviour, for example if wealthier people are impacted, then they might influence institutions and political will more effectively than poorer populations affected. It is possible that social change and behaviour could be driven by information on climate risks in a risk response scenario. Leadership, including political leadership, could help drive social change and behaviour, as well as movements for example youth-led movements (Sections 6.2.2 and 6.2.4). Enlightenment could either happen in advance of climate change impacts or as a response to climate change impacts (Section 6.2.3). It is also possible that enlightenment might be influenced or driven by other issues, for example responding effectively to some other crisis, as well as climate change.

In various scenarios, social change and behaviour influenced political will and policy as well as business and economic activity. For example, social change and behaviour included changes in consumption, demanding accountability of businesses and their activities in relation to climate change, as well as other influences on businesses and economic activity for example through shareholders or changes in business culture reflecting societal changes (Section 6.2.5).

With regards to behaviour and consumption, from Section 6.2.5 it was noted that behaviourally virtuous acts (with positive contributions to the global response) tend to be followed by wicked acts (with negative contributions), undermining the global response. Furthermore, any money that is saved by these virtuous behaviours may be spent on things that emit GHGs, hence the "rebound" effect discussed in economic literature (e.g., Wei and Liu 2017). As such, there was a view that individual behaviours have limited influence on global responses to climate change and scepticism that behavioural change can lead to systemic change at scale.

An important set of behaviours influencing political will and policy were also found in the sample of possible futures, that could help create the systemic changes needed to complement behavioural changes. Specifically, social change and behaviour also included changes in public opinion, demand for accountability of political leaders, or mandates for institutions including governments, for example through voting in democracies (Section 6.2.5). Such changes can influence political will and policy, drive policies and interventions, and could create systemic changes towards climate resilient low-emissions development.

For there to be effective global responses to climate change, social change and behaviour, coupled with political will and policy, and changes to business and economic activity requires a fundamental change in what is acceptable and what is not, at an institutional or constitutional level following the distinctions made by Ostrom (1990). From Section 6.2.9, a notional social contract was identified as being an essential part of the global response to climate change and it was noted that social permissions for an effective global response don't exist yet. Interestingly,

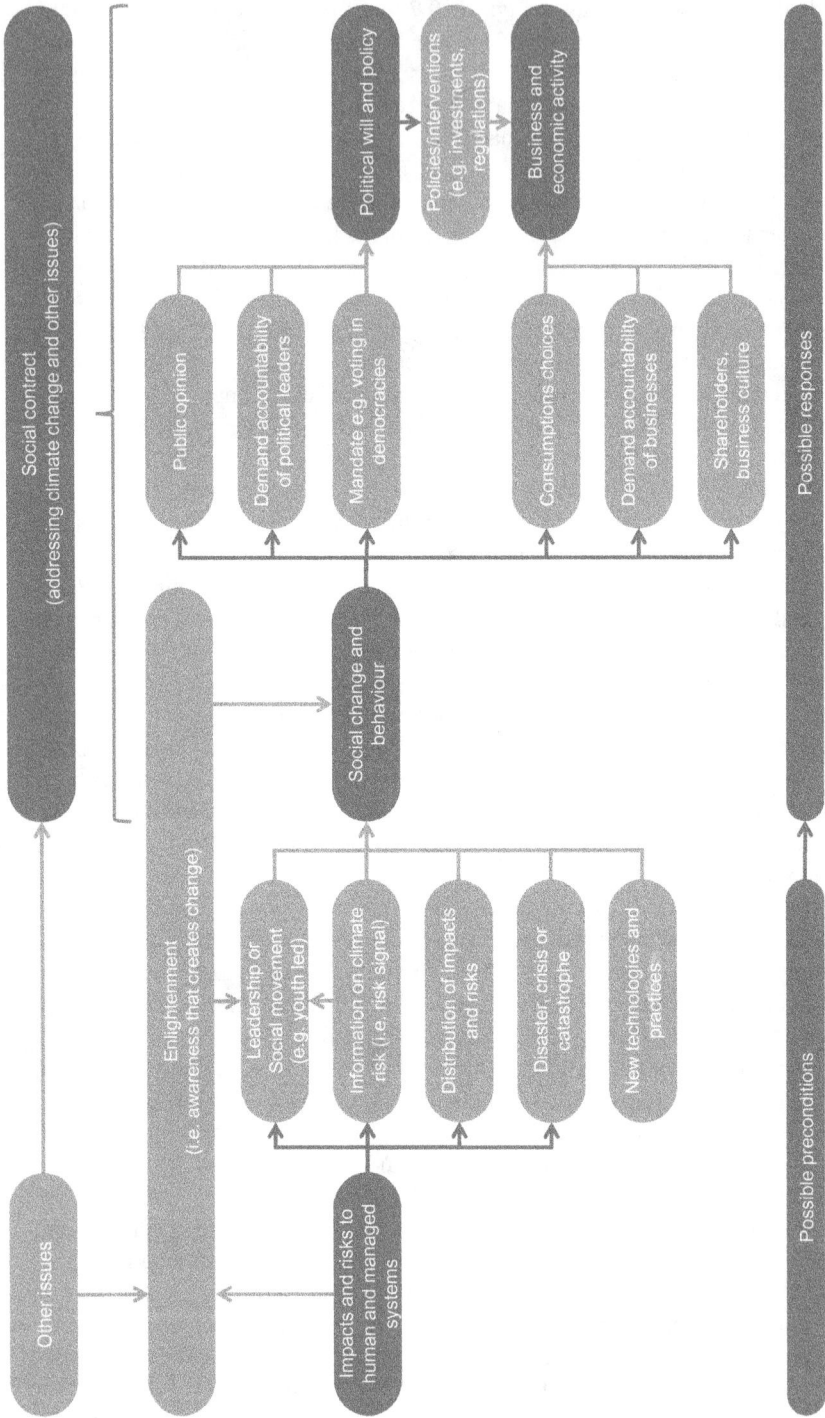

Figure 10.7 Preconditions for social change and behaviour, as well as the influences social change and behaviour might have on political will and policy, business and economic activity.

during the course of the research presented in this book, the Green New Deal was published in the United States (Friedman 2019) and could be said to constitute a proposal for a new social contract.

10.3.2 *Politically driven responses*

Political will and policy regard the ambition level of government leaders, and others in government when it comes to positions on climate change and the interventions they make. Figure 10.8 maps political will and policy-related themes and scenarios from Section 6.3, including climate change signals and preconditions for political will and policy, as well as follow-on conditions influenced by political will and policy responses.

Political will and policy include social change and behaviour as possible preconditions along with business and economic activity-related lobbying, as well as impacts on human and managed systems and related disasters, crises, or some catastrophes (Sections 6.3.1–6.3.3). With regards to social change and behaviour, institutional capacity and processes may be important, including having ways for social change to translate into governance and policy (Section 6.3.4). Social change and behaviour can also create mandates as well as public pressure, that can drive political will and policy (Section 6.3.3). New technologies and practices can also influence political will and policy, for example changes in costs and prices may make some options more politically viable (Section 6.3.6).

The capacity and level of development within a jurisdiction could also influence political will and policy in particular the ability to deliver climate resilient low-emissions development and related interventions (Section 6.3.4).

With regards to political will and policy, there are many options identified in the sample of possible futures (Sections 6.3.7–6.3.13). These include the possibility of having a stringent policy regime, governments creating technology-related incentives, investments, and supporting research (Section 6.3.7). Policies might include the forced retirement of GHG emitting technologies (Section 6.3.8) or investments in climate-resilient low emissions infrastructure (Section 6.3.9).

At the subnational level, local government might invest in climate-resilient low emissions infrastructure. Governance might include regulations on land use and related practices, as well as land tenure (Section 6.3.12), legislation supporting localisation (Section 6.3.11), or even public service withdrawal from impacted areas (Section 6.3.10). Stress testing of adaptation plans could also be important, contributing to improved resilience (Section 6.3.13).

10.3.3 *Business-driven responses*

Business and economic activity regard actions, or inaction, by individuals or groups undertaking productive activities, in many cases driven by a profit motive. Figure 10.9 maps business and economic activity-related themes and scenarios from Section 6.4, including climate change signals and preconditions for business

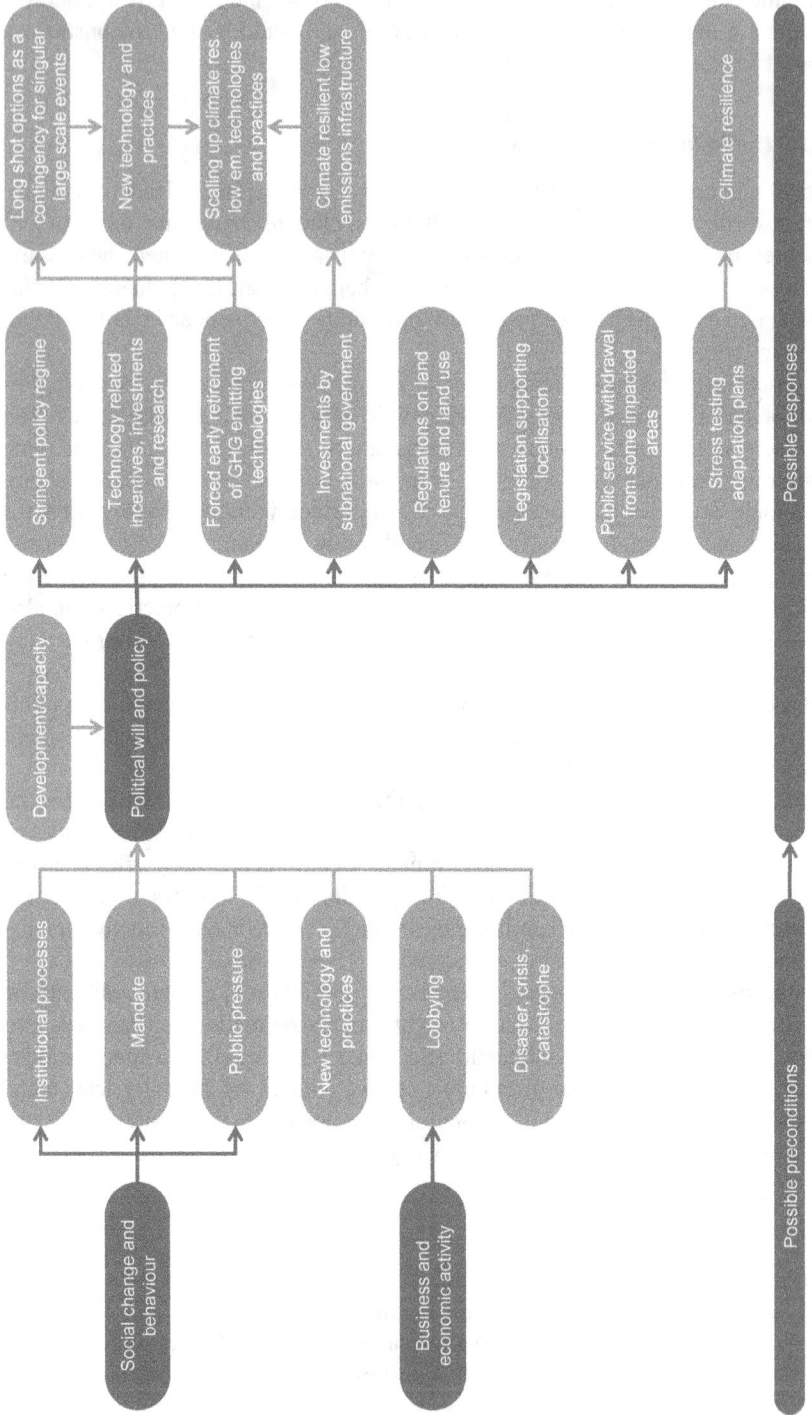

Figure 10.8 Preconditions for climate change related political will and policy and the influence political will and policy might have on responses.

and economic activity, as well as business and economic activity-related responses and subsequent conditions.

Possible preconditions for business and economic activity to respond to climate change and related impacts, include social and change and behaviour, political will and policy as well as impacts and risks to human and managed systems affecting business interests such as supply chains or stock valuations (Section 6.4.1). Social change and behaviour could include changes in consumption. Political will and policy could include subsidies, incentives, regulations, or legislative support for localisation (Section 6.4.7). Meanwhile climate finance can change the incentive structure for businesses looking for sources of finance, while new climate-resilient, low-emissions technologies and practices that are proven to work or become cheaper than alternatives might also drive business and economic activity to be climate resilient and have low emissions. Alternatively, there may be enlightenment of shareholders or those who work in a business resulting in changes in business ethos (Section 6.4.5).

Business and economic activity could also include coalitions of big business or others lobbying governments for political leadership, especially if businesses are directly affected by climate change. Businesses could lobby for a carbon price or world trade rules that provide a basis for addressing climate change and providing certainty regarding regulations (Section 6.4.1). It is even possible that business and economic activity could end up having a re-industrial revolution (Section 6.4.5).

10.3.4   *Interactions between actors and interests*

The global response to climate change relies on contributions from all types of actors in society. Importantly political will and policy as well as business and economic activity are parts of social change and behaviour. Based on the signals and responses mapped in Sections 10.3.1–10.3.3, there are many possible interactions between social change and behaviour, political will and policy as well as business and economic activity and these are summarised in Figure 10.10. These interactions include public opinion, special interests, laws and regulations, investments, consumption, goods services, and prices.

In Figure 10.10, political will and policy have an important role in shaping the social contract, for example through laws and regulations. However, climate change is not the only problem on the policy agenda. As such, climate change cannot be addressed in isolation, but needs to be addressed with other problems. This includes coalitions of actors and interests, linking different problems with common responses, for example following the "garbage can" method of policy making (Enserink et al. 2013), or "growing the pie" when it comes to international cooperation (Wetlaufer 1996). For such coalitions to form, actor interests and incentives are important.

10.3.5   *Actor incentives*

From the sample of possible futures, the extent to which options are commercially viable, and the extent to which actors have ambition has a bearing on social,

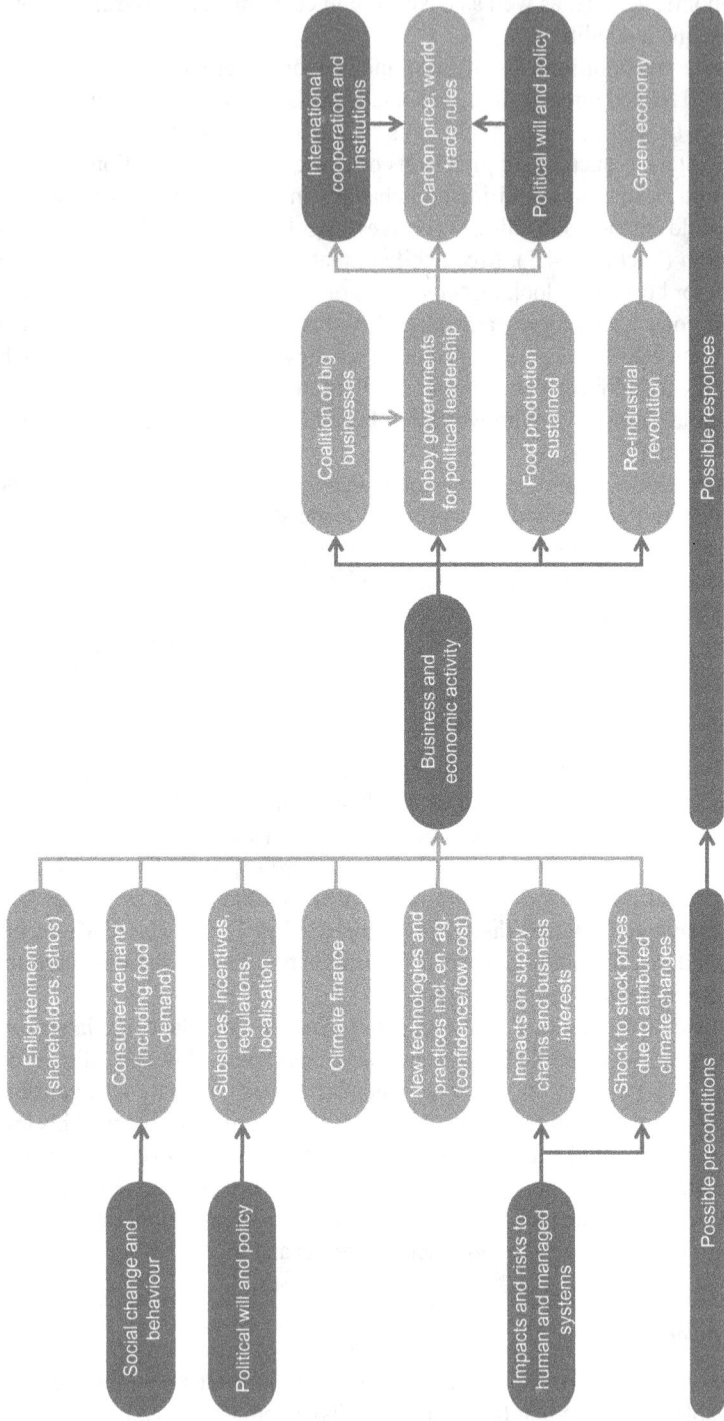

Figure 10.9 Preconditions for business and economic activity-related responses to climate change as well as possible influences business and economic activities might have on the wider response to climate change.

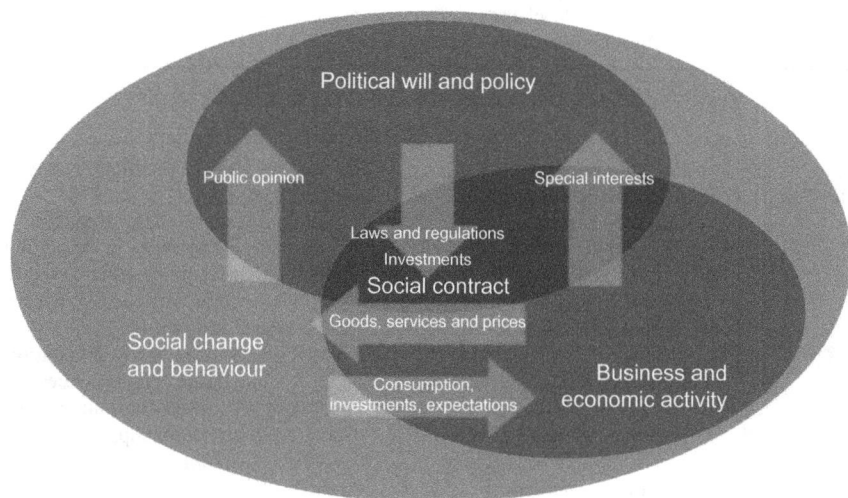

Figure 10.10 Venn diagram showing political will and policy as well as business and economic activity as a subset of social change and behaviour, including important
interactions between these sets of actors and interests.

political, and business responses. Table 10.7 presents a matrix of possible responses
where climate ambition is high or low, and the options available are commercially
viable or non-commercial (i.e., not commercially viable without government or
other support). Non-commercial options are immediate response risks meanwhile
commercial options are immediate response opportunities.

Arguably, under the Kyoto Protocol climate change was a problem without
commercially viable options (Section 6.3), hence the need to assign responsibility
and determine which governments would support climate action and the global
response. With commercially viable renewable energy options, electricity storage
and transport options (IRENA 2019), the response regime has changed dramatically from the time of the Kyoto Protocol. In many cases, mitigation technologies
are commercially viable and the most cost-effective option. However, for scale to
be achieved, it is essential to have high-ambition responses, creating competition
and strategic interests.

In high-ambition scenarios where commercial options are available, response
options are an opportunity that can generate strategic interests, competition, and
"winning teams" (i.e., coalitions) as described in Sections 5.5.3, 6.4.4 and 7.2.2. If
there is low ambition and commercial options are available, then commercial opportunities are forfeited for political or social reasons, for example due to a cynical
response (Section 5.4.1). This suppresses competition for market share and profits
which are essential drivers for rapid scaling of commercially viable response options, including for example renewable energy generation or technologies related
to energy storage and the electrification of transport. For the global response to
climate change, it would be ideal if there was high ambition competition between
the United States and China for market share and value when it comes to renewable

Table 10.7 Responses to climate change depend on ambition levels and the availability of commercially viable options

Climate ambition	Non-commercial response options	Commercial response options
High ambition	**The climate change problem is too difficult to ignore but responding is a risk** Responses could include governments providing unilateral, bilateral or multilateral investments, grants, and subsidies. Responses could also include coordinated policy interventions. Business and social leaders might also undertake research and development of long-shot technologies or practices.	**The climate change problem is an opportunity** Commercial response opportunities create strategic interests in expanding markets, gaining market share, establishing industries, and generating employment and profits. Business competition drives innovation in new technologies and practices, accelerated with subsidies and other forms of government support. The options available meet the needs of society.
Low ambition	**The climate change problem is too difficult to solve** Limited cooperation on research and development of technology and practices Information sharing on research, and experimentation with technologies, markets and policies.	**The climate change problem is politically or socially unattractive** Commercial opportunities are forfeited for political or social reasons. Business competition drives innovation and the development of technologies and practices with little or no government support.

energy technologies, the electrification of transport and energy storage options, including research and development. Meanwhile leadership on sustainable land use practices and related research could also be helpful.

However, commercial interests are not the only incentive or possible drivers for responses to climate change. From the sample of possible futures, it's possible there could be climate change related catastrophes, disasters or crises (Section 5.2.2). The intensity of loss and damage constitutes climate stress to actors and their interests. Once loss and damage become perceptible (i.e., attributable to climate change), then there may be an influence on behaviour at the societal level. With increased stress, climate change may be severe but endurable or terminal to various actors and their interests. Catastrophes can happen at different scales from personal to global or transgenerational, and affect specific regions or geographic groupings such as SIDS. As such, Table 10.8 adapts Figure 2.1 (Section 2.2.3) addressing geographic categories relevant to climate change and the global response. States with similar geographies and hazards may also share national interests and be potential allies when it comes to international cooperation and negotiations.

However, one area of the global response to climate change where it is difficult to imagine how actors' incentives might align to creating cooperation is GHG removals.

Table 10.8 Qualitative categories of risk

Scope	Intensity of losses and damage (i.e., climate stress)			
	Imperceptible	Perceptible	Severe but endurable	Crushing/ Terminal
Trans-generational	Not relevant	Influence on behaviour	Global existential risks	
Global	Not relevant	Influence on behaviour	Global catastrophic risks	
Regional or similar national geographies (e.g. SIDS)	Not relevant	Influence on behaviour	Shared catastrophic risks to national interests	
National	Not relevant	Influence on behaviour	Catastrophic risks to national interests	
Local	Not relevant	Influence on behaviour	Catastrophic risks to non-state actors	
Personal	Not relevant	Influence on behaviour	Catastrophic risks to individual interests	

Source: Author adapted from Bostrom (2013).

10.4   Greenhouse gas removals

A key question from Section 1.5 regarded the preconditions for actors to undertake atmospheric GHG removals. From the analysis of scenarios involving atmospheric GHG removals in Section 7.3, several themes emerged. The first was the need for GHG removals. It was noted that GHG removals should be a supplement to mitigation of the large quantities of GHG emitted into the atmosphere (Figure 10.11). Once these quantities have been reduced then GHG removals become more viable. Some respondents suggested GHG removals should be kept as a contingency option in case climate sensitivity is higher than expected or some threshold is passed, and a feedback mechanism is activated.

There are a range of practical limitations to atmospheric GHG removals for example related to the energy required to power technology-based options or the amount of land required to support natural removal options. If energy became sufficiently cheap, it was suggested that maybe other uses of the same energy could be found that help with adaptation rather than removing GHGs from the atmosphere (Section 7.3.4). There were also concerns around the possibility of trade-offs or unintended consequences, for example related to land or where captured carbon might be stored.

The need for political will and policies supporting GHG removals was noted, including land policies, or research and development support for atmospheric GHG removal technologies. The timing of the GHG removals is an important issue, with one respondent suggesting it would not be until the end of the century before atmospheric GHG removals would be viable. However, this would most likely lead to a very significant overshoot, assuming the budget for GHG emissions is used up in the first half of this century.

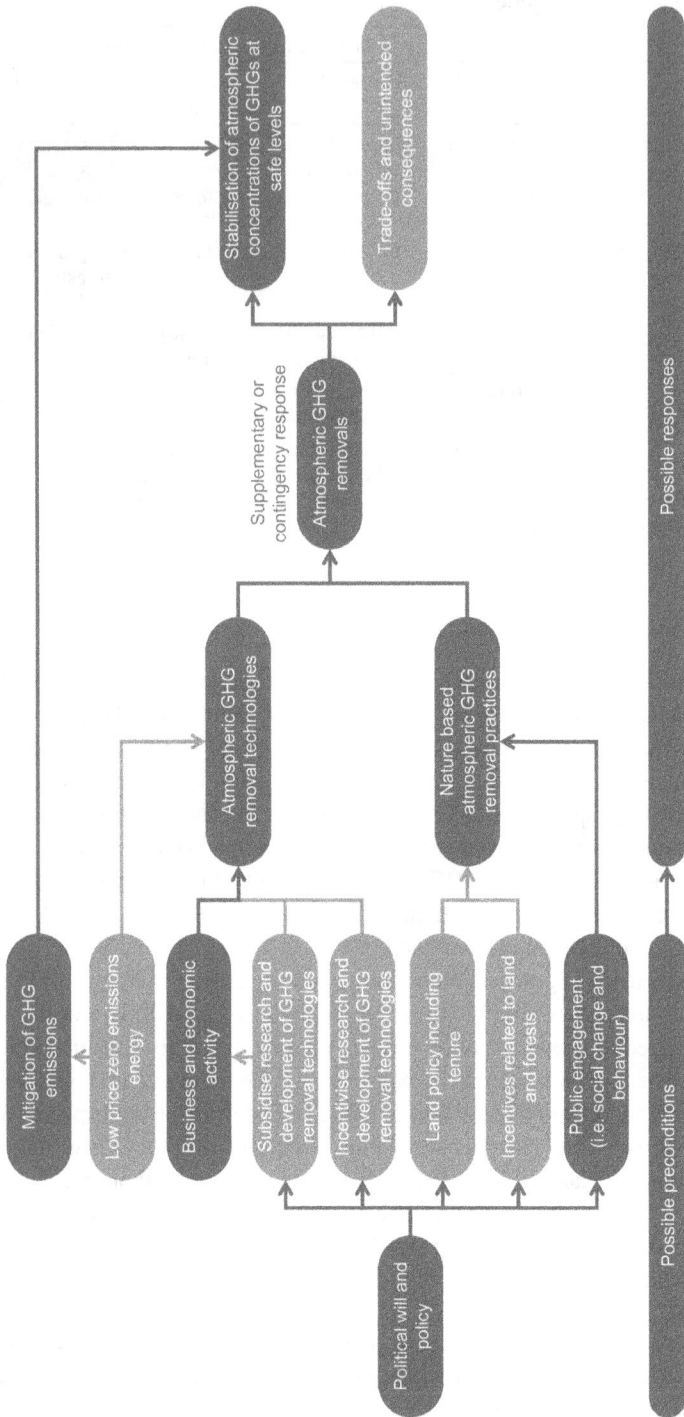

Figure 10.11 Preconditions for GHG removals from the atmosphere.

10.5 International cooperation

One of the key questions for this book regarded preconditions for effective international cooperation on climate change. However, from the scenarios and themes in Chapter 8, it can't be taken for granted that there will be a rules-based international order. It was noted that there is tension between nationalism on the one hand and internationalism on the other. It was also noted that some leaders are sceptical of the rules-based international order and whether it is in their national interests or domestic political interests, to support or undermine the rules-based international regime.

From the sample of possible futures, as the climate change signal increases it is possible that climate change could become the premier international issue and a focus of international cooperation (Section 8.2.2). This might include an emergency response, with climate change elevated to the United Nations Security Council. In such scenarios, there is a stringent enforced global response to climate change but the UNFCCC is not the lead institution. The UNFCCC would likely also have an advisory and coordination role along with an information gathering and verification role, including the possibility of county inspections like the IAEA. Meanwhile, institutions such as the World Bank and other multilateral development banks could provide climate finance while institutions such as the World Trade Organisation could help facilitate trade penalties and sanctions related to climate change.

Other triggers and drivers for international cooperation on climate change could include business interests being adversely affected by climate change, international leadership, and enlightenment (Figure 10.12). It is also possible that some States advocate for international cooperation on climate change due to an interest in sustaining a rules-based international regime.

The effectiveness of international cooperation on climate change depends upon the power and influence of coalitions of states. From Section 8.3.4, it is possible that powerful states could make offers that cannot be refused for example in a scenario where climate change is elevated to the level of the United Nations Security Council. In an emergency response scenario where there is very high ambition and powerful states form a coalition, then it is likely there would be a stringent enforced international agreement and global response to climate change (Figure 10.13). Low to high ambition cost-benefit-based agreements would be unlikely to result in a global response with sufficient scale and timeliness given the current situation and the dramatic GHG emissions reductions needed between 2020 and 2030 (Section 2.2.4). Even with commercially viable response options increasingly available to governments (Section 10.3.5) it would take very high levels of ambition for governments to shift investment patterns sufficiently (Section 5.3.2), abandon infrastructure that otherwise locks in GHG emissions (Sections 2.3.2, 6.3.9 and 6.4.6), and have businesses respond to climate change rather than treating effective responses as being optional (e.g. Shell's Sky Scenario in Section 2.3.1). These factors all influence the cost-benefit analyses that governments make meanwhile the costs of future climate change impacts (Section 2.2.3) are discounted. Given lags in the climate change system (Section 2.2.1), overshoot (Sections 2.2.4 and 5.3.1)

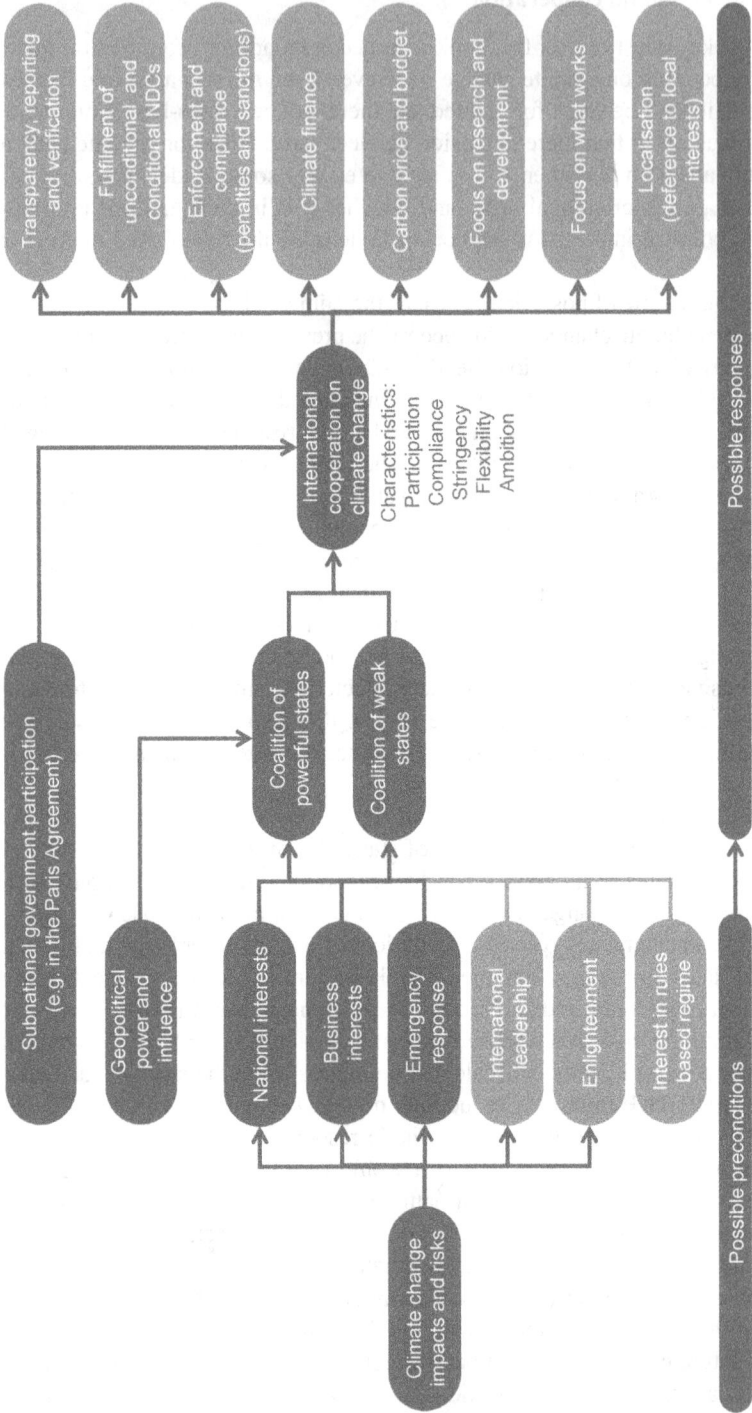

Figure 10.12 Preconditions for international cooperation on climate change and related international cooperation responses.

is more likely than not when states make decisions based on cost-benefit analyses. It is debatable whether cost-benefit analyses are even the right type of analysis for addressing potentially catastrophic risks such as climate change (Section 13.4). Regardless of what analyses are made, coalitions of weak states would struggle to generate a global response of sufficient scale and timeliness.

With regards to areas and approaches for international cooperation, the importance of transparency, reporting, and verification was highlighted by respondents. Some scenarios have the fulfilment of unconditional and conditional NDCs. The possibility of enforcement and compliance-related measures were explored too. Climate finance served to help facilitate, or at least encourage, the global response to climate change. Carbon prices and budgets were discussed, but would be complicated given the range of distribution issues, including winners and losers under any scheme. The possibility of transfer payments and compensation were also discussed, but the criteria for such transfers would likely be contentious (Section 7.4.3).

Some respondents suggested international cooperation could focus on things such as what works when it comes to policies, practices or technologies. Likewise, it could be possible to focus on research and development, for example monitoring research and development expenditure or the development of contingency options in the form of geo-engineering technologies.

The possibility of having local government involved and formally becoming a part of the Paris Agreement including self-determined contributions (akin to NDCs) and related monitoring and reporting and verification processes was raised. Another approach suggested, was localisation where deference is given to local interests when it comes to the movement of goods, services, and wastes (Section 6.3.11). Essentially, local communities and areas would be empowered, including through horizontal diplomacy.

10.6 Conclusions

As noted in Chapter 1, decision making is central to the global response to climate change. Drawing from the scenarios and themes in previous chapters, a signal response model was developed. Impacts and risks on human and managed systems were shown to be an important signal meanwhile ambition levels, response triggers, attitudes, power and capacity, were all important elements influencing actions and their contribution to the global response to climate change (Section 10.2). Another important influence was the extent to which options available are seen as an opportunity or threat by actors.

Important categories of actors in the global response to climate change were social, political, and business actors (Section 10.3). Based on the scenarios and themes from previous chapters, a range of pathways were mapped involving social change and behaviour, political will and policy, as well as business and economic activity. In practice these pathways are interconnected, and the social system includes political and business actors. Social permissions to act on climate change

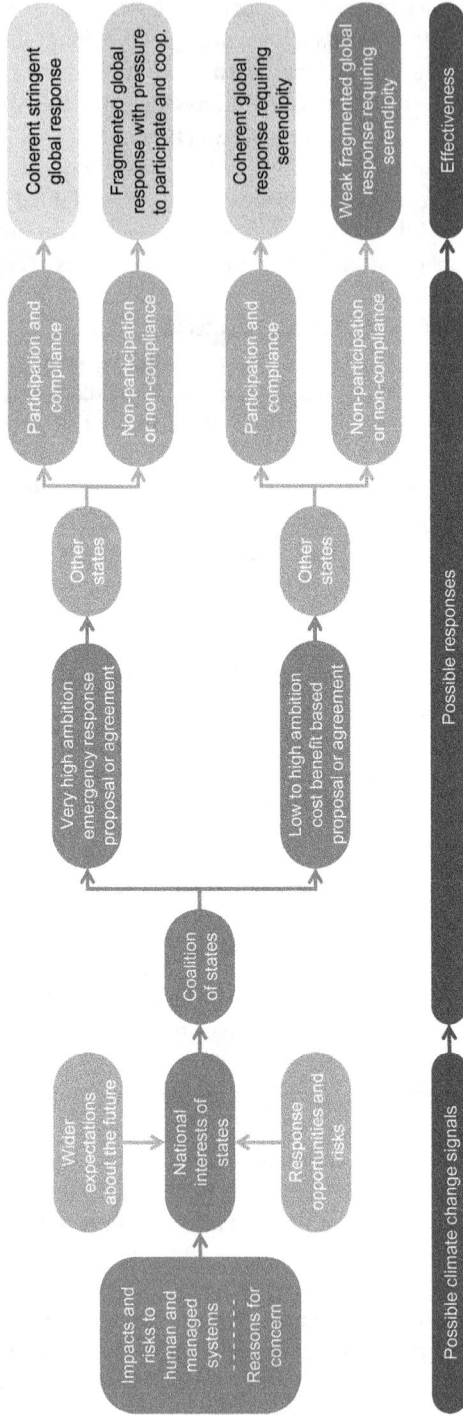

Figure 10.13   The effectiveness of international cooperation depends on the power of coalitions and their levels of ambition.

were also identified as being important, involving a notional "social contract" between all these actors.

Other pathways were also mapped involving GHG removals (Section 10.4) as well as international cooperation on climate change (Section 10.5). The effectiveness of international cooperation on climate change depends on the power of the countries involved. While GHG removals are essential to limiting atmospheric concentrations of GHGs, risks were noted as well as practical limitations around the cost of energy for technology-based GHG removals, and the durability of removals for natural removals.

These models, pathways, and insights form an important part of the analysis of preconditions for effective global response in Chapter 11.

11 Effective global responses to climate change

11.1 Introduction

In this chapter we bring together the scenarios and themes, models and pathways from previous chapters, identifying the "conditions" (i.e., preconditions) necessary for effective global responses to climate change. This chapter also answers the five key questions from Chapter 1, specifically:

What influence might climate change have on actors and the global response to climate change?

What are the preconditions for effective international cooperation on climate change?

What are the preconditions for actors to remove greenhouse gases (GHGs) from the atmosphere at a scale required to limit climate change to safe levels?

And most importantly:

What are the preconditions for effective global responses to climate change? Under what conditions would actors act on effective response options?

So, how is this discussion organised?

Drawing from the literature, we start with preconditions for effective global responses to climate change from the Paris Agreement in Section 11.2, and preconditions from the IPCC's 1.5 Degree Report in Section 11.3.

Section 11.4 discusses gaps in our knowledge of preconditions, for example not addressed by the Paris Agreement or IPCC's 1.5 Degree Report. Section 11.4 then identifies elements that make up the global response system, of which there are many. These elements are addressed in the following sections.

Section 11.5 looks at climate change signals and the scale of the climate change problem, noting distributional issues can limit the effectiveness of responses.

The question of how climate change might influence actors and their responses is discussed in Section 11.6, including the distinction between problems (people think should be solved) and conditions (people are willing to live with) (Section 11.6.1) and possible signal response combinations (Section 11.6.2). There is a risk that people treat climate change as a condition and accept climate impacts breaching the UNFCCC objective.

DOI: 10.4324/9781003465911-14

Actors, interests and social permissions are discussed in Section 11.7, highlighting coalitions of actors are essential to get the social permission needed for climate action and effective global responses.

The question of preconditions for effective international cooperation is discussed in Section 11.8 including the influence coalitions, power, and capacity have on the effectiveness of global responses.

Section 11.9 takes on the question of preconditions for effective GHG removals. This section highlights the need for cheap low emissions energy to support technology-based removals and discusses nature-based removals including concerns around the duration of these removals.

Given the long period of time being considered, the influence of other possible changes and scenarios on climate and the global response are discussed in Section 11.10. Future events, that are not widely anticipated, could influence climate or our response to climate change.

Effective global responses are plotted in Section 11.11. Two key variables for effective global responses are climate stress signals and serendipity. If our ambition is inadequate, we need serendipity.

The question of preconditions for effective global responses to climate change is addressed in Section 11.12. Section 11.12 shows effective global responses to climate change depend on a mix of climate change related, response-related and other scenario-related preconditions.

Knowing the preconditions for effective global responses is all very well and good, but we need people to act on effective response options. Section 11.13 looks at the question of conditions under which actors would form coalitions and act on effective response options. This includes a discussion on the extent to which society might treat climate change as a problem including the possibility that society changes the problem definition away from the UNFCCC objective.

Lastly, Section 11.14 notes that failing to fulfil the UNFCCC objective creates existential risks for some people, groups or even states. The existential risk associated with climate change means we are playing an infinite game with profound questions of survival and justice to be considered.

Supporting the discussion in this chapter, are a series of figures analysing relationships between climate change, possible responses, and the effectiveness of these responses. Tables also support the discussion including Table 11.6 which presents the preconditions for effective global responses (i.e., fulfilling the UNFCCC objective).

Interestingly, it was only towards the end of this research that I realised some preconditions for effective global responses to climate change already existed in the literature, starting with the Paris Agreement.

11.2   Preconditions for effective responses from the Paris Agreement

An important part of the global response to climate change is the Paris Agreement. The Paris Agreement operationalises the UNFCCC, sets out actions for the global

Table 11.1 The Paris Agreement purpose includes preconditions for an effective global response to climate change

Paris Agreement actions	*Preconditions for effective global responses*
Strengthen the global response to the threat of climate change Pursue efforts to limit the temperature increase to 1.5 °C above pre-industrial levels	A stronger global response to climate change through the pursuit of efforts to limit global warming to 1.5°C
Increasing the ability to adapt to the adverse impacts of climate change	Capability to adapt to the adverse impacts of climate change
Making finance flows consistent with a pathway towards low greenhouse gas emissions and climate- resilient development	Finance flows consistent with climate resilient and low greenhouse gas emissions development
Foster climate resilience and low greenhouse gas emissions development	Development is climate resilient with low greenhouse emissions
In a manner that does not threaten food production	Food production not threatened by the global response to climate change
Equity in the light of different national circumstances	The global response is equitable with common but differentiated responsibilities

Source: Paris Agreement   *Source: Author*

response, and by extension the preconditions for effective global responses to climate change. From Section 3.3.2, the Paris Agreement's purpose includes seven "actions" for the global response to be effective and achieve the "outcome" of limiting global warming to well below 2°C from pre-industrial times. From these actions, six conditions for an effective global response are derived in Table 11.1. For example, "increasing the ability to adapt to the adverse impacts of climate" implies "capability to adapt to the adverse impacts of climate change" is a precondition for effective global responses to climate change.

Table 11.1 shows that effective global responses must be transformational, including for example: all financial flows being consistent with climate resilient and low GHG emissions development; and, development being climate resilient with low GHG emissions. Furthermore, there should be the capability to adapt to the adverse impacts of climate change. The global response should be equitable with common but differentiated responsibilities and the response should be strengthened through efforts to limit global warming to 1.5°C from preindustrial times. Importantly, food production should not be threatened by the global response to climate change.

11.3   Preconditions for effective responses from the literature

In addition to international agreements, there is a vast body of literature regarding the climate change problem, options for addressing climate change and required global responses. For example, the IPCC's Fifth Assessment included over 9,200 references regarding the physical science basis (IPCC 2014e), over 12,000 references regarding impacts, adaptation, and vulnerability (IPCC 2014f), and close to 10,000 references regarding the mitigation of climate change (IPCC 2014g). From Chapter 2, Project Drawdown provided a list of 100 options for addressing climate change (Hawken 2017), meanwhile Figueres et al. (2017) addressed the timing and scale required of the global response to climate change based on a carbon budget. Blok et al. (2018) and van Vuuren et al. (2018) each made assessments of options for limiting global warming to 1.5°C. Blok et al. (2018) focused on energy-related mitigation options and highlighted the need for massive investments, meanwhile van Vuuren et al. (2018) provided a set of scenarios and highlighted the need for carbon dioxide removals. Shell (2018), the oil and gas company, prepared a "Sky" Scenario that highlighted the need for energy and development to support the sustainable development goals and carbon dioxide removals to limit global warming to 2°C. All of Shell's other scenarios exceeded 2°C of global warming. In many ways each of these publications is like a pair of hands feeling the proverbial elephant from Section 1.4, each describing something different, but collectively help describe the global response system (i.e., the elephant) and possible futures (i.e., how the elephant might move).

IPCC reports stand out from the literature, as they synthesise knowledge and describe the main features of the global response system including possible responses. The IPCC's 1.5 Degree Report is especially important, as it brought together the results of over 6,000 publications to assess global responses capable of limiting global warming to 1.5°C and compared these with responses that limit global warming to 2°C. Importantly, Chapter 2 of the IPCC's 1.5 Degree Report addressed mitigation pathways compatible with 1.5°C and noted "There is a diversity of potential pathways consistent with 1.5°C, yet they share some key characteristics" (Rogelj et al. 2018, p. 129). By definition, these "key characteristics" are preconditions for effective global responses to climate change, that limit global warming to 1.5°C. Table 11.2 presents key characteristics (i.e., preconditions) of 1.5°C pathways in the left-hand column, while the right-hand column shows related preconditions for effective global responses from the Paris Agreement (Section 11.2 above).

From Table 11.2, the key characteristics of 1.5°C pathways focused on things that help limit GHG emissions, which makes sense given that GHGs drive global warming. Key characteristics also included shifts in investment which is consistent with eventually having economies where finance flows are consistent with climate-resilient low GHG emissions development. Low emission development also requires a rapid and profound decarbonisation of energy supply and comprehensive emissions reductions in the coming decade, greater mitigation efforts on the demand side including switching from fossil fuels to electricity (Rogelj et al. 2018).

Importantly, GHG removals in the form of CDR are needed at scale before mid-century but such removals were not in the Paris Agreement purpose (Section 11.2).

Table 11.2 Key characteristics of 1.5°C pathways with related Paris Agreement preconditions for effective global responses

Key characteristics of 1.5°C pathways (i.e., preconditions)	Related preconditions from the Paris Agreement
Considerable shifts in investment patterns	Finance flows consistent with climate resilient and low greenhouse gas emissions development
Rapid and profound near-term decarbonisation of energy supply	Development is climate resilient with low greenhouse gas emissions
Greater mitigation efforts on the demand side	
Switching from fossil fuels to electricity in end-use sectors	
Comprehensive emission reductions will be implemented in the coming decade	
CDR at scale before mid-century	NA – Not mentioned in the Paris Agreement purpose

Source: IPCC 1.5 Degree Report (Rogelj et al. 2018)   *Source: Author*

11.4   Other system elements and possible conditions

Despite the large and growing body of literature, there remain gaps in the literature when it comes to the global response to climate change especially at the systems level. For example, the term "global response to climate change" had not been defined in the literature prior to the research presented in this book (Section 1.4). From the analysis of the sample of possible futures in Chapter 10, there are many elements to the global response system that might have a bearing on effective global responses, including:

- Climate change signals e.g., climate change impacts
- Risks and concerns as well as the extent to which response options are perceived as opportunities or threats
- Ambition levels
- The actors and interests involved
- Coalitions of actors
- Their power to influence other actors and capacity to develop and apply response options
- Response triggers, drivers and attitudes
- The extent to which various options are considered
- The actions taken including adaptation, mitigation, GHG removals or other actions
- The contributions these actions make to the global response
- System dynamics including unintended consequences and other social, economic or environmental changes
- The extent to which other events influence climate or affect the global response.

For each element of the global response system, different conditions are possible, for example, the themes and scenarios from Chapters 5 to 9 each represent different combinations of conditions.

The preconditions from the Paris Agreement purpose addressed issues of ambition, resilience, low emissions development, the avoidance of unintended consequences on food production, and fairness, meanwhile the IPCC's 1.5 Degree Report focused on actions, technologies, and practices that contribute towards climate resilient low emissions development. In each case, only part of the global response system is addressed. The sections below discuss global response system elements and related conditions needed for effective global responses.

11.5   Climate change signals and scale of the problem

The extent to which climate change is a problem that might generate a response is uncertain. From Section 5.2.2, many respondents said climate change and related impacts on human and managed systems could be a "disaster", "crisis", "catastrophe", and "emergency" or "threat multiplier". However, not all respondents expected such severe climate change impacts and risks. It was noted that there are a mix of positive and negative climate change impacts, and at an aggregate level climate change is only estimated to limit global economic growth by only 2–4% by 2100. This estimate is consistent with the literature (Tol, 2009) however most estimates of economic costs and benefits were made before impacts on human and managed systems became discernible (Section 10.2.1). More recently Burke et al. (2015) have shown temperature effects on productivity are non-linear, meanwhile there are concerns that economic estimates fail to adequately address the full range of climate change and related risks, (e.g. DeFries et al. 2019).

With regards to limiting the impact on Gross Domestic Product or Gross World Product, this doesn't mean climate change related disasters, crises or emergencies will be avoided. For example, these aggregate economic indicators obscure the distribution of costs and benefits (Tol et al. 2004, Tol 2009). The distribution of impacts and risks is important as it creates issues of equity (Dietz et al. 2007). In some scenarios, the distribution of impacts raises the possibility of actors needing to migrate due to climate change (Section 7.4.6). It was noted that climate-driven migration could create domestic and international tensions and in some scenarios conflict if other actors respond to migration defensively (Sections 7.4.6 and 7.4.7). Meanwhile, attempting to balance costs and benefits, or limit the need for migration, through transfers or compensation is very difficult (Section 7.4.3). In short, limiting aggregate economic impacts of climate change at the global level, or national level, may not be sufficient to limit social and political tensions or conflict due to distributional issues.

11.6   The influence of climate change on responses

From Chapter 1, a key question was: What influence might climate change have on actors and the global response to climate change? From Section 10.2.1, the emergence of an attributable climate change stress signal represents a fundamental shift in the "global response system", where climate change can less easily be ignored because costs and benefits are more apparent. However, the divergence in climate change scenarios from Section 11.5 highlights uncertainty regarding the extent to which climate change is a "problem" that needs to be solved versus a "condition" society can live with (see Section 11.6.1). Meanwhile, from the sample of possible futures a range of climate change signal and response combinations were identified, including likely ambition levels and contributions to the global response (see Section 11.6.2).

11.6.1   *Problems versus conditions*

Kingdon (1995) noted that there are many conditions, including "bad weather" that while undesirable, society lives with. Only some conditions are considered problems where people believe something should be done about the condition (Section 2.2). Applying this to climate stress signals from Section 10.2.1, it is possible that some climate stress signals may be considered conditions society can live with, while other climate change signals may be considered problems society needs to solve (Figure 11.1). For example, the fact that respondents did not anticipate effective global responses to impacts on natural systems (Section 5.2.3) suggests that impacts on natural systems, including unique and threatened systems (RFC1), are conditions society is collectively willing to live with, even though these impacts and risks are a problem to some actors, for example indigenous peoples living in the arctic (IPCC 2018). It is an important question as to which impacts and risks to human and managed systems might be considered conditions versus problems.

With regards to extreme weather events (RFC2) from Section 2.2.3, Kingdon (1995) noted such events typically only generate periods of interest in a problem, although these periods may constitute political or policy windows for responses. From the sample of possible futures, it may take a succession of distributed events

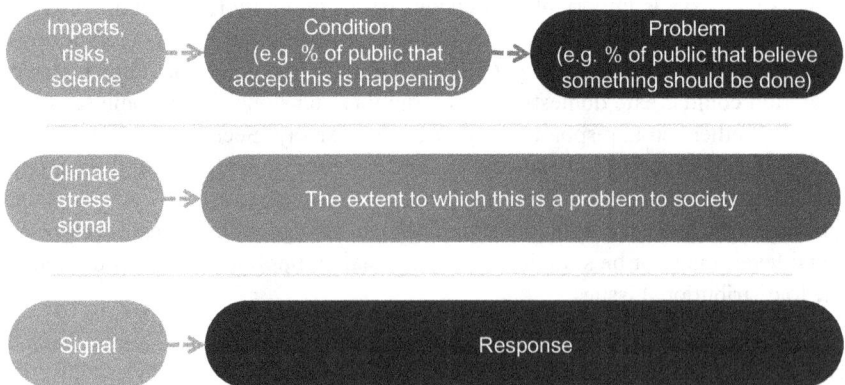

Figure 11.1 The extent to which climate change is a problem to society, applying distinctions from Kingdon (1995).

to generate effective global responses (Section 5.2.4). The distribution of impacts (RFC3) in relation to the power and capacity of actors to respond is also important (Sections 10.2.1 and 5.2.4).

Global aggregate impacts (RFC4) such as "global monetary damage, global-scale degradation and loss of ecosystems and biodiversity" (IPCC 2018, p. 11) would constitute a "disaster", "crisis", "catastrophe" or "emergency". However, an emergency response consisting of "containment and defence" (Grubb et al. 2014) would involve adaptation to the climate change problem and living with some of the hazards and impacts, essentially treating them as conditions. An emergency response consisting of "containment and fix" on the other hand, involves a mix of adaptation to immediate impacts, with mitigation and removals to solve the long-term climate change problem in a way that gets closer to fulfilling the UNFCCC objective (Section 10.2.2).

Large-scale singular events (RFC5), include large, abrupt or irreversible changes in systems caused by global warming, such as tipping points, feedback mechanisms or cascade effects, can also be characterised as a "disaster" or "catastrophe". However, the extent to which large-scale singular events constitute a "crisis" or "emergency" depends on the speed of change. For example, an important distinction can be made between fast feedbacks and other feedbacks (Section 9.10.3). Given the magnitude of large-scale singular events, such as passing a tipping point that makes the collapse of ice sheets in Antarctica or Greenland inevitable, an apathetic or hopeless non-response is possible (Section 5.4.1). In such a scenario, there may be acceptance that sea levels will rise and a scheduled retreat from coastal areas is required (Sections 6.3.10 and 6.4.9). Tipping points, slow feedbacks or cascade effects would constitute inevitable "conditions" rather than "problems" that are reversible. Even fast feedbacks or cascade effects may constitute "conditions" that are accepted if the scale is too large to be solved with available technologies and practices. The problem could become one of adaptation to immediate impacts and risks, while mitigation and removals are used to limit other long-term climate change related risks rather than reversing the sea level rise.

11.6.2 Signals and responses

Conceptually, the global response to climate change, including levels of ambition, are influenced by the climate change signal (Section 10.2.1) including climate stresses along with response risks and opportunities (Figure 11.2). From the sample of possible futures, climate change signals may or may not influence contributions to the global response to climate change (Section 10.2.2). For example, there are scenarios (Figure 11.2) where there are high levels of climate change stress followed by non-responses due to apathy or a sense of hopelessness. Scenarios also include cynical responses for example a fossil fuel push where climate change stress is ignored for the benefit of special interests (Section 5.4.1). Given the current situation, with regards to technologies reliant on fossil fuels and land use practices, non-responses and cynical responses constitute negative contributions to the global response (Figure 11.2).

From the sample of possible futures, not all non-responses make negative contributions to the global response. In some scenarios, where renewable energy and storage become cheap much quicker than expected, these technologies are taken up at scale and in a timely manner, for commercial reasons rather than climate change related reasons (Section 7.4.2). As such, actors can make incidentally positive contributions to the global response to climate change (Figure 11.2), although in such a scenario land use-related emissions would also need to be addressed along with aviation and marine GHG emissions sources.

From Section 10.2.2, climate change signals could trigger low to high ambition cost-benefit-based responses, high to very high ambition enlightened responses or very high-ambition emergency responses (Figure 11.2). However, responding to climate change with ambition does not necessarily mean positive contributions will be made to the global response in the form of adaptation, mitigation, and GHG removals. Response attitudes also influence the options considered, actions taken and contributions to the global response. For example, a very high ambition emergency response involving a defensive response attitude could result in containment and defence with adaptation to impacts and limited mitigation or GHG removals (Sections 10.2.2 and 5.5.1). It is also possible that a cost-benefit response coupled with a defensive attitude could result in the selection of defensive adaptation responses while largely ignoring mitigation or GHG removal options. As such, Figure 11.2 shows defensive response attitudes are likely to make very negative contributions to the global response to climate change. Meanwhile, if actors don't act on mitigation or GHG removal options, it is expected they will adapt to some extent (Sections 5.4.2 and 6.3.4).

Other response attitudes identified from the sample of possible futures include cooperative, practice-focused, technological and competitive attitudes. Given that climate change can be framed as a collective action problem, it would seem cooperative response attitudes are essential to make positive contributions to the global response. However, waiting for global decisions on how to cooperate may not be timely (Section 5.5.1). Importantly, when options that make positive contributions to the global response are anticipated to be commercially viable, then competitive attitudes can help rapidly scale up relevant technologies and practices making a positive contribution to the global response (Section 5.5.3). Technological response attitudes can also make a positive contribution to the global response (Section 5.5.2). However, from the sample of possible futures, technological attitudes also include risks with regards to the possibility of unintended consequences, trade-offs, levels of consumption, resource use, and related wastes (Section 7.2.4). Practice-based responses include challenges related to human behaviour and the management of ecosystems (Section 5.5.2) and are generally not as scale-able as technological options, but are still needed as part of effective global response scenarios especially when it comes to land use (Section 7.4.2).

Table 11.3 simplifies Figure 11.2, and groups response triggers, drivers and attitudes into three categories, based on likely contributions to the global response to climate change. With regards to signals in Table 11.3, climate change signal strength also depends on the actors and interests affected including the extent to

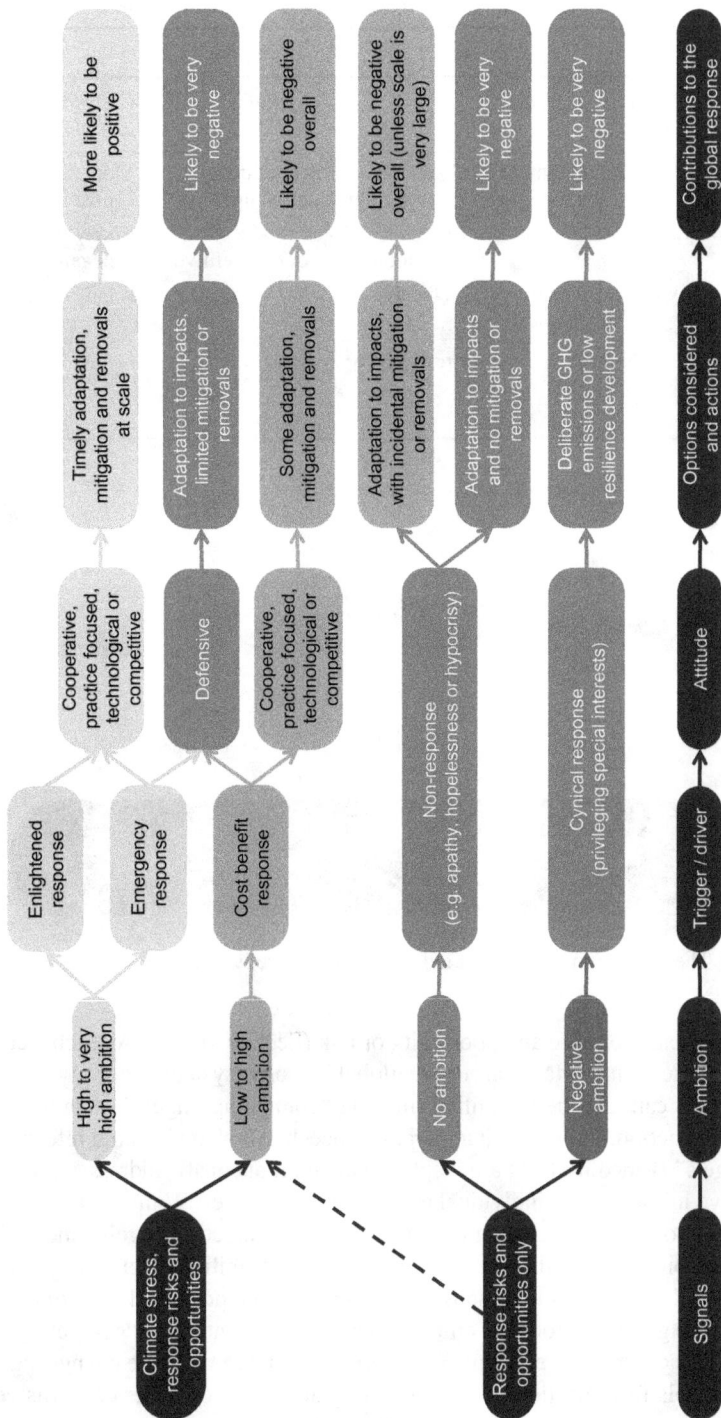

Figure 11.2 Climate change signals and many other factors influence contributions to the global response.

Table 11.3 Summary of climate change related signals, responses, and contributions

Signals	Response		
	Trigger or driver	*Attitude*	*Contribution to global response*
Climate stress, response risks and opportunities	Enlightenment or emergency	Cooperative, practice-focused, technological, or competitive	More likely to be positive
	Cost-benefit	Cooperative, practice-focused, technological, or competitive	Likely to be negative overall
	Cost-benefit or emergency	Defensive	Likely to be very negative
Response risks and opportunities only	No trigger e.g. non-response or cynical		

Figure 11.3 The "competing angels" of individual and collective action.

which response options are an opportunity or risk (Section 10.2.1). As such, actors and interests are an important part of the global response system.

Figure 11.2 can also be simplified into the "competing angels" of individual and collective action. In his first inauguration speech, Abraham Lincoln referred to "better angels" (Lincoln 1861) a metaphor that simultaneously addresses positive and negative influences on individual and collective choices. With this in mind, it is possible to borrow the metaphor and consider the "competing angels" and in doing so better communicate the influence that interests, ambition levels, triggers and drivers, as well as attitudes and inclinations, can have on individual and collective actions and the global response to climate change. Importantly, the "competing angels" highlight contrasting elements of Figure 11.2, and the climate change signal response models from Section 10.2.2, including the angels of wider concerns versus self-interest, ambitious and unambitious angels, enlightened and cynical angels as well as cooperative and defensive angels (Figure 11.3).

11.7   Actors, interests, and social permissions

From the analysis in Chapter 6, three broad groups of actors, interests, and actions were identified consisting of social change and behaviour, political will and policy as well as business and economic activity. While there are many other more detailed breakdowns of actors and interests, including demographic, political or sectoral breakdowns, this book focused on the main features of the global response system (Section 1.4), and as such limits the breakdown of actors, interests, and related actions to three categories.

From the sample of possible futures, political actors and business actors are part of society and as such, are influenced by social change (Section 10.3.4). Furthermore, business and political actors need social permissions if they are to make timely responses at the scales needed to address climate change (Section 6.2.9). Social permissions and related social contracts are central to interactions between domestic social, political, and business actors (see Figure 10.10 in Section 10.3.4) forming the rules of the game that guide responses to climate change in combination with other issues and norms. Several respondents noted that the global response cannot be done in isolation but must be part of a wider set of considerations. An important part of defining and redefining social contracts, national interests and the global response to climate change, are coalitions. This includes coalitions in support of international cooperation.

11.8   Coalitions and effective international cooperation on climate change

It's been acknowledged by the IPCC and others that international cooperation is an important part of the global response to climate change (e.g. Stavins et al. 2014). As such, a key question from Chapter 1, was: What are the preconditions for effective international cooperation on climate change?

From Section 5.2.4, coalitions of both state and non-state actors were identified as being important for international cooperation. However, for these coalitions to be effective, the actors participating need the capacity to develop and apply necessary technologies and practices, and the power to influence other actors and institutional arrangements i.e., "the rules of the game". Having this capacity and power means the coalition can influence the options available and limit the chances of other actors undermining the global response.

From Section 5.2.4, the distribution of climate change signals within states as well as between states was identified as being important for international cooperation. Applying Putnam's (1988) two-level game (Section 3.4.2) the distribution of climate changes and related impacts and risks within a state, coupled with the extent to which response options are perceived as risks or opportunities by various actors, can affect coalitions of domestic actors and hence the national interests of the state, for example when it comes to international cooperation on climate change.

From Section 8.2.2, if the climate stress signal, in the form of impacts on human and managed systems, gets strong enough, then it is possible that climate change could become the premier international issue and an issue of international security.

As such, climate change could be addressed in the United Nations Security Council, in which case the permanent members with veto power, which currently consist of China, France, Russia, United Kingdom, and United States of America, would need to agree if there is to be an effective global response. From Section 8.3.3, it is anticipated that if powerful states backed a stringent international climate agreement, then other states would comply rather than dare upset the powerful states.

An important question when it comes to the effectiveness of international cooperation is whether capable and powerful coalitions of States will respond to climate change risk signals (e.g., respond to climate change as a "threat multiplier") or whether such coalitions will only form once there are attributable climate change and related impacts on human and managed systems. If capable and powerful coalitions do not form until there are impacts, then the climate change and related hazards precipitating a response will likely be part of the new range of normal conditions. In such a scenario, climate change will need to be reversible for GHG mitigation and removal responses to be effective. All other things being equal, a response to climate change risk signals would be timelier than a response to impacts (Section 5.4.2).

Table 11.4 builds on Table 11.3 from Section 11.6.2, and includes the distribution of climate change signals on actors and coalitions with power and capacity versus actors and coalitions with limited power and capacity. As in Table 11.3, actors and coalitions can either make positive or negative contributions to climate change, but actors and coalitions with power and capacity have much more influence on whether the global response is likely to be effective or not. For example, if actors and coalitions with power and capacity engage in responses that make a

Table 11.4 The likely effectiveness of the global response to climate change based on signals, actor volved, their responses and contributions made

Signals	Actors and coalitions	Possible responses			Effectivenes global resp
		Trigger or driver	Attitude	Contribution	
Climate stress, response risks and opportunities	With power and capacity	Enlightenment or emergency	Cooperative, practice-focused, technological, or competitive	More likely to be positive	More likely be effecti
		Cost-benefit	Cooperative, practice-focused, technological, or competitive	Likely to be negative overall	Unlikely to effective
		Cost-benefit or emergency	Defensive	Likely to be very negative	Very unlike be effecti
		No trigger e.g. non-response or cynical			
	With limited power and capacity	Cost-benefit, enlightenment or emergency	Cooperative, practice-focused, technological, competitive, or defensive	Positive or negative contributions	Limited influence effectiver Unlikely to effective
		No trigger e.g. non-response or cynical			

positive contribution to the global response, then the global response is more likely the be effective. Conversely, if actors and coalitions with power and capacity engage in responses that make negative contributions to the global response, then the global response is much more likely to be ineffective. Actors and coalitions with limited power and capacity have limited positive or negative influence on the effectiveness of the global response.

11.9   Preconditions for atmospheric greenhouse gas removals

From Chapter 1, a key question was: What are the preconditions for actors to remove GHG from the atmosphere at a scale required to limit climate change to safe levels? This is important, because the budget available for more GHG emissions to accumulate in the atmosphere, while limiting climate change, is rapidly running out (Section 2.2.4). From Section 11.3, the IPCC's 1.5 Degree Report noted carbon dioxide removals are needed at scale before 2050 if the global response to climate change is to be effective.

From Section 7.3.1, an important precondition for GHG removals and stabilising atmospheric concentrations of GHGs at safe levels is mitigation. Given the large quantity of GHG emissions being emitted when interviews were being conducted, it was generally considered unfeasible to use GHG removals alone to stabilise atmospheric concentrations of GHGs at safe levels. Instead, mitigation is required to dramatically lower GHG emissions and as such, reduce the quantities of GHGs that need to be removed from the atmosphere to stabilise concentrations at safe levels. Meanwhile, from Section 10.4, the availability of cheap low emissions energy is a very important factor for both mitigation and technology-based atmospheric GHG removals. One thing that was not addressed in the sample of possible futures is the possibility that if some actors are emitting GHGs into the atmosphere, other actors may not want to remove GHGs.

Given that there are large stores of carbon in the biosphere, including forests, soils, and the sea life, practices are going to be an important part of limiting GHG emissions and could be an important part of GHG removals, in particular carbon dioxide removals. This includes community engagement, and institutional arrangements especially when it comes to land tenure.

From the sample of possible futures, atmospheric GHG removals are not without risks or possible unintended consequences, including for example the large amounts of energy or land required by some GHG removal methods (Section 7.3.4). Other concerns include the duration of GHG storage and the risk of releases of carbon from the biosphere or geosphere back into the atmosphere (Section 7.3.5). An important condition for effective GHG removals is that GHGs remain out of the atmosphere.

11.10   Other changes that could influence climate or the
global response

The questions from Chapter 1 did not include: What are the other changes that could influence climate or the global response to climate change? This question is important because climate change is a long-term problem. In any given decade, or century there will be unexpected "historic" events. From Chapter 9, there are a range of possible social, economic or environmental changes that could help or hinder the global response to climate change.

Unexpected events can interact with political will and policy in unexpected ways, for example in Section 5.4.2 it was noted that a global pandemic might demonstrate the value of international cooperation, and once the pandemic is addressed, actors might also work together to address climate change (see enlightened response). Another scenario is a volcanic eruption emitting large quantities of sulphates creating a natural experiment informing solar radiation management policies (Section 9.5). As such, any response, or response strategy, should be aware that other things could happen and acknowledge the need for contingencies in case of other scenarios rather than relying on serendipity[1] or chance.

From the sample of possible futures, it is arguable that people interested in climate change focus too much on climate change, meanwhile there are many problems and conditions that affect people (Section 9.1), some of which may have more influence on the global response to climate change than so-called "climate action" where climate change is a central concern. From the sample of possible futures, scenarios include volcanic eruptions, environmental feedbacks and catastrophic cooling events. Meanwhile, social and economic scenarios include cyborgization of humanity exacerbating inequities, enlightenment, population decline, conflict including nuclear war, as well as the possibility of solar radiation management technologies being deployed and possible geopolitical ramifications.

11.11   Effective global responses

From Sections 3.3.1 and 4.3.2, an effective global response is one that fulfils the UNFCCC objective of stabilising atmospheric concentrations of GHGs at levels that allow ecosystems to adapt naturally, food production not be threatened and economic development to proceed in a sustainable manner. Given the legitimacy of the UNFCCC and the fact that the UNFCCC is the framework convention under which the global response to climate change is broadly organised, the UNFCCC objective was used to define the conditions that constitute effective global responses to climate change.

With regards to scenarios, Table 11.4 from Section 11.8 only addressed the influence actors might have on the effectiveness of global responses to climate change. Based on the CCNIIC Model in Section 3.4.2, the environment is an important part of the global response system (also see Appendix B). However, one of the complicating issues when attempting to identify effective response scenarios, is uncertainty in our understanding of natural systems and the environment, especially when it comes to radiative forcing and climate sensitivity. From Section 2.2.1 there's a considerable range of uncertainty when it comes to levels of anthropogenic radiative forcing. Similarly, from Section 2.2.3 there is a very remote chance that global warming could be limited below 1.5°C or 2°C even with much higher atmospheric GHG concentrations (Section 2.2.3). Meanwhile, from Section

Effective response scenarios

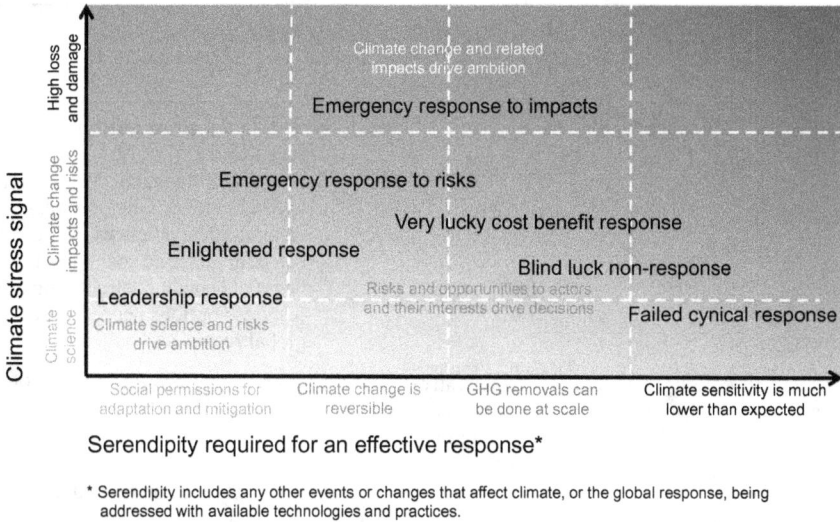

Figure 11.4 Schematic plot of effective response scenarios including climate stress levels and levels of serendipity required.

7.4.2, it is conceivable that technologies change much quicker than expected and serendipitously make it commercially viable to scale up mitigation or even remove GHGs from the atmosphere. As such, it is conceivable, but very unlikely that a blind luck non-response, or even a failed cynical response could still result in limited climate change and related impacts. However, this would require incredible levels of serendipity (Figure 11.4).

Figure 11.4 schematically plots the strength of the climate stress signal (vertical axis) against the level of serendipity required for effective global responses to climate change (horizontal axis). These responses are divided into two categories, those that rely on ambition and high levels of serendipity (lower left quarter to upper middle), and those that require incredible levels of serendipity relative to ambition (lower right quarter).

Table 11.5 builds on Figure 11.4 as well as Table 11.4 from Section 11.8, and presents the conditions under which responses relying on ambition (i.e., leadership, enlightened or emergency responses with cooperative, practice-based, technological, or competitive attitudes) and responses relying on serendipity (i.e., cost-benefit responses, defensive responses, blind luck non-responses, and failed cynical responses) could fulfil the UNFCCC objective. Importantly, leadership responses can help create the conditions needed for serendipity to arise, as do capable institutions. For example, leadership on technologies and practices including policies, means there are experiences to draw upon, and options available, when there is the collective ambition to scale up responses to climate change. Meanwhile levels of

Table 11.5 The characteristics of effective global responses to climate change (including effective r responses and serendipity)

Signals	Types of global responses		Serendipity
	Trigger or driver	*Attitude*	
Climate science, risks and impacts	Leadership response		**Creating the preconditions for serendipity** Leadership influencing social change and behaviour, policy, business, technologies, practices, and coalitio can help generate social permissions negotiate social contracts and create response options needed for effectiv global responses
Climate stress, response risks and opportunities	Enlightenment or emergency	Cooperative, practice-focused, technological, or competitive	**High levels of serendipity needed,** including: Climate sensitivity is no higher than anticipated Climate change and related impacts ca be halted GHG removals can be done at scale if needed Other unexpected changes do not hind the global response to climate chang
	Cost-benefit	Cooperative, practice-focused, technological, or competitive	**Very high levels of serendipity need** including: Climate sensitivity is lower than anticipated Climate change and related impacts ar reversible Climate resilient low emissions technologies and practices become cheap quicker than expected GHG removals technologies and pract are available and done at scale Other unexpected changes help the glc response to climate change
	Cost-benefit or emergency	Defensive	**Incredible levels of serendipity need** including: Climate sensitivity is much lower thar anticipated Climate change and related impacts ar reversible Climate resilient low emissions technologies and practices become cheap very quickly GHG removals technologies and pract are available and are eventually app at scale Other unexpected changes help the glc response to climate change
Response risks and opportunities	Blind luck non-response Failed cynical response		

development and capable institutions make it possible to act on response options (Section 6.3.4). Note: For more on the role of serendipity in effective responses, see Appendix H.

Armed with a knowledge of possible effective global responses to climate change, it is possible to identify preconditions for effective global responses to climate change.

11.12   Preconditions for effective global responses

The primary question from Section 1.5 is: What are the preconditions for effective global responses to climate change? Rather than mapping reference pathways and discussing milestones towards effective global responses, preconditions for fulfilling the UNFCCC objective are organised into interrelated themes and tiers. From Table 11.5 and the descriptions of serendipity in Section 11.11, characteristics of effective global responses can be divided into three themes, consisting of:

- Climate change related preconditions
- Global response related preconditions
- Other scenario related preconditions.

Of these, the Paris Agreement purpose and IPCC's 1.5 Degree Report addressed response-related preconditions. While it is important to understand response-related preconditions over which actors and coalitions have an influence, given the importance of serendipity in effective global responses to climate change (Section 11.11), it is essential to consider climate change related preconditions as well as other scenario-related preconditions.

With regards to tiers (i.e., the organisation of preconditions into rows), essential preconditions are described in Tier 1, meanwhile other preconditions needed to support Tier 1 preconditions, are described in Tier 2. Different combinations of Tier 2 preconditions could contribute to Tier 1 preconditions and fulfilment of the UNFCCC objective. In addition to preconditions, other factors are highlighted that could influence preconditions and the global response.

Climate change related preconditions regard the scale and reversibility of climate change. A Tier 1 precondition for effective global responses to climate change is that the scale of climate change, and related impacts, be limited (Table 11.6). At Tier 2, this could be due to mitigation of GHG emissions limiting the accumulation of GHGs in the atmosphere, GHG removals, or climate sensitivity being much lower than expected. Given the likelihood of overshoot scenarios, another Tier 1 precondition is that climate change is reversible or can be quickly stabilised. At Tier 2 this means not passing thresholds such as tipping points or cascade effects.

Response-related preconditions have been identified from the sample of possible futures as well as the Paris Agreement and IPCC's 1.5 Degree Report. Tier 1 response-related preconditions derived from the sample of possible futures consist of the global response being timely and at scale most likely involving a transformation; and contingencies being available to address any extreme climate changes or other scenarios that might influence the global response. The likely need for a transformational global response is consistent with the Paris Agreement and IPCC 1.5 Degree Report's preconditions (Table 11.7). The need for finance and development that is climate resilient with low GHG emissions constitutes a transformation of finance and development as we know it.

Tier 2 response-related preconditions from the sample of possible futures include leadership, research and development, technologies and practices (including policies), and for effective ambition-driven responses, social permissions. Leadership includes social, political, and business leadership. This leadership can generate response options and opportunities including technologies, practices including policies (Section 11.11). Cost-benefit-based responses and blind luck non-responses rely on climate sensitivity being much lower than expected (see climate change related preconditions, Table 11.6) coupled with relevant technologies and practices becoming cheaper much quicker than expected. In this scenario, climate resilient, low emissions, and GHG removal technologies and practices are adopted at scale for commercial reasons with limited consideration of climate (i.e., a limited social cost of carbon applied) or no consideration of climate change. Even in blind luck scenarios, it is important to have contingencies available, in the form of technologies and practices, in case of other changes.

Effective (i.e., non-defensive) ambition-driven responses include enlightened or emergency responses with cooperative, practice-focused, technological and competitive attitudes. To a lesser extent, cost-benefit responses with the same attitudes might also be included in this category. Ambition-driven responses need social permissions allowing climate action and any necessary trade-offs. Social permissions are based on local and domestic coalitions of actors (Section 11.7) which then translate into national interests, coalitions, and agreements at the international level (Section 11.8). For such coalitions to be effective, the actors participating need the capacity to develop and apply technologies and practices and the power to influence other actors and institutions (Section 11.8) so they also participate and comply with the global response. Institutions are an important factor contributing to the capacity and power of government actors.

Importantly, the IPCC's 1.5 Degree Report provides some guidance on the specific technologies and practices needed for limiting global warming to 1.5°C. This includes switching from fossil fuels to electricity in end-use sectors coupled with greater mitigation efforts on the demand side (Table 11.7). With regards to timeliness, decarbonisation of energy supply needs to be rapid and profound (i.e., in the near term) and part of a wider set of comprehensive emission reductions to be implemented in the coming decade. Investment priorities and patterns will need to change even quicker to finance this transformation (Table 11.7). With regards to GHG removals, the 1.5 Degree Report stated CDR needs to be at scale before 2050.

With regards to the levels of global warming that might be allowed while fulfilling the UNFCCC objective, scenarios included stabilisation of global warming at levels greater than 1.5°C or even 2°C from pre-industrial times (Section 5.3.1). In such scenarios, the question of what fulfilling the UNFCCC objective looks like

Objectives	Success criteria		
UNFCCC objective	Stabilising atmospheric concentrations of greenhouse gases at levels that allow ecosystems to adapt naturally, food production is not threatened, and economic development can proceed in a sustainable manner.		
Tiers	*Preconditions for effective global responses*		
	Climate change related	*Response related*	*Other scenario related*
Tier 1	Scale of climate change is limited.	The global response is timely and at scale (i.e., constitutes a transformation). This includes adaptation, mitigation and atmospheric GHG removals.	Contingencies are available for addressing extreme climate change or other scenarios.	Other social, economic, or environmental changes don't negatively influence climate or the global response to climate change at scale.
Tier 2	The scale of climate change can be addressed by technologies and practices. The accumulation of GHGs in the atmosphere is limited. Climate sensitivity is much lower than expected, or not much higher than expected.	Not passing thresholds such as negative tipping points or cascade effects.	Social, political, and business leadership generating response options and opportunities. For non-defensive ambition-reliant responses, preconditions include social permissions for climate action; as well as, domestic and international coalitions with sufficient capacity and power to ensure responses are timely and at scale.	Research and development of technologies, practices, and policies, so these things can be deployed quickly and at scale if needed.	The scale of other negative social, economic, or environmental changes can be addressed by technologies, practices, institutions, or policies. Other social, economic, or environmental changes limiting climate change or helping the global response to climate change.
Factors	*Climate change related*	*Response related*	*Other scenario related*
Other factors	Ecological and environmental processes and dynamics, e.g., the carbon cycle and carbon balance between the atmosphere, biosphere, hydrosphere, and geosphere.	Social change and behaviour, political will and policy, business, and economic activity; distribution of impacts and risks; response triggers, drivers, and attitudes of actors; domestic, national, and international actors; prices, taxes and subsidies, markets, support for research, development of technologies and practices; and institutions.	Ecological, environmental, and cosmological processes and dynamics, social, political, and business processes, and dynamics.

Table 11.7 Preconditions for effective global responses from the Paris Agreement and IP-CC's 1.5 Degree Report

Tier	Response related preconditions
Tier 1	**Ambition** A stronger global response to climate change through the pursuit of efforts to limit global warming to 1.5 degrees[a] **Resilience (including avoidance of unintended consequences)** Capability to adapt to the adverse impacts of climate change[a] Food production not threatened by the global response to climate change[a] **Transformation (finance and development)** Finance flows consistent with climate resilient and low greenhouse gas emissions development[a] – Considerable shifts in investment patterns[b] Development is climate resilient with low greenhouse emissions[a] – Rapid and profound near-term decarbonisation of energy supply[b] – Greater mitigation efforts on the demand side[b] – Switching from fossil fuels to electricity in end-use sectors[b] – Comprehensive emission reductions are implemented in the coming decade[b] **GHG removals** – CDR at scale before mid-century[b] **Fairness** The global response is equitable with common but differentiated responsibilities[a]

[a]   Paris Agreement.
[b]   IPCC 1.5 Degree Report.

in an overshoot scenario becomes important. These scenarios involve losing ecosystems and it is an open question as to whether society attempts to limit climate change or focuses on adaptation (Sections 11.13 and 13.5).

Response-related preconditions from the sample of possible futures do not always align with preconditions from the Paris Agreement purpose. For example, the Paris Agreement's purpose highlights fairness, equity, and common but differentiated responsibilities as important conditions. From the sample of possible futures, enlightenment scenarios also included issues of humanity and economic distribution at the same time as addressing climate change (Section 5.4.2). In enlightenment scenarios these issues cannot be addressed separately. However, enlightenment scenarios were not the only scenarios that could lead to effective global responses and it is conceivable that powerful business interests might drive political coalitions and effective emergency responses while privileging business interests (Section 6.4.1). In these scenarios, fairness is not a precondition for an effective global response but rather, coalitions of powerful and capable actors are the defining precondition.

While the UNFCCC objective includes making sure food production is not threatened by climate change, the Paris Agreement highlights food production should not to be threatened by the global response (Section 11.2). The risk of unintended consequences comes with any social, political or business changes and as such it is important to be mindful of these response risks (e.g. see Sections 11.6.2 and 11.9) in the same way it is important to ensure resilience and have contingencies.

The Paris Agreement highlighted the need for resilience (Table 11.7), while analysis of the sample of possible futures demonstrated the need for contingencies related to climate change as well as other social, economic or environmental changes (Section 11.10). Contingencies and capacity are important elements of resilience. Adaptation also contributes to resilience and from the analysis in Section 11.6.2, adaptation is included in non-responses, as actors are expected to make changes when impacts arise, regardless of whether they attribute them to climate change or not (Figure 11.2). Of course, there is always the chance of maladaptation (Section 2.2.3).

Other scenario-related preconditions rely on serendipity (see Table 11.5 in Section 11.11). Tier 1 other scenario preconditions consist of other changes, including unexpected social, economic, and environmental changes, not exacerbating climate change and related impacts, or hindering the global response to climate change. For serendipity-reliant scenarios, any other social, economic or environmental changes need to help the global response to climate change rather than hinder it. Even ambition-driven responses need to avoid social, economic or environmental changes that negatively affect climate or the global response to climate change. With regards to the literature, Kerr (2007) noted "Serendipity is not a strategy" in the title of their paper regarding national climate programmes from 21 countries. Worryingly, Kerr (2007) found serendipity was a key factor towards improving GHG emissions trends, rather than policy or some deliberative response.

See Appendix I for more information on the role of serendipity with regards to Tier 1 preconditions.

11.13   Preconditions for action

A key question is: Under what conditions would actors act on effective response options? While it is good to know the preconditions for effective global responses to climate change, it would be reassuring to know the conditions under which actors would actually act on effective response options.

Collective action depends, in part, on the extent to which society treats climate change and related impacts as a problem that needs to be solved or a condition that can be lived with (Section 2.2). Kingdon (1995) highlighted that the problem definition has politically high stakes, as it has implications regarding the options considered, with some actors and their interests potentially "helped" while others may be "hurt". The extent to which climate change is defined as a problem depends in part on the climate change signal and climate stress levels, as well as the extent to which response options are an opportunity (Section 11.6.1). These factors also influence actor ambition and how high climate change is ranked on the public policy agenda (Figure 11.5). However, individual and collective responses to climate change also depend on the availability of options that simultaneously address climate change and actor interests.

The availability of commercially viable technology options fundamentally changes the climate change problem. It was noted in Section 6.3 that the climate change was a problem without technological solutions, and Section 10.3.5 noted these conditions characterised the Kyoto era. Now climate change is a problem requiring political ambition (Section 6.3) because commercially viable technologies and practices exist (Section 10.3.5). With ambition it is possible that zero emissions technologies and practices could be scaled up through political and business coalitions driven by competitive and technological response attitudes (e.g. Sections 5.5.3, 6.4.4 and 7.2.2). For such a scenario to happen, it is important that capable and powerful actors do not pursue cynical responses privileging special interests. Furthermore, cooperative attitudes and high ambition are needed on a mix of low and negative emissions agriculture, forestry, and land use practices (policies included) for there to be an effective global response that is not overly reliant of serendipity.

Enlightened responses featured strongly in the sample of possible futures. As such, it is possible that awareness creates change in society, including social change and behaviour, political will and business responses that make positive contributions to the global response to climate change. Such a scenario might be driven by an unambiguous climate risk signal or some other global event such as a pandemic that generates awareness of the value of collective action (Section 5.4.2). However, the extent to which global events such as a pandemic might contribute to an enlightened response is an open question.

Emergency responses involve very high ambition and urgency, which Beck (2006) described as being "involuntary enlightenment" where climate change impacts and risks force awareness and change. Like with enlightened responses, it is difficult to envisage scenarios where an unambiguous climate risk signal would generate collective action capable of limiting climate change and its impacts. Meanwhile, emergency impact responses may not be effective at limiting climate change and impacts due to locked-in of climate change and related physical hazards.

Given the scale of the climate change problem and the long-term lock-in of climate change and related hazards, it is conceivable that societies collectively treat impacts and risks to natural systems as "conditions" that can be lived with. From the sample of possible futures it seemed unlikely that actors with power and capacity would willingly let food production be threatened (Section 5.3.2), however defensive responses to climate change and other social, economic or environmental scenarios could still result in food insecurity, for example due to food exports being banned in some countries. Likewise, it seems unlikely actors with power and capacity will willingly forgo economic development (e.g., see Section 8.3.3). Hence actors are anticipated to adapt to climate change and related impacts, even if climate change is not acknowledged as a driver.

If climate change impacts and risks are primarily addressed through adaptation, then the climate change problem will have changed from one of stabilising atmospheric concentrations of GHGs to a problem of resilience (i.e., reducing vulnerability and exposure to climate change and related hazards). In such a scenario, the UNFCCC objective is essentially abandoned.

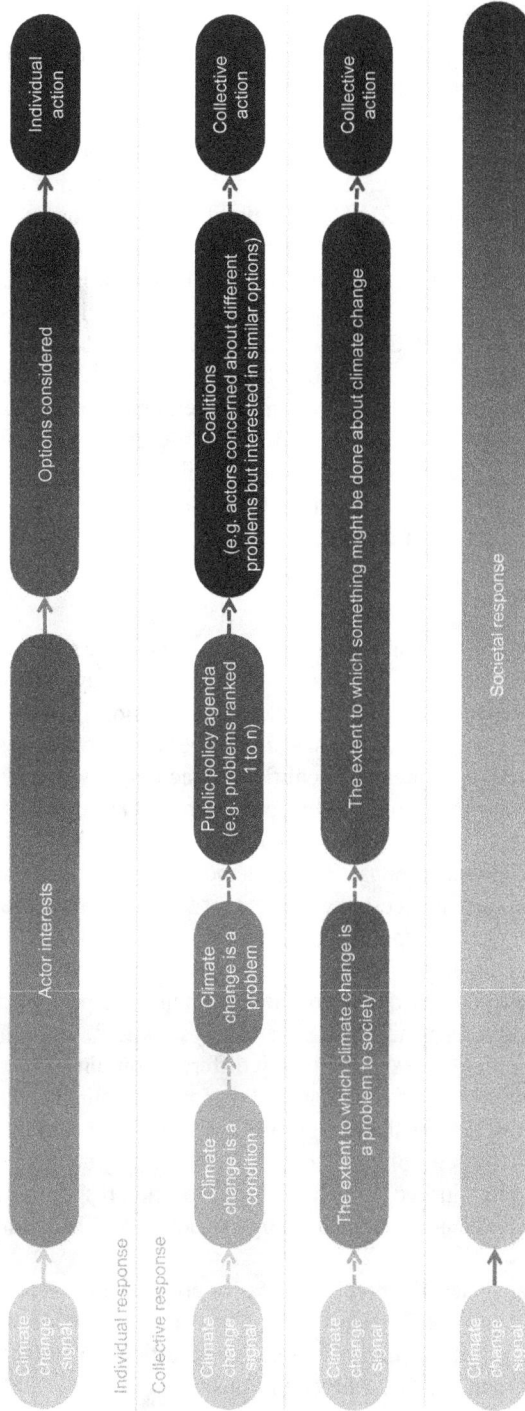

Figure 11.5 Preconditions for individual and collective action on options.

11.14   Existential risk

If the UNFCCC objective is abandoned, or fails to be achieved, then for some actors climate change could be catastrophic at personal, local or national levels (Section 10.3.5). For example, actors in unique and threatened systems may not be able to continue cultural practices (Section 2.2.3), meanwhile low-lying small island developing states may not remain viable as states with further climate change and sea level rise (Section 5.2.2). This highlights the fundamental nature of the climate change problem, as an "infinite" game, where the aim is to stay in the game i.e., "continuing the play" (Carse 1986). So, while this book implicitly framed the climate change problem as a "finite game" where it is possible to solve climate change by fulfilling the UNFCCC objective, the UNFCCC objective has the characteristics of an "infinite game", e.g. ecosystems being able to adapt naturally, food production not being threatened and economic development being able to proceed in a sustainable manner (Section 3.3.1). An effective global response to climate change is not a one-off event, and there is no finish line, instead it is something that requires vigilance and effort. Failure on the other hand creates existential risks and questions of climate justice for all of us.

11.15   Conclusions

By looking deep into the crystal ball (i.e., sample of possible futures) in Chapters 5–9, and then analysing signals, responses, actors, and pathways in Chapter 10, elements of the global response system were identified (Section 11.4). These elements consisted of climate change signals; risks and concerns; opportunities and threats; ambition levels; actors and interests; coalitions of actors; power and capacity; response triggers, drivers and attitudes; the options considered by actors; the actions taken including adaptation, mitigation, GHG removals; and the contributions these actions make to the global response. Other elements include system dynamics including unintended consequences; and, the extent to which other events influence climate or affect the global response.

Like a jigsaw puzzle, each of these elements were brought together, looking for links. Unlike a jigsaw puzzle, for each of these elements it is possible to have more than one set of scenarios or conditions. So, an analysis was made on these elements and conditions, including the extent to which different combinations contribute towards more effective, or less effective, global responses to climate change. Further complicating the analysis was the recognition that there is uncertainty including a very small chance that serendipity may play a role in helping achieve the UNFCCC objective, if ambition is limited. Even with high ambition responses serendipity is needed because we have emitted such a large quantity of GHGs into the atmosphere already.

Based on these analyses, the preconditions for effective global responses to climate change were identified (Section 11.12). These include climate change related, response-related and other scenarios-related preconditions. In total there are five Tier 1 preconditions and ten Tier 2 preconditions.

Tier 1 preconditions consist of

- The scale of climate change is limited
- Climate change and related impacts are reversible or can be quickly stabilised
- The global response is timely at scale and transformational, including adaptation, mitigation, and atmospheric GHG removals
- Contingencies are available for addressing extreme climate change or other scenarios
- Other social, economic, or environmental changes don't negatively influence climate or the global response to climate change at scale.

Tier 2 preconditions consist of

- The accumulation of GHGs in the atmosphere is limited
- Climate sensitivity is much lower than expected, or not much higher than expected
- The scale of climate change can be addressed by technologies and practices
- Not passing thresholds such as negative tipping points or cascade effects
- Social permissions for climate action
- Social, political, and business leadership generating response options and opportunities
- Domestic and international coalitions with sufficient capacity and power to drive timely responses at scale
- Research and development of contingent technologies, practices and policies, so these technologies and practices can be deployed quickly and at scale if needed
- Other negative social, economic, or environmental changes can be addressed by institutions, technologies, or practices (including policies)
- Other events and social, economic, or environmental changes, help the global response to climate change or limit climate change.

The chapter finished by questioning how we might define the climate change problem in the future, and notes that we cannot "solve" climate change because there is no finish line, but through vigilance and effort we can improve our collective situation.

What does all this mean for you and me? Chapter 12 looks at what this study has learnt, how we can use this information, its limitations, and whether we are up to the challenge of limiting climate change and its impacts.

Note

1 The Oxford Dictionary defines serendipity as "the occurrence and development of events by chance in a happy or beneficial way." In the context of this study, serendipity refers to: the occurrence and development of events that by chance, rather than intent, contribute towards fulfilling the UNFCCC objective.

12 How to limit climate change and its impact

12.1 Introduction

At the beginning of this book, a climate veteran said, "We are all actors". The same veteran highlighted we need to "break the box in which we are located."

When I started working on this book, I wanted to "break the box" and "solve" the climate change problem.[1] Initially, I wanted to find options for limiting climate change and its impacts; fulfilling our internationally agreed objectives of stabilising atmospheric concentrations of greenhouse gases (GHGs) while allowing ecosystems to adapt naturally, ensuring food production is not threatened and economic development can proceed in a sustainable manner. I quickly realised there are adaptation, mitigation, and GHG removal options, they have been peer-reviewed, published, and assessed by the IPCC (Section 2.3). The question is not whether we have options, it is under what conditions would we act on these options? These "conditions" are "preconditions" for effective global responses.

So, how do we find preconditions for effective global responses? It quickly became apparent the past is not representative of the future (Sections 1.2 and 10.2.1). Instead of looking to the past, we needed to take lessons from the future. And instead of exploring a few reference scenarios, we needed to explore all possible futures (Section 4.3.3).

By looking deep into the crystal ball and searching the "sample of possible futures," this book explored: the influence climate change might have on us and our responses; the preconditions for effective international cooperation by state and non-state actors; the conditions under which governments, businesses or others would remove GHGs from the atmosphere at scale; and, the conditions under which we would actually do what it takes to limit climate change and the related impacts. In short, the book "solves" the climate change puzzle, in much the same way you can solve a riddle. And like any riddle, there is a twist (spoiler alert, turns out we are playing the infinite game), and it is important to share the "solution" (i.e., preconditions for effective responses), so these lessons can be applied.

12.2 Preconditions for effective global responses to climate change

In public and academic discourse, the expression "climate crisis" has been used to describe the current climate change situation (Section 2.2.3). This book finds

DOI: 10.4324/9781003465911-15

the climate issue to date is largely a "crisis of response" rather than a "crisis of impacts". For example, impacts on people, property, and livelihoods were largely undetectable until the period 2010–2020 (Section 10.2.1) unless you were living in "unique and threatened systems"[2] (Section 11.6.1). However, climate change lock-in means global warming and other climatic changes will persist for long periods (e.g., centuries, Section 2.2.3), meanwhile the risk of impacts increases markedly with global warming above 1°C (Section 10.2.1), making the need for timely climate responses critical.

Timely responses at scale are preconditions for effective global responses (Section 5.3). Our response to date is anything but timely or at scale. Underscoring the "crisis of response" is the fact that atmospheric GHG removals are almost certainly required to limit climate change to well below 2°C of global warming (Section 2.2.4).

In addition to timeliness and scale, effective global responses to climate change involve a combination of serendipity and ambition (Section 11.11). Ambition-driven responses consist of leadership, enlightened responses, and emergency responses. Serendipity-driven responses consist of very lucky cost-benefit-based responses, blind luck non-responses, and failed cynical responses.

Leadership is a prerequisite for any effective global response. Social, political, and business leaders create the conditions for effective global responses, for example by researching and developing technologies and practices when others don't, developing policies and business models when others won't, and forming coalitions in support of climate action and wider social change when others resist. Leadership creates the options and coalitions needed for effective global responses (Section 11.12).

Enlightened responses (i.e. awareness that creates change) and emergency (i.e. security) responses involve high to very high ambition levels where the decision to solve the climate change problem is made and the main economic question regards "cost effectiveness" i.e. how to best use available resources to limit climate change and related impacts (Section 10.2.2)? By comparison, cost-benefit-based responses, non-responses, and cynical responses each rely on very high to incredible levels of serendipity if they are to be effective (Table 11.5, Section 11.11). Of all the options required to limit climate change and related impacts, cost-benefit-based responses only consider the options that have a positive net present value (Section 10.2.2). Only some of these will actually be implemented. Blind luck non-responses involve incidental adaptation, mitigation, and GHG removals driven by commercial or other considerations rather than climate change related considerations (Section 10.2.2). Failed cynical responses involve the failure of special interests to inhibit the global response to climate change.

It would be reassuring to think ambition-driven responses (e.g., leadership, enlightened, and emergency responses) could drive effective global responses without serendipity, but this is not true. Due to current levels of climate change and GHG emissions, even ambition-driven responses rely on high levels of serendipity if they are to limit climate change and related impacts (Section 11.11). So, while Kerr (2007) noted "serendipity is not a strategy", serendipity is a precondition for effective global responses to climate change.

Serendipitous conditions required for effective global responses include: climate sensitivity being the same or lower than expected; not breaching climate system thresholds, tipping points or triggering cascade effects (including so-called "large scale singular events"); the accumulation of GHGs being kept below what can be addressed with available technologies and practices; other unexpected social, economic or environmental changes helping limit climate change and related impacts; and, other unexpected changes helping (or at least not hindering) the global response. These are "climate change" and "other scenario" related preconditions that are essential for any "response" to be effective (Section 11.12).

According to the IPCC's 1.5 Degree Report, "response" related preconditions consist of timeliness and scale nothing short of a transformation including adaptation, mitigation, and atmospheric GHG removals (Rogelj et al. 2018, Table 11.7). From the scenarios compiled as part of this book, preconditions for such a transformation include social permissions for adaptation, mitigation and GHG removals at scale; domestic and international coalitions forming around climate resilient low emissions technologies and practices (including policies); coalitions forming with the capacity and power to ensure these technologies and practices are adopted at scale and in a timely manner; social, political and business leadership on technologies and practices, social, political and business leadership creating domestic and international coalitions in support of the global response to climate change; and, availability of contingencies in case of extreme climate change or unexpected social, economic or environmental events (given the long periods of time being considered) (Section 11.12).

From the literature, "response" related preconditions derived from the Paris Agreement purpose include ambition in the form of a stronger global response to climate change through the pursuit of efforts to limit global warming to 1.5°C; resilience including the capability to adapt to the adverse impacts of climate change; and changes in finance and development that are nothing short of what's required to achieve climate resilient low GHG emissions development (Table 11.7, Section 11.12). While fairness featured in the Paris Agreement-based preconditions, it did not feature in scenarios collected as part of the research presented in this book. Instead, the capacity and power of coalitions featured as preconditions (Section 11.12). Preconditions from the Paris Agreement and IPCCs 1.5 Degree Report focused on technologies and practices including finance for example, rather than the social, political or business actors and interests that might develop or apply these technologies and practices.

12.3   Other conclusions

IPCC assessments coupled with the searchable sample of possible futures provide rich datasets for exploring preconditions for effective global responses to climate change. As such, this book includes a number of other conclusions. For example, there have been two fundamental shifts in the global response system in the period 2010–2020. The first is the climate change signal has shifted from a mix of "climate science" and "risks" in the future, to a signal where there are "impacts" on people, property, and livelihoods (i.e., human and managed systems) (Section 10.2.1).

An important question addressed by this book regarded the influence climate change itself might have on our global response to climate change (Section 1.5). From the sample of possible futures, there were no effective global response scenarios based on concerns about the environment, for example regarding impacts and risks to "natural systems". As such, an important finding of this book is only impacts and risks to people, property, and livelihoods drive effective global responses (Section 5.2.3). At the same time, the extent to which response options are a risk (i.e., transition risk) or an opportunity also influences actors and their decision making (Section 10.2.2).

The second fundamental shift regards incentives and the availability of commercially viable response options (Section 10.3.5). In the past, many mitigation options were not commercially viable and as such the development of these options required a mix of social, political or business leadership, and adoption at scale required political will and government support. Thanks to this leadership, there is now a suite of commercially viable renewable energy, energy storage, and transportation technologies that mitigate GHG emissions. These technologies can drive the formation of coalitions i.e., "winning teams" involving social, political, and business actors. Importantly, competition between businesses and between governments can drive research and development as well as production and adoption of these technologies at scale. However, the United States government's low ambition under President Trump has dampened national competition for market share or dominance in these technology categories (Section 10.3.5). It should also be noted that even with these technologies, there still needs to be changes in land use and marine practices, in support of adaptation, mitigation and GHG removals (Section 7.4.2).

When stepping back and looking at the main features of the global response system (Section 10.2), several categories of actors were identified as being important when it comes to decision making and the structure of the global response system (Section 10.3). With regards to international cooperation, the distinction between "state" (i.e., central government) and "non-state" actors is very important, as state actors have primacy defining the national interests of "sovereign states" and can enhance or inhibit international cooperation between non-state actors (Section 3.4.2). At the domestic level (i.e., within states), three overarching categories of actors were identified consisting of social, political, and business actors. Each of these actors influences the global response to climate change differently, for example social actors drive social change and behaviour, political actors decide on policy, meanwhile business actors are focused on economic activities (Figure 10.6, Section 10.3). Importantly, each of these actors has interests influencing decision making and their contributions to climate change.

When it comes to decision making, this book borrows Abraham Lincoln's metaphor of "better angels" and identified "the competing angels of individual and collective action" based on the climate change signal response models developed earlier in this book (Section 10.2.2). The competing angels include among others, the angels of self-interest and wider concerns, unambitious and ambitious angels, cynical and enlightened angels, defensive and cooperative angels. These "angels" compete within us whenever we make decisions. Meanwhile, each angel is competing and cooperating with other angels when it comes to our collective response to climate change. The extent to which any "angel" influences the effectiveness of the global response to climate change depends on the actors and coalitions being influenced, the options available, and the capacity and power of actors and coalitions. From the sample of possible futures, the angels of cooperative, practice-focused, technological, and competitive attitudes can all have a role in effective global responses (Section 11.11). However, scenarios where defensive angels prevail fail to limit climate change and related impacts (Section 11.6.2).

From the literature, an important "reason for concern" in successive IPCC assessments has been the "distribution of impacts" (Section 2.2.3). From the sample of possible futures, the distribution of impacts was also important. For example, preconditions for effective international cooperation on climate change include the distribution of climate change and related impacts. If powerful countries such as China, France, Russia, the United Kingdom, and the United States of America each suffer a "crisis of impacts" then it is possible climate change could become the premier international issue, elevated from being an issue of "environment and development" to one of "security" for example in the United Nations Security Council (Section 11.8). Given the power and capacity of these countries, the expectation is that other states would participate in the global response and comply with whatever agreements these powerful countries reach. The distribution of climate change and related impacts within states and on different groups can also influence the extent to which climate change is seen as a "condition" that can be lived with versus a "problem" that needs to be solved in the national interest (Section 11.8).

Preconditions for actors to remove GHGs from the atmosphere start with mitigation at scale limiting the quantities of GHGs that need to be removed. Land management practices, policies, and governance processes need to be in place for carbon storage in terrestrial ecosystems meanwhile very cheap renewable energy is essential, simultaneously contributing to mitigation while making technology-based GHGs removals economically viable (Section 11.9). From the literature, the IPCC's 1.5 Degree Report stated removals would need to start before 2050 if global warming is to be limited to 1.5°C (Rogelj et al. 2018, Section 11.3).

Institutions (i.e., the "rules of the game") are an important part of the global response to climate change (Section 3.3). This includes formal institutions such as laws, regulations, and contractual arrangements as well as informal institutions for example notional "social contracts" (Section 10.3.1). As already noted, social permissions are a precondition for government actors and business actors to do what it takes to limit climate change and related impacts (Section 12.2). Many effective global response scenarios involve social change and behaviour as a factor (Sections 10.3.2 and 10.3.3), however dystopian business-driven scenarios are also possible where the global response is driven by special interests rather than public opinion (Section 8.2.5). Institutional capacity, particularly in less developed countries or jurisdictions, can be a limiting factor when it comes to adopting and applying technologies and practices including policies (Section 10.3.2). Other institutional factors that can influence the global response to climate change include the availability of finance to scale up technologies and practices as well as legal precedent limiting the availability of options harmful to climate and the global response (Section 10.2.2).

With regards to the conditions under which you, me, and everyone else (i.e., actors) would actually do what it takes to limit climate change and related impacts, there was no definitive finding. While it is expected we will adapt to climate change as impacts become prevalent (Section 11.6.2), it is unclear whether we will mitigate or remove GHGs at scale. It is even possible that apathy or hopelessness could drive the global response (Section 5.4.1). In such scenarios, climate change is redefined from being a "problem" that needs to be solved, to being a "condition" that we collectively decide to live with (Section 11.13). Already the climate change "problem" has been simplified by the Paris Agreement, from being a problem of stabilising atmospheric concentrations of GHGs at levels that allow ecosystems to adapt naturally, food production not to be threatened and economic development to proceed in a sustainable manner, to a problem of limiting global warming to well below 2°C from pre-industrial times (Section D.2.1.1). Meanwhile, there is evidence that with 1°C of global warming, we have already failed to fulfil the UNFCCC objective, given that risk of impacts to coral ecosystems is already high (Section 2.2.3).

The extent to which we (i.e., societies, political leaders, and businesses) collectively treat climate change impacts on natural, managed or human systems as "conditions" that can be lived with versus "problems" that need to be solved, will be an important part of the global response (Section 11.13). Complicating the situation is the fact that climate change has the characteristics of an "infinite game" where the aim of actors, including Small Island Developing States for example, is "continuing play" and "staying in the game" i.e., survival (Section 11.14). Climate change is not "finite game" that can be won or solved, but rather effective responses involve continuous effort limiting risks at individual and collective levels including existential risks (Section 10.3.5).

12.4   Practical applications, limitations, and future research

There is a mix of practical and academic applications for the conceptual models, substantive findings, qualitative methods, and scenarios compiled in this book.

This book includes conceptual models that we can use to describe the global response. The CCNIIC Model allows us to see where we fit in the global response system. Meanwhile, the climate change signal response models allow us to see how our decisions might contribute to the global response and provide a basis for understanding how other people, organisations or states with different interests might respond to climate change.

With regards to the substantive findings of this book, whenever we prepare a climate change strategy or response (e.g. an NDC), we should be clear about our assumptions when it comes to "preconditions for effective global responses" including the scale of climate change relative to technologies and practices required to limit climate change and its impacts; the reversibility of climate change and related impacts; the global response in terms of other actors and their contributions; contingencies in case of unexpected events; and the other things that may or may not affect the strategy. The extent to which our strategies rely on serendipity ahead of ambition is also something to consider. In addition to these things, our climate change response strategies should consider coalitions, and working with other actors with common interests. Coalitions can help address issues of capacity, for example to develop and implement technologies and practices, as well as power to influence institutions and the rules of the game. Importantly, this book provides models and scenarios that can help us think about climate change and our strategies. It is also possible for us to "stress test" our strategies by considering scenarios from the sample of possible futures described in Chapters 5–9.

The climate change signal response model provides an overarching frame for organising information and monitoring climate change and responses. Much of the information needed to fill the framework is already reported. As trends change, we will be able to distinguish which scenarios appear to be playing out, and as such, adjust response strategies.

Further research on each of the sub-sectors in this book would be valuable, including scenarios for social change and behaviour, political will and policy, business and economic activity, GHG removals and international cooperation. It is important to note here that the sample for the study presented in this book lacked respondents with business backgrounds, and more people could have been interviewed from developing countries. Any new data collected can be processed using the methods from this book and results from new studies can be integrated using the overarching models of the global response system.

Interestingly, the most comparable scenarios in the literature are from non-peer-reviewed sources such as blogs, opinion pieces and popular media. For anyone that is interested, there is scope for research of this literature for example using qualitative methods developed in this book to identify themes and analyse conditions that contribute to effective global responses. Important scenarios that are missing from the sample of possible futures include momentum scenarios where the Paris Agreement drives momentum and ratcheting of climate action (Rockström 2017), or a "Marshall Plan" for climate change scenarios where vast sums of money, expertise, and resources are thrown at the climate change problem, with the intention of "solving" it (Carlin 2020, Massara 2020). There is scope for more scenarios to be added to the sample of possible futures.

Before completing this chapter, three sources of bias need to be acknowledged, consisting of the framing in relation to the UNFCCC objective rather than human development or some other overarching frame (Section 10.2.2); the selection of respondents and the extent to which the sample of possible futures they provided reflect the full range of possible scenarios (Sections 4.4 and D.2.2.1); and lastly the fact that failure scenarios, while thematically analysed, did not feature in the analysis due to limited space (Section D.4). Framing biases the results presented in this book including the language and models used to describe results. The selection of respondents may bias the sample of possible futures collected, and it is possible that with a wider set of respondents, an even wider set of scenarios might have been identified. Meanwhile there could be further analysis of failure scenarios. These are all things I hope other researchers consider pursuing.

Related to the issue of framing, the biggest weakness of the signal response model is that it focused on decision making from a climate change perspective, rather than the perspective of the actors making the decisions. Most of us are not focused on climate change when we make most of our decisions, but rather we are focused on fulfilling needs, meeting expectations or aspirations with the means and options available to us. Climate change is just one of many signals or concerns we may or may not take into account when making decisions. In many cases, other factors will generate incentives and disincentives influencing our decisions. As such, an important area of research is to understand decision-making processes from an actor perspective, including the processes and considerations that go into economic, political, and social decisions.

With these limitations in mind, the qualitative models developed in this book could be developed into quantitative models. For example, the CCNIIC Model and climate change signal response models could be used as a basis for developing agent-based models. This could include modelling multiple agents, each with different situations and interests when it comes to climate change impacts, options available, response risks and opportunities, triggers and drivers, attitudes and inclinations, capacities to develop and apply technologies and practices, and levels of power to influence other actors and the rules of the game.

An important issue that deserves more research is the extent to which climate change might collectively be treated as a "condition" that can be lived with versus a "problem" that needs to be solved, drawing on concepts from Kingdon (1995). If the problem definition changes from stabilising atmospheric GHG concentrations at levels that allow ecosystems to adapt naturally, food production not being threatened and economic development being able to proceed in a sustainable manner, then this raises questions of climate justice including which of these things we are collectively willing to abandon and who will be impacted.

At the same time, it is important to remember climate change is not the only challenge that affects people, our planet or collective prosperity (Section 9.1). Climate change needs to be addressed simultaneously with other challenges facing individuals, communities, and states (Section 11.10). When first drafting these conclusions, the COVID-19 pandemic was sweeping the world, and like climate change, effective responses required a mix of collective action coupled with technologies and practices applied at scale in a timely manner. The CCNIIC Model, climate change signal response model, and competing angels of individual and collective action can each be adapted and used as frames to explore effective global pandemic responses by substituting COVID-19 hazards and signals into these models (Section 13.3).

The extent to which models developed in this book are generalisable and can be applied to other global response problems will be an interesting area of future research. It may be that the conceptual models developed in this book can be generalised and used to rapidly assess issues and options when new global hazards arise requiring coherent responses from all actors.

Lastly, with regards to "solving" climate change and other challenges, it is important to recognise that the world we live in today has changed dramatically from 10, 20, or 30 years ago, and the world in 10, 20, or 30 years time will be very different from the world today. The question is not whether the world will change, but rather, what do you want the world to be like in 10, 20, or 30 years' time? What are you going to do to make this a reality? Which technologies and practices, including policies, will you and others support? What coalitions will you form? And perhaps most importantly, which of the competing angels of individual and collective action will you listen to? Hopefully, it's your better angels and serendipity follows.

Notes

1 Wanting to "solve the climate change problem" was wilfully naïve but regardless, this aspiration triggered the research and drove important choices throughout, for example broadening the focus from effective "international cooperation" options to effective "global responses." Both are research worthy, but the focus on effective global responses allowed for a first principles driven search for climate change "solutions" in the form of effective response scenarios and related preconditions, rather than restricting research to a limited set of scenarios that happen to involve international cooperation.
2 For people living in unique and threatened systems, climate change may already be a "crisis of impacts."

Part IV

Epilogue

Part IV is an epilogue and consists of one chapter.

Chapter 13 makes the case for considering global response problems as a category, applies lessons from Chapters 10 and 11 to COVID-19, and then reflects on what constitutes success after failure.

DOI: 10.4324/9781003465911-16

13 Lessons for other global response problems

13.1   Introduction

The research for this book was conceived in the shadows of the Paris Agreement and was drafted in the midst of a global pandemic. COVID-19, like climate change, turns out to be a global response problem (Section 13.2). In Chapter 1 it was said solving climate change is like solving a riddle. The value of solving a riddle is not the answer itself, but the ability to take the answer and apply it to similar problems. In this epilogue I want to share lessons from climate change and look at how they might be applied to COVID-19 (Section 13.3) and other global response problems. This includes scanning the signals actors might respond to, and the types of responses different actors might have.

13.2   Global response problems as a category

As noted previously (Sections 1.4 and 3.3), the IPCC has characterised climate change as a collective action problem at the global scale (IPCC 2014a). This characterisation served as a point of departure for this book, but as research progressed, this characterisation appeared problematic. On the one hand, collectively, all of our actions contribute to the global response. This makes climate change a global response problem because all human actions and inaction, influence the achievement of the UNFCCC objective (Section 1.4). But on the other hand, the term "collective action" gives the impression that the global response to climate change depends on "collectively" deciding on a response, for example through some governance arrangement or process such as the UNFCCC and climate negotiations. As noted by Respondent 1, collective decision making through the UNFCCC might even inhibit progress (Section 5.5.1), although Appendix C suggests the UNFCCC creates a signal, leading to wider engagement in climate change and the search for solutions (see Figure C.3).

An alternative characterisation of climate change could be as a "global response problem". So how do we define a "global response problem" as a category?

A global response problem is: **An issue that requires coherent actions by actors around the world for the issue to be resolved with minimal negative impacts.**

DOI: 10.4324/9781003465911-17

"Global problems" can be either global in scale, due to interconnections around the world, or are localised but common to people in many countries (adapted from Tanter 2008). "Global response problems" are a subset of global problems, restricted to problems that are due to interconnections around the world, including social, economic and environmental connections. These problems require coherent responses by actors, and the actions of one actor can affect many other actors around the world.

Effective global responses depend on "coherence" i.e., the extent to which something is logical and consistent, forming a unified whole (adapted from the Oxford Dictionary). This involves a mix of social change and behaviour, political will and policy as well as business and economic activity (Section 10.3.4) including relevant technologies and practices.

The phrase "minimal negative impacts" is included in the definition because some global response problems will resolve themselves without intervention, but in ways that adversely affect people. The aim is to limit the problem, and its adverse impacts, to the extent possible.

The extent to which a global response problem is a threat is important (Section 2.2.3). Using climate change as an example, catastrophic climate change could have the characteristics of a severe but endurable catastrophe (Section 10.3.5); where a worst-case scenario could involve climate change reducing civilisation down to the level of "medieval warlords" or even the level of "Neolithic hunter gatherer" (Section 9.8).

Some problems may be resolved by running their course (e.g., a pandemic) but this is far from ideal in terms of preserving human life. Compared to climate change, COVID-19 is much less persistent as a "problem" that needs to be solved, taking just over three years to go from being declared a pandemic (11 March 2020, Ghebreyesus 2020) to having the public health emergency being declared over (5 May 2023, WHO 2023a), essentially becoming a "condition" that is lived with (Section 11.6.1) despite continued waves of infection and COVID-19 being fatal or "crushing" for some people. Meanwhile, climate change and related impacts will persist for a long time (Section 2.2.3), requiring adaptation actions even after atmospheric levels of greenhouse gases have been stabilised at safe levels.

The extent to which actions are coherent will have a bearing on the effectiveness of global responses to climate change, pandemics (WHO 2023b), and other global response problems. As witnessed during the COVID-19 pandemic, just because a threat is potentially catastrophic (Figure 2.1 in Section 2.2.3), this does not mean there will be social permissions or the political will to act, for example due to special interests or other reasons (Figure 10.10 in Section 10.3.4).

From Sections 5.4.2 and 9.3, there was a pandemic scenario, where the world came together to successfully address the pandemic and then took on other challenges together, such as climate change. It is fair to say the scenario was naïve and the COVID-19 response did not unfold as described pre-pandemic. This highlights the limitations of scenarios. That said, when the COVID-19 pandemic was getting underway, I was able to compare what I was seeing globally with the scenario, identify deviations, and quickly gauge that COVID-19 was not going to result in unified global action on climate change.

Perhaps more importantly, I was able to compare the global response to COVID-19 with the models I had developed for understanding the global response to climate change, quickly forming a picture of what was happening, how various actors might respond, and what this meant for the effectiveness of the global response to COVID-19.

13.3   COVID-19 and preconditions for effective global responses

In the section below, I compare COVID-19 with the definition of what constitutes a global response problem, then apply the signal response model (from Chapter 10), and the framework of preconditions for an effective global response (from Chapter 11).

COVID-19 as an issue was global in nature due to interconnections around the world (Szocik 2021, Ali et al. 2022). COVID-19 was "terminal", "severe but endurable", or "perceptible" to different people. For most jurisdictions COVID-19 was "severe but endurable" politically, economically, and socially (Szocik 2021, Ali et al. 2022). COVID-19 required coherent actions by many actors around the world for the problem to be resolved with minimal loss of life (Boyd and Wilson 2021). The extent to which COVID-19 was addressed with minimal loss of life varied greatly between regions and jurisdictions (Sachs et al. 2022).

From Figure 13.1, signals include epidemiology, impacts on health and the future risks to health. For some actors, COVID-19 represented a response opportunity, for example pharmaceutical companies, or the producers of hand sanitiser and medical masks (Peters et al. 2021). For many other businesses, COVID-19 responses represented a risk especially at the beginning of the pandemic when cities and countries were going into "lock-down" (Pitterle and Niermann 2021).

The ambition level of responses around the world varied greatly (Boin et al. 2020). For example, many political leaders and countries demonstrated emergency responses, with high ambition only once COVID-19 had arrived and it was clear death rates were rising exponentially. Other leaders and countries demonstrated enlightened responses, where awareness of the risk meant they acted with urgency immediately, for example Prime Minister Jacinda Ardern from New Zealand, locking down borders and the country when the first cases were detected, until COVID-19 was eliminated, then waited for vaccines to become available, immunising the population and only re-opening the country to travellers once herd immunity had been achieved (Gauld 2023). Characterising the enlightened response, New Zealand Prime Minister Jacinda Ardern stated, "We must go hard, and go early, and do everything we can to protect New Zealanders' health." (Ardern, March 14, 2020). The New Zealand government also put in measures to protect businesses from revenue losses while retaining employees (OECD 2022, Gauld 2023).

Other leaders attempted to balance economic costs and benefits at various stages during the pandemic (Sachs 2022). In the case of the UK and the USA, this approach proved to be inadequate due to the highly infectious nature of the virus and mortality rates associated with the virus. The US and UK lost many lives (Sachs 2022), especially prior to the availability of vaccines.

At various stages through the COVID-19 pandemic there were non-responses and arguably cynical responses where special interests were put ahead of the public good meanwhile evidence and professional advice were disregarded (Ringe and Rennó 2023).

With regards to response attitudes, there was cooperation and competition around the development, production and distribution of vaccines, including geopolitically driven competition on vaccines (Sachs 2022). The focus on vaccines as a solution also serves as an example of a technological attitude. Practice-focused attitudes included authorities promoting changes in behaviours for example mask wearing and physical distancing. Defensive attitudes included the closing of borders, and restricting travel and movements of people especially from outside a jurisdiction (Sachs 2022).

Important constraints on the options available and responses of countries to COVID-19 included the technologies and practices available to countries as well as institutions and their capacity. For some countries, pursuing the unprecedented rapid development of a vaccine was something they were able to support for example the United Kingdom or the United States of America, as they had the technologies available to do so as well as institutions capable of adapting processes to allow for rapid deployment of the vaccines after initial trials (Sachs 2022). Other countries had to rely on practices, the strength of their institutions as well as social change and behaviour (for a period), until vaccines were available, for example New Zealand and Vietnam.

In addition to political leadership, social change and behaviour had a large influence on responses to COVID-19 and the extent to which lives and livelihoods were maintained including social distancing and the use of masks (Sachs 2022). Likewise, the decisions made by businesses included both positive and negative contributions to the effectiveness of local and domestic responses to COVID-19, and ultimately the global response to COVID-19.

Table 13.1 complements Figure 13.1, to provide more detail on COVID-19 response ambition levels, triggers, decision criteria and actions. Importantly, Table 13.1 adapts Table 10.4 from Chapter 10, demonstrating the value of the framework for assessing not only climate change, but other global response problems such as COVID-19.

While Figure 13.1 and Table 13.1 apply well to COVID-19, they are not a perfect fit and require some adapting. For example, an element missing from the models in Figures 10.4 and 10.5 is the influence of social, political and economic filters, biases and amplifiers (Pers Comm Sylvia Frean 2023). In the case of COVID-19, social media had a large influence on people's responses, including misinformation (Sachs 2022). These filters influence decision making, the information that is considered legitimate and relevant, as well as perceived interests. Likewise, these filters and biases influence response attitudes and ultimately the options considered by people individually and collectively (Sachs 2022) including the misuse of hydroxychloroquine and ivermectin as COVID-19 treatments. Figure 13.1 is adapted to include filters, biases and amplifiers.

Table 11.6 (from Chapter 11) presented preconditions for effective global response to climate change. Table 13.2 adapts these preconditions, applying them to COVID-19.

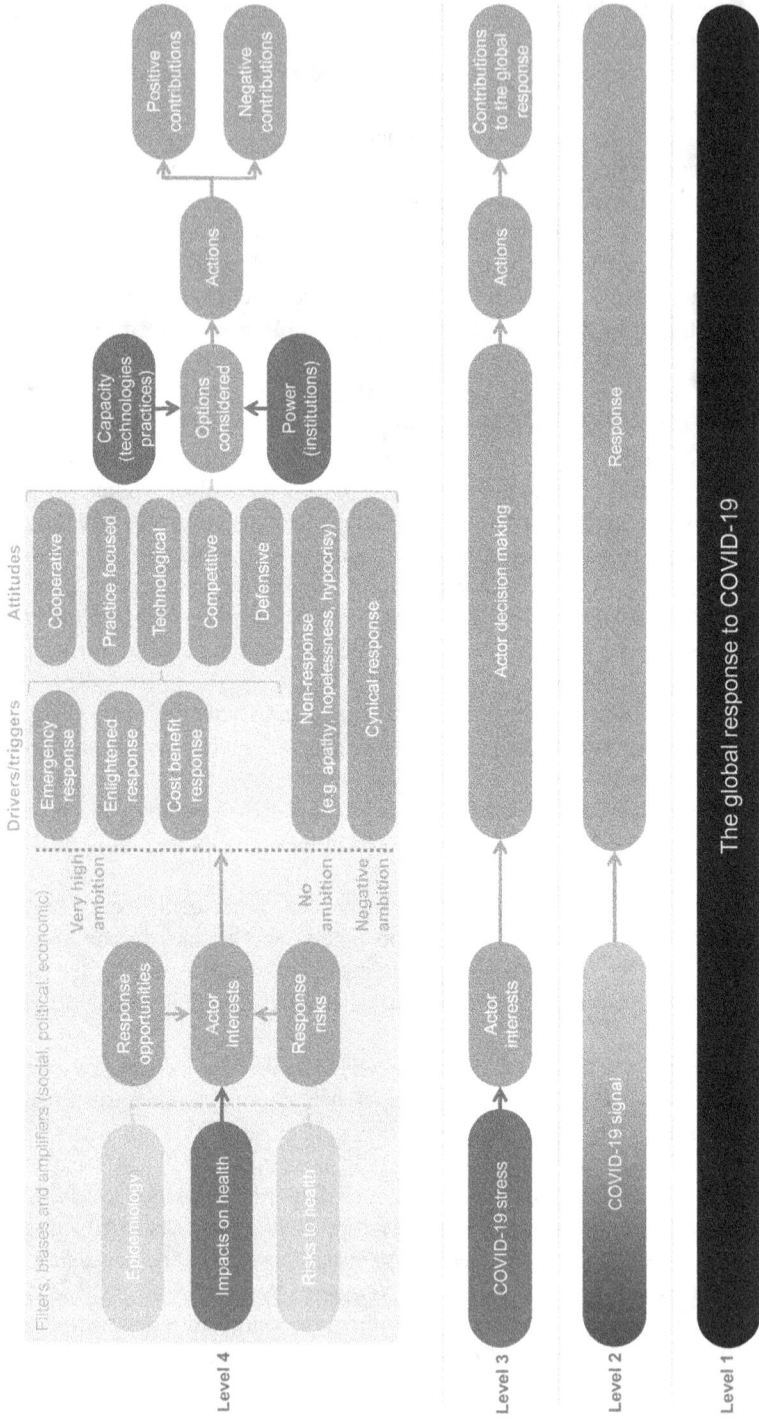

Figure 13.1 Signal response model applied to COVID-19.

Ambition	Trigger/driver (strategy)	Justification and framing	Decision criteria and types of options considered	Actions
Negative	Cynical response (non-cooperation)	**Special interests**: the extent to which special interests benefit.	Special interests and the extent to which these can be enhanced regardless of COVID-19.	Misinformation and a disregard for facts, e.g., inject or drink disinfectant
None	Non-response (free riding and no cost contributions)	**None**: Denial, apathy or hopelessness leading to inaction or incidental helpful contributions to the global response to COVID-19.	Behavioural and other options not considered. COVID-19 illness acted upon for reasons of self-preservation.	Incidental negative contributions to the global response, "ignorance is bliss" Incidental positive contributions to the global response
Low to high	Cost-benefit response (competition and cooperation)	**Cost-benefit analysis**: Decisions to act on COVID-19 are based on costs and benefits of the options being considered	Low ambition criteria $NPV\ (COVID\text{-}19) > 0$ Moderate ambition criteria $NPV\ (COVID\text{-}19) \sim 0$ High ambition criteria $NPV\ (COVID\text{-}19) < 0$	Act at costs up to social cost of COVID-19
High to very high	Enlightened response (cooperation)	**Evidence** of impacts, risks and need to limit COVID-19 and its impacts	The extent to which collective action, including behavioural changes, will fulfil the objective of limiting COVID-19 and its impacts.	Collective action on behavioural responses, policies and technologies including vaccines e.g., "go hard go early" (Ardern 2020)
Very high	Security response (cooperation and coercion)	**Security**: COVID-19 is a clear and present danger. It has been decided that COVID-19 needs to be acted upon and now it is a question of what can be achieved with the resources available	A decision is made on a defensive outcome or approach (e.g., lockdowns) taking into account willingness to pay for (and afford) the options available. Once an outcome or approach is decided, the most cost-effective options for achieving this within an acceptable level of risk are assessed and selected.	Containment and defence
			A decision is made on an outcome or approach including containment (e.g., lock-downs) and a fix (e.g. vaccines) taking into account willingness to pay (and afford) the options available. Once an outcome or approach is decided, the most cost-effective options for achieving this within an acceptable level of risk are assessed and selected.	Containment and fix

Source: Author applying Table 10.4 to COVID-19.

Table 13.2 Preconditions for effective responses to COVID-19

Objectives	Success criteria		
COVID-19 objective	"save lives and minimize impact" WHO Director, 11 March 2020. Pandemic Declaration. (Ghebreyesus 2020)		

Preconditions for effective global responses

Tiers	*COVID-19 related*	*Response related*	*Other scenario related*
Tier 1	Scale of COVID-19 is limited.	The global response is timely and at scale.	Other social, economic or environmental changes don't negatively influence COVID-19 or the global response to COVID-19 at scale.
	COVID-19 and related impacts are reversible or can be quickly stabilised.	Contingencies are available for addressing extreme COVID-19 or other scenarios.	The scale of other negative social, economic or environmental changes can be addressed by technologies, practices, institutions or policies.
Tier 2	The scale of COVID-19 can be addressed by technologies and practices (including behavioural changes).	Social, political and business leadership generating response options and opportunities. For non-defensive ambition-reliant responses, preconditions include social permissions for COVID-19 action, as well as domestic and international coalitions with sufficient capacity and power to ensure responses are timely and at scale.	Other social, economic or environmental changes limiting COVID-19 or helping the global response to COVID-19.
	Not passing thresholds such as negative tipping points or cascade effects.	Research and development of technologies, practices and policies, so these things can be deployed quickly and at scale if needed.	

Factors	*COVID-19 related*	*Response related*	*Other scenario related*
Other factors	Virological, ecological and environmental processes and dynamics.	Social change and behaviour, political will and policy, business and economic activity; distribution of impacts and risks; response triggers, drivers and attitudes of actors; domestic, national and international actors; prices, taxes and subsidies, markets, support for research, development of technologies and practices;	Ecological, environmental and cosmological processes and dynamics; social, political and business processes and

The process of adapting the Table and preconditions to COVID-19, started by identifying an authoritative objective against which the effectiveness of the response could be assessed. For COVID-19 the words of the World Health Organisation Director General were used when he declared the pandemic and stated the need to "save lives and minimize impact" (Ghebreyesus 2020).

In Table 13.2, COVID-19-related preconditions included the Tier 1 precondition that the scale of COVID-19 should be limited, but this was not the case due to the "r" factor resulting in the exponent spread of the disease (Sachs 2022). However, the Tier 2 precondition that "The scale of COVID-19 can be addressed by technologies and practices (including behavioural changes)" was useful, because at the outset of the pandemic, technologies such as vaccines were lacking, but practices such as lockdowns and social distancing were effective at limiting the transmission of the disease (Sachs 2022). Likewise, the response-related precondition that contingencies are available was not true at the outset of the pandemic, but the Tier 2 precondition that "Research and development of technologies, practices and policies, so these things can be deployed quickly and at scale if needed" did eventuate.

Other response-related preconditions included the extent to which responses were timely and at scale. This varied across jurisdictions, sometimes due to the Tier 2 precondition that there needs to be social permissions for such actions (Sachs 2022), and coalitions with sufficient power and capacity form to deliver the required actions.

With regards to other scenarios, there were other natural disasters in the midst of the COVID-19 pandemic, for example wildfires in Australia, earthquakes in Croatia, floods in China, a Cyclone Harold affected Vanuatu, Fiji, and Tonga (Sohrabizadeh et al. 2021). Research identified the need for risk reduction strategies and policies, given that once the pandemic is in place, other disasters may also arise (Ishiwatari et al. 2020, Sohrabizadeh et al. 2021).

Taken together, the models and frameworks developed for climate change provided a template readily applied to COVID-19, yielding insights. These models and frameworks may also be useful for other new global response problems, but this requires further testing.

13.4   Conclusions

Climate change and COVID-19 can both be defined as global response problems that require coherent actions by actors around the world to be resolved with minimal negative impacts. To effectively address global response problems, coherence is needed, a principle the WHO highlights (WHO 2023b).

In this epilogue we have demonstrated that models developed for understanding climate change, signals, and responses, and preconditions for effective global responses, can be applied to COVID-19. Still much more research is needed to validate these findings and assess the extent to which these models might be useful for understanding other new global response problems and what we can do about them. Hopefully we can do this before the next global response problem arises.

References

Ali, M.J., Bhuiyan, A.B., Zulkifli, N. and Hassan, M.K., 2022. The COVID-19 pandemic: conceptual framework for the global economic impacts and recovery. In Hassan, M.K., Muneeza, A. and Sarea, A.M. (eds.), *Towards a Post-Covid Global Financial System*, Emerald Publishing Limited, Leeds, pp. 225–242.

Aldy J.E., and Stavins, R.N., 2010. Introduction. In: Post-Kyoto International Climate Policy: Implementing Architectures for Agreement. Aldy, J. E. and Stavins, R. N. (eds.), Cambridge University Press, Cambridge, UK, pp. 1 – 28.

Allen, M.R., O.P. Dube, W. Solecki, F. Aragón-Durand, W. Cramer, S. Humphreys, M. Kainuma, J. Kala, N. Mahowald, Y. Mulugetta, R. Perez, M. Wairiu, and K. Zickfeld, 2018. Framing and context. In: *Global Warming of 1.5°C. An IPCC Special Report on the Impacts of Global Warming of 1.5°C Above Pre-industrial Levels and Related Global Greenhouse Gas Emission Pathways, in the Context of Strengthening the Global Response to the Threat of Climate Change, Sustainable Development, and Efforts to Eradicate Poverty* [online] [Masson-Delmotte, V., P. Zhai, H.-O. Pörtner, D. Roberts, J. Skea, P.R. Shukla, A. Pirani, W. Moufouma-Okia, C. Péan, R. Pidcock, S. Connors, J.B.R. Matthews, Y. Chen, X. Zhou, M.I. Gomis, E. Lonnoy, T. Maycock, M. Tignor, and T. Waterfield (eds.)]. In Press. Available from: https://www.ipcc.ch/site/assets/uploads/sites/2/2019/05/SR15_Chapter1_Low_Res.pdf [Accessed 6 December 2019].

Anand, S., Gupta, A. and Tyagi, S.K., 2015. Solar cooling systems for climate change mitigation: A review. *Renewable and Sustainable Energy Reviews* [online], *41*, pp. 143–161. Available from: https://www.sciencedirect.com/science/article/abs/pii/S1364032114007187 [Accessed 3 May 2020].

Aplin, P., 2006. On scales and dynamics in observing the environment. *International Journal of Remote Sensing* [online], *27*(11), pp. 2123–2140. Available from: https://www.tandfonline.com/doi/full/10.1080/01431160500396477 [Accessed 1 March 2020].

Archer, C., 2015. *International Organizations* [Kindle Reader, Ipad Mini]. Routledge, Abingdon Oxon.

Ardern, J. March 14, 2020. Major steps taken to protect New Zealanders from COVID-19. Prime Minister of New Zealand [online]. Available from: https://www.beehive.govt.nz/release/major-steps-taken-protect-new-zealanders-covid-19 [Accessed 13 December 2023].

Attenborough, D., 2018. *Transcript of the Speech by Sir David Attenborough* [online]. COP24, Katowice, Poland. 3rd December 2018. United Nations Framework Convention on Climate Change. Available from: https://unfccc.int/sites/default/files/resource/The%20People%27s%20Address%202.11.18_FINAL.pdf [Accessed 30 October 2019].

Barrett, S., 2005. *Global Climate Change and the Kyoto Protocol: Environment and State-craft: The Strategy of Environmental Treaty-Making* [online]. Oxford University Press.

Available from: http://www.oxfordscholarship.com/view/10.1093/0199286094.001.00 01/acprof-9780199286096-chapter-15. [Accessed 1 October 2018].

Basadur, M., Pringle, P., Speranzini, G. and Bacot, M., 2000. Collaborative problem solving through creativity in problem definition: Expanding the pie. *Creativity and Innovation Management* [online], *9*(1), pp. 54–76. Available from: https://onlinelibrary.wiley.com/doi/epdf/10.1111/1467-8691.00157 [Accessed 8 December 2019].

Beck, U., 2006. Living in the world risk society: A Hobhouse Memorial Public Lecture given on Wednesday 15 February 2006 at the London School of Economics. *Economy and Society* [online], *35*(3), pp. 329–345. Available from: https://www.tandfonline.com/doi/abs/10.1080/03085140600844902 [Accessed 8 August 2020].

Bethke, I., Outten, S., Otterå, O.H., Hawkins, E., Wagner, S., Sigl, M. and Thorne, P., 2017. Potential volcanic impacts on future climate variability. *Nature Climate Change* [online], *7*(11), pp. 799–805. Available from: https://doi.org/10.1038/nclimate3394 [Accessed 21 December 2020].

Bindoff, N.L., P.A. Stott, K.M. AchutaRao, M.R. Allen, N. Gillett, D. Gutzler, K. Hansingo, G. Hegerl, Y. Hu, S. Jain, I.I. Mokhov, J. Overland, J. Perlwitz, R. Sebbari and X. Zhang, 2013. Detection and attribution of climate change: From global to regional. In: *Climate Change 2013: The Physical Science Basis. Contribution of Working Group I to the Fifth Assessment Report of the Intergovernmental Panel on Climate Change* [online] [Stocker, T.F., D. Qin, G.-K. Plattner, M. Tignor, S.K. Allen, J. Boschung, A. Nauels, Y. Xia, V. Bex, and P.M. Midgley (eds.)]. Cambridge University Press, Cambridge and New York. Available from: http://www.ipcc.ch/pdf/assessment-report/ar5/wg1/WG1AR5_Chapter10_FINAL.pdf [Accessed 8 August 2018].

Bishop, P., Hines, A. and Collins, T., 2007. The current state of scenario development: an overview of techniques. *Foresight* [online], *9*(1), pp. 5–25. Available from: https://www.emerald.com/insight/content/doi/10.1108/14636680710727516/full/pdf?title=the-current-state-of-scenario-development-an-overview-of-techniques [Accessed 7 December 2019].

Blok, K., van Exter, P. and Terlouw, W. 2018. *Energy Transition Within 1.5°C A Disruptive Approach to 100% Decarbonisation of the Global Energy System by 2050* [online]. ECOFYS. Available from: https://www.ecofys.com/files/files/ecofys-a-navigant-company-2018-energy-transition-within-1.5c.pdf [Accessed 21 May 2018].

Boardman, A.E., Greenberg, D.H., Vining, A.R. and Weimer, D.L. 2006. *Cost-Benefit Analysis: Concepts and Practice* (3rd edition). Pearson/Prentice Hall, Upper Saddle River, NJ, 560 pp.

Bodansky, D. and Diringer, E., 2010. *The Evolution of Multilateral Regimes: Implications for Climate Change*, p. 23 [online]. Arlington: Pew Center on Global Climate Change. Available from: http://www.environmentportal.in/files/evolution-multilateral-regimes-implications-climate-change.pdf [Accessed 12 February 2017].

Bodansky, D., December 2012. The Durban platform: Issues and options for a 2015 agreement. [online] *Center for Climate and Energy Solutions*. Available from: https://www.c2es.org/site/assets/uploads/2012/11/durban-platform-issues-and-options.pdf [Accessed 7 December 2019].

BOE, 2015. The impact of climate change on the UK insurance sector: A Climate Change Adaptation Report by the Prudential Regulation Authority. Bank of England Prudential Regulatory Authority. Available from: https://www.bankofengland.co.uk/-/media/boe/files/prudential-regulation/publication/impact-of-climate-change-on-the-uk-insurance-sector.pdf [Accessed 21 May 2018].

Böhmelt, T., Koubi, V. and Bernauer, T., 2014. Civil society participation in global governance: Insights from climate politics. *European Journal of Political Research* [online], *53*(1), pp. 18–36. Available from: https://onlinelibrary.wiley.com/doi/abs/10.1111/1475-6765.12016 [Accessed 14 September 2018].

Boin, A., Lodge, M. and Luesink, M., 2020. Learning from the COVID-19 crisis: An initial analysis of national responses. *Policy Design and Practice*, *3*(3), pp. 189–204.

Borio, C., Drehmann, M. and Tsatsaronis, K., 2014. Stress-testing macro stress testing: Does it live up to expectations? *Journal of Financial Stability* [online], *12*, pp. 3–15. Available from: https://www.sciencedirect.com/science/article/pii/S1572308913000454 [Accessed 21 May 2018].

Bostrom, N. and Cirkovic, M.M., eds., 2008. *Global Catastrophic Risks*. [Kindle Reader, Ipad Mini] Oxford University Press.

Bostrom, N., 2013. Existential risk prevention as global priority. *Global Policy*, *4*(1), pp. 15–31.

Boyd, M. and Wilson, N., 2021. Anticipatory governance for preventing and mitigating catastrophic and existential risks. *Policy Quarterly*, *17*(4), pp. 20–31.

Braun, V. and Clarke, V., 2006. Using thematic analysis in psychology. *Qualitative Research in Psychology* [online], *3*(2), pp. 77–101. Available from: https://www.tandfonline.com/doi/pdf/10.1191/1478088706qp063oa?needAccess=true [Accessed 8 December 2019].

Brown, C. and Wilby, R.L., 2012. An alternate approach to assessing climate risks. *Eos, Transactions American Geophysical Union* [online], *93*(41), pp. 401–402. Available from: http://onlinelibrary.wiley.com/doi/10.1029/2012EO410001/full [Accessed 21 May 2018].

Burke, M., Hsiang, S.M. and Miguel, E., 2015. Global non-linear effect of temperature on economic production. *Nature* [online], *527*(7577), pp. 235–239. Available from: https://www.nature.com/articles/nature15725 [Accessed 8 August 2020].

Cambridge Dictionary. *Cambridge Dictionary* [online]. Available from: https://dictionary.cambridge.org/ [Accessed 11 December 2019].

Cameron, E. and DeAngelis, K., 2012. What is ambition in the context of climate change? *Blog World Resources Institute* [online]. Available from: https://www.wri.org/blog/2012/11/what-ambition-context-climate-change [Accessed 1 November 2019].

Carlin, D., April 8, 2020. America needs a green marshall plan to fight COVID-19 and climate change. [online] *Forbes*. Available from: https://www.forbes.com/sites/davidcarlin/2020/04/08/america-needs-a-green-marshall-plan-to-fight-covid-19-and-climate-change/#2d764c037943 [Accessed 4 May 2020].

Carney, M., September 29, 2015. *Breaking the Tragedy of the Horizon–Climate Change and Financial Stability* [online]. Speech given at Lloyd's of London. Available from: http://www.futurebusinesscouncil.com/wp-content/uploads/2016/12/speech844.pdf [Accessed 21 May 2018].

Carse, J., 1986. *Finite and Infinite Games*. [Kindle reader, Ipad Mini]. Free Press, New York.

CDP, 2014. *Global Governance and Global Rules for Development in the Post-2015 Era.* [online] United Nations, New York. Available from: https://www.un.org/en/development/desa/policy/cdp/cdp_publications/2014cdppolicynote.pdf [Accessed 9 December 2019].

Clarke L., K. Jiang, K. Akimoto, M. Babiker, G. Blanford, K. Fisher-Vanden, J.-C. Hourcade, V. Krey, E. Kriegler, A. Löschel, D. McCollum, S. Paltsev, S. Rose, P. R. Shukla, M. Tavoni, B. C. C. van der Zwaan, and D.P. van Vuuren, 2014. Assessing transformation pathways. In: *Climate Change 2014: Mitigation of Climate Change. Contribution of Working Group III to the Fifth Assessment Report of the Intergovernmental Panel on*

Climate Change [online] [Edenhofer, O., R. Pichs-Madruga, Y. Sokona, E. Farahani, S. Kadner, K. Seyboth, A. Adler, I. Baum, S. Brunner, P. Eickemeier, B. Kriemann, J. Savolainen, S. Schlömer, C. von Stechow, T. Zwickel, and J.C. Minx (eds.)]. Cambridge University Press, Cambridge and New York. Available from: https://www.ipcc.ch/site/ assets/uploads/2018/02/ipcc_wg3_ar5_chapter6.pdf [Accessed 27 October 2019].

Collins, M., R. Knutti, J. Arblaster, J.-L. Dufresne, T. Fichefet, P. Friedlingstein, X. Gao, W.J. Gutowski, T. Johns, G. Krinner, M. Shongwe, C. Tebaldi, A.J. Weaver and M. Wehner, 2013. Long-term climate change: Projections, commitments and irreversibility. In: *Climate Change 2013: The Physical Science Basis. Contribution of Working Group I to the Fifth Assessment Report of the Intergovernmental Panel on Climate Change* [Stocker, T.F., D. Qin, G.-K. Plattner, M. Tignor, S.K. Allen, J. Boschung, A. Nauels, Y. Xia, V. Bex, and P.M. Midgley (eds.)]. Cambridge University Press, Cambridge and New York. Available from: http://www.ipcc.ch/pdf/assessment-report/ar5/wg1/WG1AR5_ Chapter12_FINAL.pdf [Accessed 2 July 2018].

Crist, E., 2007. Beyond the climate crisis: A critique of climate change discourse. *Telos* [online], *141*(Winter), pp. 29–55. Available from: http://biophilosophy.ca/ Teaching/2070papers/crist.pdf [Accessed 6 August 2020].

CSF and CSC 2012. *Foresight a Glossary.* [online] Centre for Strategic Studies and Civil Service College, Singapore. Available from: https://www.csf.gov.sg/files/media-centre/ publications/csf-csc_foresight--a-glossary.pdf [Accessed 9 March 2019].

Cubasch, U., D. Wuebbles, D. Chen, M.C. Facchini, D. Frame, N. Mahowald, and J.-G. Winther, 2013. Introduction. In: *Climate Change 2013: The Physical Science Basis. Contribution of Working Group I to the Fifth Assessment Report of the Intergovernmental Panel on Climate Change* [online] [Stocker, T.F., D. Qin, G.-K. Plattner, M. Tignor, S.K. Allen, J. Boschung, A. Nauels, Y. Xia, V. Bex, and P.M. Midgley (eds.)]. Cambridge University Press, Cambridge and New York. Available from: http://www.ipcc.ch/pdf/ assessment-report/ar5/wg1/WG1AR5_Chapter01_FINAL.pdf [Accessed 2 July 2018].

Davidson, K., Briggs, J., Nolan, E., Bush, J., Håkansson, I. and Moloney, S., 2020. The making of a climate emergency response: Examining the attributes of climate emergency plans. *Urban Climate* [online], *33*, p. 100666. Available from: https://www-sciencedirect-com. libproxy.ucl.ac.uk/science/article/pii/S2212095520300171 [accessed 7 August 2020].

Davies, R. April 5, 2012. Criteria for assessing the evaluability of Theories of Change. *Rick on the Road* [online]. Available from: http://mandenews.blogspot.co.uk/2012/04/criteria-for-assessing-evaluablity-of.html [Accessed 8 December 2019].

de Coninck, H., A. Revi, M. Babiker, P. Bertoldi, M. Buckeridge, A. Cartwright, W. Dong, J. Ford, S. Fuss, J.-C. Hourcade, D. Ley, R. Mechler, P. Newman, A. Revokatova, S. Schultz, L. Steg, and T. Sugiyama, 2018. Strengthening and implementing the global response. In: *Global Warming of 1.5°C. An IPCC Special Report on the Impacts of Global Warming of 1.5°C Above Pre-industrial Levels and Related Global Greenhouse Gas Emission Pathways, in the Context of Strengthening the Global Response to the Threat of Climate Change, Sustainable Development, and Efforts to Eradicate Poverty* [online] [MassonDelmotte, V., P. Zhai, H.-O. Pörtner, D. Roberts, J. Skea, P.R. Shukla, A. Pirani, W. Moufouma-Okia, C. Péan, R. Pidcock, S. Connors, J.B.R. Matthews, Y. Chen, X. Zhou, M.I. Gomis, E. Lonnoy, T. Maycock, M. Tignor, and T. Waterfield (eds.)]. In Press. Available from: https://www.ipcc.ch/site/assets/uploads/sites/2/2019/05/SR15_ Chapter4_Low_Res.pdf [Accessed 11 December 2019].

DeFries, R.S., Edenhofer, O., Halliday, A.N., Heal, G.M., Lenton, T., Puma, M., Rising, J., Rockström, J., Ruane, A., Schellnhuber, H.J. and Stainforth, D., 2019. *The Missing*

Economic Risks in Assessments of Climate Change Impacts [online]. The Grantham Research Institute on Climate Change and the Environment, The Earth Institute, and The Potsdam Institute for Climate Impact Research. Available from: https://academiccommons. columbia.edu/doi/10.7916/d8-0brx-2380/download [Accessed 4 May 2020].

Dietz, S., Hepburn, C.J. and Stern, N., December 2007. Economics, ethics and climate change. *Ethics and Climate Change* [online]. Available from: https://papers.ssrn.com/ sol3/papers.cfm?abstract_id=1090572 [Accessed 8 August 2020].

Dryzek, J.S., Norgaard, R.B. and Schlosberg, D., 2011. Climate change and society: Approaches and responses. *The Oxford Handbook of Climate Change and Society* [online], Oxford University Press, Oxford pp. 3–17. Available from: https://doi.org/10.1093/ oxfordhb/9780199566600.003.0001 [Accessed 9 August 2020].

Duffield, J., 2007. What are international institutions? *International Studies Review* [online], *9*(1), pp. 1–22. Available from: https://onlinelibrary-wiley-com.libproxy.ucl.ac.uk/doi/ epdf/10.1111/j.1468-2486.2007.00643.x [Accessed 18 September 2018].

Duverger, M., 1972. *The Study of Politics*. Springer Netherlands, Dordrecht.

Engerman, S.L., 1986. Slavery and emancipation in comparative perspective: A look at some recent debates. *The Journal of Economic History* [online], *46*(2), pp. 317–339. Available from: https://www.cambridge.org/core/journals/journal-of-economic-history/ article/slavery-and-emancipation-in-comparative-perspective-a-look-at-some-recent-debates/F3536659F6ED2F54EF7435674AFD2DC6 [Accessed 3 May 2020].

Enserink, B., Koppenjan, J.F.M. and Mayer, I.S., 2013. A policy sciences view on policy analysis. In W.A.H. Thissen and W.E. Walker (eds.), *Public Policy Analysis: New Developments*. Springer, New York, pp. 11–40.

Evans, R.G., 2006. The blind men, the elephant and the CT Scanner. *Healthcare Policy* [online], *1*(3), p. 12. Available from: https://www.ncbi.nlm.nih.gov/pmc/articles/ PMC2585342/ [Accessed 1 March 2020].

Fajardy, M., Köberle A., MacDowell N. and Fantuzzi, A., 2019. BECCS deployment: A reality check. *Grantham Institute Briefing Paper* [online], *28*. Available from: https:// www.imperial.ac.uk/media/imperial-college/grantham-institute/public/publications/ briefing-papers/BECCS-deployment---a-reality-check.pdf [Accessed 3 May 2020].

Farber, D.A., 2011. Issues of scale in climate governance. In: Dryzek, J.S., R.B. Norgaard and D. Schlosberg (eds.) *The Oxford Handbook of Climate Change and Society* [online], Oxford University Press, Oxford, pp. 479–498. Available from: https://doi.org/10.1093/ oxfordhb/9780199566600.003.0032 [Accessed 9 August 2020].

Field, C.B., V.R. Barros, K.J. Mach, M.D. Mastrandrea, M. van Aalst, W.N. Adger, D.J. Arent, J. Barnett, R. Betts, T.E. Bilir, J. Birkmann, J. Carmin, D.D. Chadee, A.J. Challinor, M. Chatterjee, W. Cramer, D.J. Davidson, Y.O. Estrada, J.-P. Gattuso, Y. Hijioka, O. Hoegh-Guldberg, H.Q. Huang, G.E. Insarov, R.N. Jones, R.S. Kovats, P. Romero-Lankao, J.N. Larsen, I.J. Losada, J.A. Marengo, R.F. McLean, L.O. Mearns, R. Mechler, J.F. Morton, I. Niang, T. Oki, J.M. Olwoch, M. Opondo, E.S. Poloczanska, H.-O. Pörtner, M.H. Redsteer, A. Reisinger, A. Revi, D.N. Schmidt, M.R. Shaw, W. Solecki, D.A. Stone, J.M.R. Stone, K.M. Strzepek, A.G. Suarez, P. Tschakert, R. Valentini, S. Vicuña, A. Villamizar, K.E. Vincent, R. Warren, L.L. White, T.J. Wilbanks, P.P. Wong, and G.W. Yohe, 2014. Technical summary. In: *Climate Change 2014: Impacts, Adaptation, and Vulnerability. Part A: Global and Sectoral Aspects. Contribution of Working Group II to the Fifth Assessment Report of the Intergovernmental Panel on Climate Change* [online] [Field, C.B., V.R. Barros, D.J. Dokken, K.J. Mach, M.D. Mastrandrea, T.E. Bilir, M. Chatterjee, K.L. Ebi, Y.O. Estrada, R.C. Genova, B. Girma, E.S. Kissel, A.N. Levy, S. MacCracken, P.R.

Mastrandrea, and L.L. White (eds.)]. Cambridge University Press, Cambridge and New York, pp. 35–94. Available from: http://www.ipcc.ch/pdf/assessment-report/ar5/wg2/WGIIAR5-TS_FINAL.pdf [Accessed 2 July 2018].

Figueres, C., Schellnhuber, H.J., Whiteman, G., Rockström, J., Hobley, A. and Rahmstorf, S., 2017. Three years to safeguard our climate. *Nature News* [online], *546*(7660), p. 593. Available from: https://www.nature.com/articles/546593a [Accessed 11 July 2018].

Forster, P., T. Storelvmo, K. Armour, W. Collins, J.-L. Dufresne, D. Frame, D.J. Lunt, T. Mauritsen, M.D. Palmer, M. Watanabe, M. Wild, and H. Zhang, 2021. The earth's energy budget, climate feedbacks, and climate sensitivity. In: *Climate Change 2021: The Physical Science Basis. Contribution of Working Group I to the Sixth Assessment Report of the Intergovernmental Panel on Climate Change* [online] [Masson-Delmotte, V., P. Zhai, A. Pirani, S.L. Connors, C. Péan, S. Berger, N. Caud, Y. Chen, L. Goldfarb, M.I. Gomis, M. Huang, K. Leitzell, E. Lonnoy, J.B.R. Matthews, T.K. Maycock, T. Waterfield, O. Yelekçi, R. Yu, and B. Zhou (eds.)]. Cambridge University Press, Cambridge and New York, pp. 923–1054. Available from: Chapter 7: The Earth's Energy Budget, Climate Feedbacks and Climate Sensitivity (ipcc.ch) [Accessed 9 December 2023].

Friedman, L., February 21, 2019. What is the green new deal? A climate proposal, explained. *The New York Times* [online]. Available from: https://www.nytimes.com/2019/02/21/climate/green-new-deal-questions-answers.html [Accessed 4 May 2020].

Füser, K., Hein, B. and Somma, M., 2012. Inverse stresstests: Neue Perspektiven auf ein relevantes Thema (2). *Bank-Zeitschrift für Bankpolitik und Bankpraxis* [online], *5*, p. 45. Available from: http://www.die-bank.de/news/neue-perspektiven-auf-ein-relevantes-thema-2-4710/ [Accessed 21 May 2018].

Gauld, R., 2023. A review of public policies on Covid-19: The New Zealand experience. *Public Administration and Policy*, *26*(1), pp. 10–20.

Ghebreyesus, T. A., March 11, 2020. WHO Director-General's opening remarks at the media briefing on COVID-19 [online]. Available from: WHO Director-General's opening remarks at the media briefing on COVID-19 - 11 March 2020 [Accessed 13 December 2023].

Go, K. and Carroll, J.M., 2004. The blind men and the elephant: Views of scenario-based system design. *Interactions* [online], *11*(6), pp. 44–53. Available from: https://dl.acm.org/doi/10.1145/1029036.1029037 [Accessed 01 March 2020].

Goodwin, J., Gillenwater, M., Romano D., and Radunsky, K., 2019. Introduction to national GHG inventories. In: Calvo Buendia, E., Tanabe, K., Kranjc, A., Baasansuren, J., Fukuda, M., Ngarize S., Osako, A., Pyrozhenko, Y., Shermanau, P. and Federici, S. (eds.), *Refinement to the 2006 IPCC Guidelines for National Greenhouse Gas Inventories Volume 1: General Guidance and Reporting* [online]. Switzerland: Intergovernmental Panel on Climate Change. Available from: https://www.ipcc-nggip.iges.or.jp/public/2019rf/pdf/1_Volume1/19R_V1_Ch01_Introduction.pdf [Accessed 3 May 2020].

GOS, 2016. *Foresight for Cities: A Resource for Policy-Makers* [online]. London: Government Office for Science, Future of Cities. Available from: https://assets.publishing.service.gov.uk/government/uploads/system/uploads/attachment_data/file/516443/gs-16-5-future-cities-foresight-for-cities.pdf [Accessed 8 August 2020].

Graichen, J., Healy, S., Siemons, A., Höhne, N., Kuramochi, T., Gonzales Zuñiga, S. and Wachsmuth, J., 2016. *Climate Initiatives, National Contributions, and the Paris Agreement* [online]. Berlin: Öko-Institute. Available from: https://newclimate.org/wp-content/uploads/2017/09/161031_technical-annex_allsectors_clean-version.pdf [Accessed 08 July 2018].

Grubb, M., Hourcade, J.C. and Neuhoff, K., 2014. *Planetary Economics: Energy, Climate Change and the Three Domains of Sustainable Development*. [Kindle Reader, Ipad Mini]. Routledge Earthscan, Abingdon.

Harries, E., Hodgson, L. and Noble, J., 2014. *Creating Your Theory of Change: NPC's Practical Guide* [online]. New Philanthropy Capital. Available from: https://www.thinknpc.org/wp-content/uploads/2018/07/Creating-your-theory-of-change1.pdf [Accessed 8 December 2019].

Hawken, P. ed., 2017. *Drawdown: The Most Comprehensive Plan Ever Proposed to Reverse Global Warming*. [Kindle Reader, Ipad Mini] Penguin, New York.

Hekkert, M.P., Suurs, R.A., Negro, S.O., Kuhlmann, S. and Smits, R.E., 2007. Functions of innovation systems: A new approach for analysing technological change. *Technological Forecasting and Social Change* [online], *74*(4), pp. 413–432. Available from: https://doi.org/10.1016/j.techfore.2006.03.002 [Accessed 25 December 2020].

Hermwille, L., 2018. Making initiatives resonate: How can non-state initiatives advance national contributions under the UNFCCC? *International Environmental Agreements: Politics, Law and Economics* [online], *18*(3), pp. 447–466. Available from: https://link.springer.com/article/10.1007/s10784-018-9398-9 [Accessed 14 September 2018].

Hochrainer-Stigler, S., Linnerooth-Bayer, J. and Lorant, A., 2017. The European Union Solidarity Fund: An assessment of its recent reforms. *Mitigation and Adaptation Strategies for Global Change* [online], *22*(4), pp. 547–563. Available from: https://link.springer.com/article/10.1007/s11027-015-9687-3 [Accessed 21 May 2018].

Hodgson, G. M., March 2006. What are institutions? *Journal of Economic Issues* [online], *XL*(1), pp. 1–25. Available from: http://www.geoffrey-hodgson.info/user/bin/whatareinstitutions.pdf [Accessed 1 February 2017].

Hofmann, B., 2019. Policy responses to new ocean threats: Arctic warming, maritime industries, and international environmental regulation. In: Harris, P. (ed.), *Climate Change and Ocean Governance: Politics and Policy for Threatened Seas* [online]. Cambridge University Press, Cambridge, pp. 215–235. Available from: https://doi.org/10.1017/9781108502238.014 [Accessed 20 December 2020].

IEGL, 2017. *Mapping the Climate Regime: Interactive Climate Regime Map* [online]. Griffith University's Institute for Ethics, Governance and Law. Available from: https://climateregimemap.net/hierarchical [Accessed 11 October 2017].

IPCC, 1995. *IPCC Second Assessment Climate Change 1995* [online]. Intergovernmental Panel on Climate Change. Available from: http://www.ipcc.ch/pdf/climate-changes-1995/ipcc-2nd-assessment/2nd-assessment-en.pdf [Accessed 9 August 2018].

IPCC, 2001. *Climate Change 2001: Synthesis Report. A Contribution of Working Groups I, II, and III to the Third Assessment Report of the Intergovernmental Panel on Climate Change* [online] [Watson, R.T. and the Core Writing Team (eds.)]. Cambridge University Press, Cambridge, and New York, 398 pp. Available from: https://www.ipcc.ch/site/assets/uploads/2018/05/SYR_TAR_full_report.pdf [Accessed 5 December 2019].

IPCC, 2007. *Climate Change 2007: Synthesis Report - Summary for Policymakers. Contribution of Working Groups I, II and III to the Fourth Assessment Report of the Intergovernmental Panel on Climate Change* [online]. [Core Writing Team, Pachauri, R.K. and Reisinger, A. (eds.)]. IPCC, Geneva, Switzerland, p. 104. Available from: http://www.ipcc.ch/pdf/assessment-report/ar4/syr/ar4_syr_spm.pdf [Accessed 9 August 2018].

IPCC, 2012. *Managing the Risks of Extreme Events and Disasters to Advance Climate Change Adaptation. A Special Report of Working Groups I and II of the Intergovernmental Panel on Climate Change* [online] [Field, C.B., V. Barros, T.F. Stocker, D. Qin, D.J. Dokken, K.L. Ebi, M.D. Mastrandrea, K.J. Mach, G.-K. Plattner, S.K. Allen, M. Tignor, and P.M. Midgley (eds.)]. Cambridge University Press, Cambridge, United Kindom, and New York, 582 pp. Available from: https://www.ipcc.ch/pdf/special-reports/srex/SREX_Full_Report.pdf [Accessed 8 August 2018].

IPCC, 2013. Annex III: Glossary [Planton, S. (ed.)]. In: *Climate Change 2013: The Physical Science Basis. Contribution of Working Group I to the Fifth Assessment Report of the Intergovernmental Panel on Climate Change* [online] [Stocker, T.F., D. Qin, G.-K. Plattner, M. Tignor, S.K. Allen, J. Boschung, A. Nauels, Y. Xia, V. Bex, and P.M. Midgley (eds.)]. Cambridge University Press, Cambridge and New York. Available from: http://www.ipcc.ch/pdf/assessment-report/ar5/wg1/WG1AR5_AnnexIII_FINAL.pdf [Accessed 9 July 2018].

IPCC, 2014a. *Climate Change 2014: Synthesis Report-Summary for Policy Makers. Contribution of Working Groups I, II and III to the Fifth Assessment Report of the Intergovernmental Panel on Climate Change* [online] [Core Writing Team, R.K. Pachauri, and L.A. Meyer (eds.)]. IPCC, Geneva, Switzerland, 31 pp. Available from: http://www.ipcc.ch/report/ar5/syr/ [Accessed 9 July 2018].

IPCC, 2014b. Annex II: Glossary [Mach, K.J., S. Planton, and C. von Stechow (eds.)]. In: *Climate Change 2014: Synthesis Report. Contribution of Working Groups I, II and III to the Fifth Assessment Report of the Intergovernmental Panel on Climate Change* [online] [Core Writing Team, R.K. Pachauri, and L.A. Meyer (eds.)]. IPCC, Geneva, Switzerland, pp. 117–130. Available from: https://www.ipcc.ch/site/assets/uploads/2018/02/AR5_SYR_FINAL_Annexes.pdf [Accessed 6 December 2019].

IPCC, 2014c. *Climate Change 2014: Synthesis Report. Contribution of Working Groups I, II and III to the Fifth Assessment Report of the Intergovernmental Panel on Climate Change* [online] [Core Writing Team, R.K. Pachauri, and L.A. Meyer (eds.)]. IPCC, Geneva, Switzerland, 151 pp. Available from: http://www.ipcc.ch/pdf/assessment-report/ar5/syr/SYR_AR5_FINAL_full_wcover.pdf [Accessed 8 August 2014].

IPCC, 2014d. Annex II: Glossary [Agard, J., E.L.F. Schipper, J. Birkmann, M. Campos, C. Dubeux, Y. Nojiri, L. Olsson, B. Osman-Elasha, M. Pelling, M.J. Prather, M.G. Rivera-Ferre, O.C. Ruppel, A. Sallenger, K.R. Smith, A.L. St. Clair, K.J. Mach, M.D. Mastrandrea, and T.E. Bilir (eds.)]. In: *Climate Change 2014: Impacts, Adaptation, and Vulnerability. Part B: Regional Aspects. Contribution of Working Group II to the Fifth Assessment Report of the Intergovernmental Panel on Climate Change* [online] [Barros, V.R., C.B. Field, D.J. Dokken, M.D. Mastrandrea, K.J. Mach, T.E. Bilir, M. Chatterjee, K.L. Ebi, Y.O. Estrada, R.C. Genova, B. Girma, E.S. Kissel, A.N. Levy, S. MacCracken, P.R. Mastrandrea, and L.L. White (eds.)]. Cambridge University Press, Cambridge and New York, pp. 1757–1776. Available from: https://www.ipcc.ch/site/assets/uploads/2018/02/WGIIAR5-AnnexII_FINAL.pdf [Accessed 11 December 2019].

IPCC, 2014e. *Working Group I Fact Sheet* [online]. Intergovernmental Panel on Climate Change. Available from: https://www.ipcc.ch/site/assets/uploads/2018/03/WG1AR5_FactSheet.pdf [Accessed 8 August 2020].

IPCC, 2014f. *Working Group II Fact Sheet* [online]. Intergovernmental Panel on Climate Change. Available from: https://www.ipcc.ch/site/assets/uploads/2018/03/IPCC_WG2AR5_FactSheet.pdf [Accessed 8 August 2020].

IPCC, 2014g. *Working Group III Fact Sheet* [online]. Intergovernmental Panel on Climate Change. Available from: https://www.ipcc.ch/site/assets/uploads/2018/03/WGIII_AR5_FactSheet-1.pdf [Accessed 8 August 2020].

IPCC, 2018a. Summary for policymakers. In: *Global Warming of 1.5°C. An IPCC Special Report on the Impacts of Global Warming of 1.5°C above Pre-industrial Levels and Related Global Greenhouse Gas Emission Pathways, in the Context of Strengthening the Global Response to the Threat of Climate Change, Sustainable Development, and Efforts to Eradicate Poverty* [online] [Masson-Delmotte, V., P. Zhai, H.-O. Pörtner, D. Roberts, J. Skea, P.R. Shukla, A. Pirani, W. Moufouma-Okia, C. Péan, R. Pidcock, S. Connors, J.B.R. Matthews, Y. Chen, X. Zhou, M.I. Gomis, E. Lonnoy, T. Maycock, M. Tignor, and

T. Waterfield (eds.)]. In Press. Available from: https://www.ipcc.ch/site/assets/uploads/sites/2/2019/05/SR15_SPM_version_report_LR.pdf [Accessed 5 December 2019].

IPCC, 2018b. Annex I: Glossary [Matthews, J.B.R. (ed.)]. In: *Global Warming of 1.5°C. An IPCC Special Report on the Impacts of Global Warming of 1.5°C above Pre-industrial Levels and Related Global Greenhouse Gas Emission Pathways, in the Context of Strengthening the Global Response to the Threat of Climate Change, Sustainable Development, and Efforts to Eradicate Poverty* [online] [Masson-Delmotte, V., P. Zhai, H.-O. Pörtner, D. Roberts, J. Skea, P.R. Shukla, A. Pirani, W. Moufouma-Okia, C. Péan, R. Pidcock, S. Connors, J.B.R. Matthews, Y. Chen, X. Zhou, M.I. Gomis, E. Lonnoy, T. Maycock, M. Tignor, and T. Waterfield (eds.)]. In Press. Available from: https://www.ipcc.ch/site/assets/uploads/sites/2/2019/06/SR15_AnnexI_Glossary.pdf [Accessed 6 December 2019].

IPCC, 2022a. Summary for policymakers [H.-O. Pörtner, D.C. Roberts, E.S. Poloczanska, K. Mintenbeck, M. Tignor, A. Alegría, M. Craig, S. Langsdorf, S. Löschke, V. Möller, and A. Okem (eds.)]. In: *Climate Change 2022: Impacts, Adaptation and Vulnerability. Contribution of Working Group II to the Sixth Assessment Report of the Intergovernmental Panel on Climate Change* [online] [H.-O. Pörtner, D.C. Roberts, M. Tignor, E.S. Poloczanska, K. Mintenbeck, A. Alegría, M. Craig, S. Langsdorf, S. Löschke, V. Möller, A. Okem, and B. Rama (eds.)]. Cambridge University Press, Cambridge and New York, pp. 3–33. Available from: IPCC_AR6_WGII_SummaryForPolicymakers.pdf [Accessed 9 December 2023].

IPCC, 2022b: Annex III: Scenarios and modelling methods [Guivarch, C., E. Kriegler, J. Portugal-Pereira, V. Bosetti, J. Edmonds, M. Fischedick, P. Havlík, P. Jaramillo, V. Krey, F. Lecocq, A. Lucena, M. Meinshausen, S. Mirasgedis, B. O'Neill, G.P. Peters, J. Rogelj, S. Rose, Y. Saheb, G. Strbac, A. Hammer Strømman, D.P. van Vuuren, and N. Zhou (eds)]. In: *IPCC, 2022: Climate Change 2022: Mitigation of Climate Change. Contribution of Working Group III to the Sixth Assessment Report of the Intergovernmental Panel on Climate Change* [online] [P.R. Shukla, J. Skea, R. Slade, A. Al Khourdajie, R. van Diemen, D. McCollum, M. Pathak, S. Some, P. Vyas, R. Fradera, M. Belkacemi, A. Hasija, G. Lisboa, S. Luz, and J. Malley, (eds.)]. Cambridge University Press, Cambridge and New York. Available from: https://www.ipcc.ch/report/ar6/wg3/downloads/report/IPCC_AR6_WGIII_Annex-III.pdf [Accessed 10 December 2023].

IPCC, 2023a. Summary for policymakers. In: *Climate Change 2023: Synthesis Report. Contribution of Working Groups I, II and III to the Sixth Assessment Report of the Intergovernmental Panel on Climate Change* [online] [Core Writing Team, H. Lee, and J. Romero (eds.)]. IPCC, Geneva, Switzerland, pp. 1–34. Available from: https://www.ipcc.ch/report/ar6/syr/downloads/report/IPCC_AR6_SYR_SPM.pdf [Accessed 10 December 2023].

IPCC, 2023b: *Workshop Report of the Intergovernmental Panel on Climate Change Workshop on the Use of Scenarios in the Sixth Assessment Report and Subsequent Assessments* [online] [Masson-Delmotte, V., H.-O. Pörtner, D.C. Roberts, P.R. Shukla, J. Skea, P. Zhai, W. Cheung, J. Fuglestvedt, A. Garg, B. O'Neill, J. Pereira, J. Portugal Pereira, K. Riahi, A. Sörensson, C. Tebaldi, E. Totin, D. van Vuuren, Z. Zommers, A. Al Khourdajie, S.L. Connors, R. Fradera, C. Ludden, D. McCollum, K. Mintenbeck, M. Pathak, A. Pirani, E.S. Poloczanska, S. Some, and M. Tignor (eds.)]. Working Group III Technical Support Unit, Imperial College London, 67 pp. Available from: ipcc.ch/site/assets/uploads/2023/07/IPCC_2023_Workshop_Report_Scenarios.pdf [Accessed 10 December 2023].

IRENA, 2019. *A New World: The Geopolitics of the Energy Transformation* [online]. International Renewable Energy Agency. Available from: https://www.irena.org/-/media/Files/IRENA/Agency/Publication/2019/Jan/Global_commission_geopolitics_new_world_2019.pdf [Accessed 6 December 2019].

Ishiwatari, M., Koike, T., Hiroki, K., Toda, T. and Katsube, T., 2020. Managing disasters amid COVID-19 pandemic: Approaches of response to flood disasters. *Progress in Disaster Science, 6*, p. 100096.

James, C., 2011. *Theory of Change Review* [online]. Comic Relief. Available from: http://www.theoryofchange.org/wp-content/uploads/toco_library/pdf/James_ToC.pdf [Accessed 8 December 2019].

Kennicutt, M.C., Chown, S.L., Cassano, J.J., Liggett, D., Massom, R., Peck, L.S., Rintoul, S.R., Storey, J.W., Vaughan, D.G., Wilson, T.J. and Sutherland, W.J., 2014. Polar research: Six priorities for Antarctic science. *Nature News* [online], *512*(7512), p. 23. Available from: https://www.nature.com/news/polopoly_fs/1.15658!/menu/main/topColumns/topLeftColumn/pdf/512023a.pdf [Accessed 26 March 2021].

Keohane, R.O. and Victor, D.G., 2011. The regime complex for climate change. *Perspectives on Politics* [online], *9*(1), pp. 7–23. Available from: https://www.cambridge.org/core/journals/perspectives-on-politics/article/regime-complex-for-climate-change/F5C4F620A4723D5DA5E0ACDC48D860C0 [Accessed 14 September 2018].

Kerr, A., 2007. Serendipity is not a strategy: The impact of national climate programmes on greenhouse-gas emissions. *Area* [online], *39*(4), pp. 418–430. Available from: https://rgs-ibg.onlinelibrary.wiley.com/doi/epdf/10.1111/j.1475-4762.2007.00773.x [Accessed 8 December 2019].

Kilavuka, M., 2013. Reverse stress testing: A structured approach. *The RMA Journal* [online], *95*(6), pp. 28–33. Available from: https://cms.rmau.org/uploadedFiles/Credit_Risk/Library/RMA_Journal/Enterprise_Wide_Risk_Management/Reverse%20Stress%20Testing_%20A%20Structured%20Approach.pdf [Accessed 21 May 2018].

Kingdon, J.W., 1995. *Agendas, Alternatives, and Public Policies* (2nd edition) [Kindle Reader, Ipad Mini] HarperCollins College Publishers, New York.

Kleineberg, M., 2013. The blind men and the elephant: Towards an organization of epistemic contexts. *Knowledge Organization* [online], *40*(5), pp. 340–364. Available from: https://www.nomos-elibrary.de/10.5771/0943-7444-2013-5-340/the-blind-men-and-the-elephant-towards-an-organization-of-epistemic-contexts-volume-40-2013-issue-5 [Accessed 1 March 2020].

Krasner, S.D., 1982. Structural causes and regime consequences: Regimes as intervening variables. *International Organization* [online], *36*(2), pp. 185–205. Available from: https://www.cambridge.org/core/services/aop-cambridge-core/content/view/19A9938FE30759F777EA8EDC38BF1227/S0020818300018920a.pdf/structural_causes_and_regime_consequences_regimes_as_intervening_variables.pdf [Accessed 10 December 2019].

Lincoln, A., 1861. *First Inauguration Address*. The Avalon Project: Documents in Law, History and Diplomacy. Yale Law School.

Mangalagiu, D., Wilkinson, A. and Kupers, R., 2011. When futures lock-in the present: Towards a new generation of climate scenarios. In: *Reframing the Problem of Climate Change: From Zero Sum Game to Win-win Solutions, Earthscan, London* [online]. Available from: https://www.researchgate.net/profile/Diana_Mangalagiu/publication/254819683_When_futures_lock-in_the_present_Towards_a_new_generation_of_climate_scenarios/links/55625b6008ae9963a11b4a55/When-futures-lock-in-the-present-Towards-a-new-generation-of-climate-scenarios.pdf [Accessed 7 December 2019].

Marchiori, C., Dietz, S. and Tavoni, A., 2017. Domestic politics and the formation of international environmental agreements. *Journal of Environmental Economics and Management, 81*, pp. 115–131. Available from: https://www.sciencedirect.com/science/article/pii/S0095069616303060 [Accessed 8 August 2018].

Martin, D.F., Cornford, S.L. and Payne, A.J., 2019. Millennial-scale vulnerability of the Antarctic ice sheet to regional ice shelf collapse. *Geophysical Research Letters* [online], *46*(3), 1467–1475. Available from: https://agupubs.onlinelibrary.wiley.com/doi/epdf/10.1029/2018GL081229 [Accessed 26 March 2021].

Massara, P., March 13, 2020. Letter: Climate change needs its own Marshall Plan [online]. *Financial Times*. Available from: https://www.ft.com/content/436e10e2-63ca-11ea-b3f3-fe4680ea68b5 [Accessed 4 May 2020].

Mearsheimer, J.J., 1994. The false promise of international institutions. *International Security* [online], *19*(3), pp. 5–49. Available from https://www-jstor-org.libproxy.ucl.ac.uk/stable/pdf/2539078.pdf [Accessed 18 August 2018].

Meltzer, J.P., 2016. *Financing Low Carbon, Climate Resilient Infrastructure: The Role of Climate Finance and Green Financial Systems.* [online] Global Economy and Development at Brookings. Available from: https://papers.ssrn.com/sol3/papers.cfm?abstract_id=2841918 [Accessed 1 October 2018].

Merriam-Webster. *Dictionary* [online]. Merriam-Webster. Available from: https://www.merriam-webster.com/

Meyfroidt, P., Lambin, E.F., Erb, K.H. and Hertel, T.W., 2013. Globalization of land use: Distant drivers of land change and geographic displacement of land use. *Current Opinion in Environmental Sustainability* [online], *5*(5), pp. 438–444. Available from: https://www.sciencedirect.com/science/article/pii/S1877343513000353 [Accessed 3 May 2020].

Millennium Ecosystem Assessment, 2005. *Ecosystems and Human Well-being: Synthesis* [online]. Island Press, Washington, DC. Available from: https://www.millenniumassessment.org/documents/document.356.aspx.pdf [Accessed 3 May 2020].

Moncel, R. and van Asselt, H., 2012. All hands on deck! Mobilizing climate change action beyond the UNFCCC. *Review of European Community & International Environmental Law* [online], *21*(3), pp. 163–176. Available from: https://onlinelibrary.wiley.com/doi/epdf/10.1111/reel.12011 [Accessed 11 December 2019].

Moncel, R., Joffe, P., Levin, K. and McCall, K., 2011. *Building the Climate Change Regime: Survey and Analysis of Approaches. Summary for Stakeholder Comment* [online]. Washington, DC: World Resources Institute. Available from: http://pdf.wri.org/working_papers/building_the_climate_change_regime.pdf [Accessed 6 February 2018].

Moss, R., Babiker, M., Brinkman, S., Calvo, E., Carter, T., Edmonds, J., Elgizouli, I., Emori, S., Erda, L., Hibbard, K., Jones, R., Kainuma, M., Kelleher, J., Lamarque, J. F., Manning, M., Matthews, B., Meehl, J., Meyer, L., Mitchell, J., Nakicenovic, N., O'Neill, B., Pichs, R., Riahi, K., Rose, S., Runci, P., Stouffer, R., van Vuuren, D., Weyant, J., Wilbanks, T., van Ypersele, J. P. and Zurek M., 2008. *Towards New Scenarios for Analysis of Emissions, Climate Change, Impacts, and Response Strategies* [online]. Technical Summary. Intergovernmental Panel on Climate Change, Geneva, 25 pp. Available from: https://www.ipcc.ch/pdf/supporting-material/expert-meeting-ts-scenarios.pdf [Accessed 9 August 2016].

Moss, R.H., Edmonds, J.A., Hibbard, K.A., Manning, M.R., Rose, S.K., Van Vuuren, D.P., Carter, T.R., Emori, S., Kainuma, M., Kram, T. and Meehl, G.A., 2010. The next generation of scenarios for climate change research and assessment. *Nature* [online], *463*(7282), pp. 747–756. Available from: http://www.nature.com/nature/journal/v463/n7282/abs/nature08823.html [Accessed 8 August 2016].

Myhre, G., D. Shindell, F.-M. Bréon, W. Collins, J. Fuglestvedt, J. Huang, D. Koch, J.-F. Lamarque, D. Lee, B. Mendoza, T. Nakajima, A. Robock, G. Stephens, T. Takemura and H. Zhang, 2013. Anthropogenic and natural radiative forcing. In: *Climate Change 2013: The Physical Science Basis. Contribution of Working Group I to the Fifth Assessment*

Report of the Intergovernmental Panel on Climate Change [online] [Stocker, T.F., D. Qin, G.-K. Plattner, M. Tignor, S.K. Allen, J. Boschung, A. Nauels, Y. Xia, V. Bex, and P.M. Midgley (eds.)]. Cambridge University Press, Cambridge and New York. Available from: http://www.ipcc.ch/pdf/assessment-report/ar5/wg1/WG1AR5_Chapter08_FINAL. pdf [Accessed 2 July 2018].

National Academies of Sciences, Engineering, and Medicine, 2016. *Attribution of Extreme Weather Events in the Context of Climate Change* [online]. National Academies Press. Available from: https://www.nap.edu/catalog/21852/attribution-of-extreme-weather-events-in-the-context-of-climate-change [Accessed 8 August 2020].

Nature, 2019. *Climate Sciences* [online]. Nature. Available from: https://www.nature.com/subjects/climate-sciences [Accessed 16 March 2019].

Naughten, K.A., Holland, P.R. and De Rydt, J., 2023. Unavoidable future increase in West Antarctic ice-shelf melting over the twenty-first century. *Nature Climate Change* [online], pp. 1–7. Available from: https://www.nature.com/articles/s41558-023-01818-x [Accessed 8 December 2023].

Nordhaus, W.D., 2014. Estimates of the social cost of carbon: Concepts and results from the DICE-2013R model and alternative approaches. *Journal of the Association of Environmental and Resource Economists* [online], *1*(1–2), pp. 273–312. Available from: https://www.journals.uchicago.edu/doi/pdfplus/10.1086/676035 [Accessed 8 December 2019].

OECD Glossary of Statistical Terms. Organisation for Economic Cooperation and Development. Available from: https://stats.oecd.org/glossary/ [Accessed 11 December 2019].

OECD, 2022. *First Lessons from Government Evaluations of COVID-19 Responses: A Synthesis* [Online]. OECD Publishing. Available from: https://read.oecd-ilibrary.org/view/?ref=1125_1125436-7j5hea8nk4&title=First-lessons-from-government-evaluations-of-COVID-19-responses [Accessed 1 December 2023].

O'Neill, B.C., Carter, T.R., Ebi, K.L., Edmonds, J., Hallegatte, S., Kemp-Benedict, E., Kriegler, E., Mearns, L., Moss, R., Riahi, K., van Ruijven, B., and van Vuuren, D., March 12, 2012. *Workshop on The Nature and Use of New Socioeconomic Pathways for Climate Change Research* [online]. National Center for Atmospheric Research (NCAR), Boulder, CO. November 2–4, 2011. Meeting Report. Final Version. Available from: https://www2.cgd.ucar.edu/sites/default/files/iconics/Boulder-Workshop-Report.pdf [Accessed 6 July 2018].

O'Neill, B.C., Kriegler, E., Ebi, K.L., Kemp-Benedict, E., Riahi, K., Rothman, D.S., van Ruijven, B.J., van Vuuren, D.P., Birkmann, J., Kok, K. and Levy, M., 2017. The roads ahead: Narratives for shared socioeconomic pathways describing world futures in the 21st century. *Global Environmental Change* [online], *42*, pp. 169–180. Available from: https://www.sciencedirect.com/science/article/pii/S0959378015000060/pdfft?isDTMRedir=true&download=true [Accessed 11 December 2019].

O'Neill, B.C., Kriegler, E., Riahi, K., Ebi, K.L., Hallegatte, S., Carter, T.R., Mathur, R. and van Vuuren, D.P., 2014. A new scenario framework for climate change research: the concept of shared socioeconomic pathways. *Climatic Change* [online], *122*(3), pp. 387–400. Available from: https://link.springer.com/content/pdf/10.1007%2Fs10584-013-0905-2.pdf [Accessed 6 December 2019].

O'Neill, B., van Aalst, M., Zaiton Ibrahim, Z., Berrang Ford, L., Bhadwal, S., Buhaug, H., Diaz, D., Frieler, K., Garschagen, M., Magnan, A., Midgley, G., Mirzabaev, A., Thomas, A., and Warren, R., 2022. Key risks across sectors and regions. In Pörtner, H.-O., Roberts, D.C., Tignor, M., Poloczanska, E.S., Mintenbeck, K., Alegría, A., Craig, M., Langsdorf, S., Löschke, S., Möller, V., Okem, A., and Rama, B. (eds.), *Climate Change 2022: Impacts, Adaptation and Vulnerability. Contribution of Working Group II to the Sixth Assessment*

Report of the Intergovernmental Panel on Climate Change [online]. Cambridge University Press, Cambridge and New York. Available from: https://www.ipcc.ch/report/ar6/wg2/downloads/report/IPCC_AR6_WGII_Chapter16.pdf [Accessed 9 December 2023].

OPIL 2011. *Framework Agreement* [online]. Oxford Public International Law. Available from: https://opil.ouplaw.com/view/10.1093/law:epil/9780199231690/law-9780199231690-e703?prd=EPIL#:~:text=A%20framework%20agreement%20can%20be,agreements%20between%20the%20parties%2C%20usually [Accessed 12 December 2020].

Oppenheimer, M., M. Campos, R. Warren, J. Birkmann, G. Luber, B. O'Neill, and K. Takahashi, 2014. Emergent risks and key vulnerabilities. In: *Climate Change 2014: Impacts, Adaptation, and Vulnerability. Part A: Global and Sectoral Aspects. Contribution of Working Group II to the Fifth Assessment Report of the Intergovernmental Panel on Climate Change* [online] [Field, C.B., V.R. Barros, D.J. Dokken, K.J. Mach, M.D. Mastrandrea, T.E. Bilir, M. Chatterjee, K.L. Ebi, Y.O. Estrada, R.C. Genova, B. Girma, E.S. Kissel, A.N. Levy, S. MacCracken, P.R. Mastrandrea, and L.L. White (eds.)]. Cambridge University Press, Cambridge and New York, pp. 1039–1099. Available from: https://www.ipcc.ch/pdf/assessment-report/ar5/wg2/WGIIAR5-Chap19_FINAL.pdf [Accessed 8 August 2018].

Ostrom, E., 1990. *Governing the Commons* [online]. Cambridge University Press. Available from: http://wtf.tw/ref/ostrom_1990.pdf [Accessed 14 February 2017].

Oura, H. and Schumacher, L., 2012. *Macrofinancial Stress Testing—Principles and Practices* [online]. International Monetary Fund. Available from: http://www.imf.org/external/np/pp/eng/2012/082212.pdf [Accessed 12 February 2017].

Oxford Dictionary. *Oxford Dictionaries* [online]. Available from: https://en.oxforddictionaries.com/

Peters, A., Guitart, C. and Pittet, D., 2021. Addressing the global challenge of access to supplies during COVID-19: Mask reuse and local production of alcohol-based hand rub. *Environmental and Health Management of Novel Coronavirus Disease (COVID-19)*, pp. 419–441.

Pitterle, I., and Niermann, L., 2021. The COVID-19 crisis: What explains cross-country differences in the pandemic's short-term economic impact? DESA Working Paper No. 174, ST/ESA/2021/DWP/174, August 2021.

Pizer, W., Adler, M., Aldy, J., Anthoff, D., Cropper, M., Gillingham, K., Greenstone, M., Murray, B., Newell, R., Richels, R. and Rowell, A., 2014. Using and improving the social cost of carbon. *Science* [online], *346*(6214), pp. 1189–1190. Available from: https://science.sciencemag.org/content/sci/346/6214/1189.full.pdf [Accessed 8 December 2019].

Prins, G., Galiana, I., Green, C., Grundmann, R., Korhola, A., Laird, F., Nordhaus, T., Pielke Jnr, R., Rayner, S., Sarewitz, D. and Shellenberger, M., 2010. *The Hartwell Paper: A New Direction for Climate Policy after the Crash of 2009* [online]. London. London School of Economics. Available from: http://eprints.lse.ac.uk/27939/1/HartwellPaper_English_version.pdf [Accessed 22 April 2020].

Purich, A. and Doddridge, E.W., 2023. Record low Antarctic sea ice coverage indicates a new sea ice state. *Communications Earth & Environment* [online], *4*(1), p. 314. Available from: https://www.nature.com/articles/s43247-023-00961-9 [Accessed 8 December 2023].

Putnam, R.D., 1988. Diplomacy and domestic politics: The logic of two-level games. *International organization* [online], *42*(3), pp. 427–460. Available from: https://www.cambridge.org/core/journals/international-organization/article/diplomacy-and-domestic-politics-the-logic-of-twolevel-games/B2E11FB757C4465C4097015BD421035F [Accessed 14 September 2018].

Randalls, S., 2010. History of the 2 C climate target. *Wiley Interdisciplinary Reviews: Climate Change* [online], *1*(4), pp. 598–605. Available from: https://onlinelibrary.wiley.com/doi/epdf/10.1002/wcc.62 [Accessed 8 December 2019].

Riahi, K., R. Schaeffer, J. Arango, K. Calvin, C. Guivarch, T. Hasegawa, K. Jiang, E. Kriegler, R. Matthews, G.P. Peters, A. Rao, S. Robertson, A.M. Sebbit, J. Steinberger, M. Tavoni, D.P. van Vuuren, 2022. Mitigation pathways compatible with long-term goals. In: *IPCC, 2022: Climate Change 2022: Mitigation of Climate Change. Contribution of Working Group III to the Sixth Assessment Report of the Intergovernmental Panel on Climate Change* [online] [P.R. Shukla, J. Skea, R. Slade, A. Al Khourdajie, R. van Diemen, D. McCollum, M. Pathak, S. Some, P. Vyas, R. Fradera, M. Belkacemi, A. Hasija, G. Lisboa, S. Luz, and J. Malley, (eds.)]. Cambridge University Press, Cambridge and New York. Available from: https://www.ipcc.ch/report/ar6/wg3/downloads/report/IPCC_AR6_WGIII_Chapter03.pdf [Accessed 9 December 2023].

Ringe, N. and Rennó, L., 2023. *Populists and the Pandemic: How Populists Around the World Responded to Covid-19*. Routledge, Abingdon, p. 321.

Risse, T., 2002. Constructivism and International Institutions: Toward Conversations Across Paradigms. In Katznelson, I., and Milner, H. (eds.), *Political Science as Discipline. Reconsidering Power, Choice, and the State at Century's End*, W. W. Norton, New York, 597–623.

Robinson, J.B., 1990. Futures under glass: a recipe for people who hate to predict. *Futures* [online], *22*(8), pp.820–842. Available from: https://doi.org/10.1016/0016-3287(90)90018-D [Accessed 1 October 2018].

Robock, A., 2000. Volcanic eruptions and climate. *Reviews of Geophysics* [online], *38*(2), pp. 191–219. Available from: https://agupubs.onlinelibrary.wiley.com/doi/epdf/10.1029/1998RG000054 [Accessed 3 May 2020].

Rockström, J., Gaffney, O., Rogelj, J., Meinshausen, M., Nakicenovic, N. and Schellnhuber, H.J., 2017. A roadmap for rapid decarbonization. *Science* [online], *355*(6331), pp. 1269–1271. Available from: http://science.sciencemag.org/content/355/6331/1269.summary [Accessed 1 October 2018].

Rogelj, J., D. Shindell, K. Jiang, S. Fifita, P. Forster, V. Ginzburg, C. Handa, H. Kheshgi, S. Kobayashi, E. Kriegler, L. Mundaca, R. Séférian, and M.V.Vilariño, 2018. Mitigation pathways compatible with 1.5°C in the context of sustainable development. In: *Global Warming of 1.5°C. An IPCC Special Report on the Impacts of Global Warming of 1.5°C above Pre-industrial Levels and Related Global Greenhouse Gas Emission Pathways, in the Context of Strengthening the Global Response to the Threat of Climate Change, Sustainable Development, and Efforts to Eradicate Poverty* [online] [Masson-Delmotte, V., P. Zhai, H.-O. Pörtner, D. Roberts, J. Skea, P.R. Shukla, A. Pirani, W. Moufouma-Okia, C. Péan, R. Pidcock, S. Connors, J.B.R. Matthews, Y. Chen, X. Zhou, M.I. Gomis, E. Lonnoy, T. Maycock, M. Tignor, and T. Waterfield (eds.)]. In Press. Available from: https://www.ipcc.ch/site/assets/uploads/sites/2/2019/05/SR15_Chapter2_Low_Res.pdf [Accessed 27 October 2019].

Rose, S.K., Diaz, D.B. and Blanford, G.J., 2017. Understanding the social cost of carbon: A model diagnostic and inter-comparison study. *Climate Change Economics* [online], *8*(02), p. 1750009. Available from: https://www.worldscientific.com/doi/pdf/10.1142/S2010007817500099 [Accessed 8 December 2019].

Rounsevell, M.D. and Metzger, M.J., 2010. Developing qualitative scenario storylines for environmental change assessment. *Wiley Interdisciplinary Reviews: Climate Change* [online], *1*(4), pp. 606–619. Available from: https://onlinelibrary.wiley.com/doi/epdf/10.1002/wcc.63 [Accessed 6 December 2019].

Sachs, J.D., Karim, S.S.A., Aknin, L., Allen, J., Brosbøl, K., Colombo, F., Barron, G.C., Espinosa, M.F., Gaspar, V., Gaviria, A. and Haines, A., 2022. The Lancet Commission on lessons for the future from the COVID-19 pandemic. *The Lancet, 400*(10359), pp. 1224–1280.

Savaresi, A., 2016. The Paris agreement: A new beginning? *Journal of Energy & Natural Resources Law* [online], *34*(1), pp. 16–26. Available from: https://www.tandfonline.com/doi/abs/10.1080/02646811.2016.1133983 [Accessed 3 May 2020].

Self, S., 2006. The effects and consequences of very large explosive volcanic eruptions. *Philosophical Transactions of the Royal Society A: Mathematical, Physical and Engineering Sciences* [online], *364*(1845), pp. 2073–2097. Available from: https://royalsocietypublishing.org/doi/pdf/10.1098/rsta.2006.1814 [Accessed 3 May 2020].

Shell, 2018. *Sky: Meeting the Goals of the Paris Agreement* [online]. Shell. Available from: https://www.shell.com/promos/meeting-the-goals-of-the-paris-agreement/_jcr_content.stream/1530643931055/d5af41aef92d05d86a5cd77b3f3f5911f75c3a51c1961fe1c-981daebda29b726/shell-scenario-sky.pdf [Accessed 11 July 2018].

Siegele, L., 2012. *Loss and Damage: The Theme of Slow Onset Impact* [online]. Loss & Damage in Vulnerable Countries Initiative, Climate Development Knowledge Network. Available from: https://germanwatch.org/sites/germanwatch.org/files/publication/6674.pdf [Accessed 8 August 2020].

Simmons, B.A. and Martin, L.L., 2002. International organizations and institutions. *Handbook of International Relations* [online], pp. 192–211. Available from: https://sk.sagepub.com/reference/hdbk_intlrelations/n10.xml [Accessed 7 August 2020].

Smith, P., Davis, S.J., Creutzig, F., Fuss, S., Minx, J., Gabrielle, B., Kato, E., Jackson, R.B., Cowie, A., Kriegler, E. and Van Vuuren, D.P., 2016. Biophysical and economic limits to negative CO_2 emissions. *Nature Climate Change* [online], *6*(1), pp. 42–50. Available from: https://www.nature.com/articles/nclimate2870 [Accessed 7 August 2020].

Sohrabizadeh, S., Yousefian, S., Bahramzadeh, A. and Vaziri, M.H., 2021. A systematic review of health sector responses to the coincidence of disasters and COVID-19. *BMC Public Health, 21*(1), pp. 1–9.

Stavins R., J. Zou, T. Brewer, M. Conte Grand, M. den Elzen, M. Finus, J. Gupta, N. Höhne, M.-K. Lee, A. Michaelowa, M. Paterson, K. Ramakrishna, G. Wen, J. Wiener, and H. Winkler, 2014. International cooperation: Agreements and instruments. In: *Climate Change 2014: Mitigation of Climate Change. Contribution of Working Group III to the Fifth Assessment Report of the Intergovernmental Panel on Climate Change* [online] [Edenhofer, O., R. Pichs-Madruga, Y. Sokona, E. Farahani, S. Kadner, K. Seyboth, A. Adler, I. Baum, S. Brunner, P. Eickemeier, B. Kriemann, J. Savolainen, S. Schlömer, C. von Stechow, T. Zwickel, and J.C. Minx (eds.)]. Cambridge University Press, Cambridge and New York. Available from: http://www.ipcc.ch/pdf/assessment-report/ar5/wg3/ipcc_wg3_ar5_chapter13.pdf [Accessed 11 July 2018].

Steenmans, I., 2019. *Integrated Analytical Practice: Tidying Up Our Thoughts* [Powerpoint Presentation]. Analytical Methods for Policy STEP0020. Session 19.

Steffen, W., 2011. A truly complex and diabolical policy problem. *The Oxford Handbook of Climate Change and Society* [online], Oxford University Press, Oxford pp. 21–37. Available from: https://doi.org/10.1093/oxfordhb/9780199566600.003.0002 [Accessed 9 August 2020].

Stein, D. and Valters, C., 2012. *Understanding Theory of Change in International Development* [online]. Justice and Security Research Programme, International Development Department, London School of Economics and Political Science, London. Available from: http://eprints.lse.ac.uk/56359/1/JSRP_Paper1_Understanding_theory_of_change_in_international_development_Stein_Valters_2012.pdf [Accessed 8 December 2019].

Stern, N., 2007. *The Economics of Climate Change: The Stern Review* [online]. Cambridge University Press, Cambridge and New York, 692 pp. Available from: https://webarchive. nationalarchives.gov.uk/20100407172811/http://www.hm-treasury.gov.uk/stern_review_ report.htm [Accessed 8 December 2019].

Stern, P.C., Ebi, K.L., Leichenko, R., Olson, R.S., Steinbruner, J.D. and Lempert, R., 2013. Managing risk with climate vulnerability science. *Nature Climate Change* [online], *3*(7), p. 607. Available from: https://www.nature.com/articles/nclimate1929 [Accessed 21 June 2018].

Swart, R., Fuss, S., Obersteiner, M., Ruti, P., Teichmann, C. and Vautard, R., 2013. Beyond vulnerability assessment. *Nature Climate Change* [online], *3*(11), p. 942. Available from: https://www.nature.com/articles/nclimate2029 [Accessed 21 May 2018].

Szocik, K., 2021. Conceptual issues in COVID-19 pandemic: An example of global catastrophic risk: A response to: The traditional definition of pandemics, its moral conflations, and its practical implications: A defense of conceptual clarity in global health laws and policies by T. De Campos, *Cambridge Quarterly of Healthcare Ethics* (CQ *29*(2)). *Cambridge Quarterly of Healthcare Ethics*, *30*(1), pp. 199–202.

Taleb, N.N., 2007. *The Black Swan: The Impact of the Highly Improbable*. Random House, New York.

Tanter, R., 2008. *Introduction to Global Problems*. Nautilus Institute for Security and Sustainability [online]. Available from: https://nautilus.org/gps/intro/ [Accessed 29 June 2021].

Taplin, D.H. and Rasic, M., 2012. *Facilitator's Source Book* [online]. ActKnowledge. Available from: https://www.theoryofchange.org/wp-content/uploads/toco_library/pdf/ ToCFacilitatorSourcebook.pdf [Accessed 8 December 2019].

Taplin, D.H., Clark, H., Collins, E. and Colby, D.C., 2013. *Theory of Change* [online]. ActKnowledge. Available from: https://www.theoryofchange.org/wp-content/uploads/ toco_library/pdf/ToC-Tech-Papers.pdf [Accessed 8 December 2019].

TCFD, 2017. *Final Report: Recommendations of the Task Force on Climate-Related Financial Disclosures* [online]. Task Force on Climate-related Disclosures. Available from: https://www.fsb-tcfd.org/wp-content/uploads/2017/06/FINAL-TCFD-Report-062817. pdf [Accessed 21 May 2018].

The Free Dictionary. Farlex. Available from: https://www.thefreedictionary.com/ [Accessed 11 December 2019].

Tol, R.S., 2009. The economic effects of climate change. *Journal of Economic Perspectives* [online], *23*(2), pp. 29–51. Available from: https://www.aeaweb.org/articles?id=10.1257/ jep.23.2.29 [Accessed 8 August 2020].

Tol, R.S., Downing, T.E., Kuik, O.J. and Smith, J.B., 2004. Distributional aspects of climate change impacts. *Global Environmental Change* [online], *14*(3), pp. 259–272. Available from: https://www.sciencedirect.com/science/article/pii/S0959378004000421 [Accessed 8 August 2020].

Tollefson, J., 2016. Antarctic model raises prospect of unstoppable ice collapse. *Nature News* [online], *531*(7596), p. 562. Available from: https://www.nature.com/news/polopoly_fs/ 1.19638!/menu/main/topColumns/topLeftColumn/pdf/531562a.pdf [Accessed 26 March 2021].

Toth, F.L., M. Mwandosya, C. Carraro, J. Christensen, J. Edmonds, B. Flannery, C. Gay-Garcia, H. Lee, K.M. Meyer-Abich, E. Nikitina, A. Rahman, R. Richels, Y. Ruqiu, A. Villavicencio, Y. Wake, J. Weyant, J. Byrne, R. Lempert, I. Meyer, A. Underdal, J. Pershing, and M. Shechter, 2001. Chapter 10 Decision-making frameworks. In: *Climate Change 2001: Mitigation: Contribution of Working Group III to the Third Assessment Report of the Intergovernmental Panel on Climate Change* [online] *(Vol. 3)* [Metz, B., O. Davidson,

R. Swart, and J. Pan (eds.)]. Cambridge University Press. Available from: https://www. ipcc.ch/site/assets/uploads/2018/03/10.pdf [Accessed 10 December 2019].

UN, 2012. *International Recommendations for Water Statistics* [online]. Department of Economic and Social Affairs, United Nations. New York. Available from: https://unstats. un.org/unsd/publication/seriesM/seriesm_91e.pdf [Accessed 11 December 2019].

UN, 2018a. *Growth in United Nations Membership, 1945-Present* [online]. United Nations. Available from: http://www.un.org/en/sections/member-states/growth-united-nations-membership-1945-present/index.html [Accessed 15 August 2018].

UN, 2018b: Chapter XXVII Environment: 7. *United Nations Framework Convention on Climate Change, New York, 9 May 1992* [Online]. United Nations. Available from: https:// treaties.un.org/Pages/ViewDetailsIII.aspx?src=TREATY&mtdsg_no=XXVII-7&chapter =27&Temp=mtdsg3&clang=_en [Accessed 14 September 2018]

UN, 2020. *The Climate Crisis – A Race We Can Win* [online]. New York: United Nations. Available from: https://www.un.org/en/un75/climate-crisis-race-we-can-win [Accessed 6 August 2020].

UNDP, 2014. *Foresight as a Strategic Long-Term Planning Tool for Developing Countries* [online]. United Nations Development Programme. Available from: https://www.undp. org/content/dam/undp/library/capacity-development/English/Singapore%20Centre/ GPCSE_Foresight.pdf [Accessed 7 December 2019].

UNECE 2011. *Framework Convention Concept: Note by the Secretariat* [online]. United Nations Economic Commission for Europe. Available from: https://www.unece.org/fileadmin/DAM/hlm/sessions/docs2011/informal.notice.5.pdf [Accessed 12 December 2020].

UNEP, 2017. *The Emissions Gap Report 2017* [online]. United Nations Environment Programme (UNEP). Available from: https://wedocs.unep.org/bitstream/handle/20.500. 11822/22070/EGR_2017.pdf [Accessed 6 February 2018].

UNEP, 2023. *Emissions Gap Report 2023: Broken Record – Temperatures Hit New Highs, Yet World Fails to Cut Emissions (Again)* [online]. Nairobi. Available from: https:// wedocs.unep.org/bitstream/handle/20.500.11822/43922/EGR2023.pdf?sequence= 3&isAllowed=y [Accessed 9 December 2023].

UNFCCC Glossary. *UNFCCC Glossary of Climate Change Acronyms and Terms* [online]. United Nations Framework Convention on Climate Change. Available from: https://unfccc.int/process-and-meetings/the-convention/glossary-of-climate-change-acronyms-and-terms [Accessed 11 December 2019].

UNFCCC, 1992. *United Nations Framework Convention on Climate Change* [online]. United Nations. Available from: http://unfccc.int/files/essential_background/background_ publications_htmlpdf/application/pdf/conveng.pdf [Accessed 15 January 2018].

UNFCCC, 2012. *Slow Onset Events: Technical Paper* [online]. United Nations Framework Convention on Climate Change. UNFCCC/TP/2012/7. Available from: http://unfccc.int/ resource/docs/2012/tp/07.pdf [Accessed 10 March 2018].

UNFCCC, 2015. Paris Agreement. [online] United Nations Framework Convention on Climate Change Secretariat. Available from: http://unfccc.int/files/essential_background/ convention/application/pdf/english_paris_agreement.pdf [Accessed 27 January 2017].

UNFCCC, 2018a. *Climate: Get the Big Picture* [online]. United Nations Framework Convention on Climate Change. Available from: https://bigpicture.unfccc.int/ [Accessed 14 September 2018].

UNFCCC, 2018b. *Statistics on Participation and In-session Engagement* [online]. United Nations Framework Convention on Climate Change. Available from: https://unfccc.int/ process/parties-non-party-stakeholders/non-party-stakeholders/statistics-on-participation -and-in-session-engagement [Accessed 14 September 2018].

UNFCCC, 2019. *Nationally Determined Contributions* [online]. United Nations Framework Convention on Climate Change. Available from: https://unfccc.int/process-and-meetings/the-paris-agreement/nationally-determined-contributions-ndcs [Accessed 6 December 2019].

UNFCCC, 2020. *Party Groupings* [online]. Bonn, United Nations Framework Convention on Climate Change. Available from: https://unfccc.int/process-and-meetings/parties-non-party-stakeholders/parties/party-groupings [Accessed 3 May 2020].

UNICEF, 2017. *Introduction to the Convention on the Rights of the Child* [online]. United Nations Children's Fund. Available from: https://www.unicef.org/french/crc/files/Definitions.pdf [Accessed 23 January 2017].

UNTC, 2019. *Glossary* [online]. United Nations Treaty Collection. Available from: https://treaties.un.org/pages/Overview.aspx?path=overview/glossary/page1_en.xml [Accessed 10 December 2019].

UNTERM, 2019a. *Government. The United Nations Terminology Database* [online]. United Nations. Available from: https://unterm.un.org/unterm/Display/record/UNDP/NA?Origin alId=imp-2015-11-11_14-16-32-222 [Accessed 10 December 2019].

UNTERM, 2019b. *Treaty. The United Nations Terminology Database* [online]. United Nations. Available from: https://unterm.un.org/unterm/Display/record/UNHQ/NA?Origi nalId=ea1364fe8faf691785256e7b006a2c26 [Accessed 11 December 2019].

UNWater 2019. *What Is Water Security?* [online]. United Nations. Available from: https://www.unwater.org/publications/water-security-infographic/ [Accessed 11 December 2019].

US Congress, 2019. *House Select Committe on the Climate Crisis* [online]. Available from: https://climatecrisis.house.gov/ [Accessed 6 August 2020].

van Vuuren, D.P., Stehfest, E., Gernaat, D.E., van den Berg, M., Bijl, D.L., de Boer, H.S., Daioglou, V., Doelman, J.C., Edelenbosch, O.Y., Harmsen, M. and Hof, A.F., 2018. Alternative pathways to the 1.5 C target reduce the need for negative emission technologies. *Nature Climate Change* [online], *8*(5), p. 391. Available from: https://www.nature.com/articles/s41558-018-0119-8 [Accessed 17 May 2018].

Vázquez, H. R., 2011. *International Cooperation for Development: A Latin American Perspective. The South-South Opportunity* [online]. Available from: http://www.southsouth.info/m/blogpost?id=3952417%3ABlogPost%3A11463 [Accessed 3 July 2017].

Vogel, I., 2012. *Review of the Use of 'Theory of Change' in International Development* [online]. Department of International Development. Available from: https://assets.publishing.service.gov.uk/media/57a08a5ded915d3cfd00071a/DFID_ToC_Review_VogelV7.pdf [Accessed 8 December 2019].

Wadhams, P., 2012. Arctic ice cover, ice thickness and tipping points. *Ambio* [online], *41*(1), pp. 23–33. Available from: https://link.springer.com/content/pdf/10.1007%2Fs13280-011-0222-9.pdf [Accessed 7 December 2019].

Warner, K. and Van der Geest, K., 2013. Loss and damage from climate change: Local-level evidence from nine vulnerable countries. *International Journal of Global Warming* [online], *5*(4), pp. 367–386. Available from: https://collections.unu.edu/eserv/UNU:2096/warner_vandergeest_2013_loss_and_damage.pdf [Accessed 10 December 2019].

Warner, K., van der Geest, K., Kreft, S., Huq, S., Harmeling, S., Kusters, K. and de Sherbinin, A., 2012. *Evidence from the Frontlines of Climate Change: Loss and Damage to Communities Despite Coping and Adaptation* [online]. Report No.9. Bonn: United Nations University Institute for Environment and Human Security (UNU-EHS). Available from: https://collections.unu.edu/eserv/UNU:1847/pdf10584.pdf [Accessed 11 December 2019].

Webb, J.W., 2021. *Exploring Preconditions for Effective Global Responses to Climate Change* (Doctoral dissertation, UCL (University College London)) [online]. Available from: https://discovery.ucl.ac.uk/id/eprint/10120531/8/Jeremy%20Winston%20Webb%20PhD%20Thesis%202021.pdf [Accessed 11 October 2023].

Wei, T. and Liu, Y., 2017. Estimation of global rebound effect caused by energy efficiency improvement. *Energy Economics* [online], *66*, pp. 27–34. Available from: https://www.sciencedirect.com/science/article/pii/S0140988317301949 [Accessed 3 May 2020].

Wessman, C.A., 1992. Spatial scales and global change: Bridging the gap from plots to GCM grid cells. *Annual Review of Ecology and Systematics* [online], *23*(1), pp. 175–200. Available from: https://www.annualreviews.org/doi/pdf/10.1146/annurev.es.23.110192.001135 [Accessed 1 March 2020].

Wetlaufer, G.B., 1996. The limits of integrative bargaining. *The Georgetown Law Journal* [online], *85*(369), p. 369. Available from: https://heinonline.org/HOL/Page?handle=hein.journals/glj85&div=18&g_sent=1&casa_token=&collection=journals [Accessed 8 December 2019].

WHO, May 5, 2023a. WHO chief declares end to COVID-19 as a global health emergency. *United Nations News.* [online]. Available from: WHO chief declares end to COVID-19 as a global health emergency | United Nations [Accessed 13 December 2023].

WHO, 2023b. *Strengthening Health Emergency Prevention, Preparedness, Response and Resilience* [online]. Geneva: World Health Organization. Available from: https://cdn.who.int/media/docs/default-source/emergency-preparedness/who_hepr_wha2023-21051248b.pdf?sfvrsn=a82abdf4_3&download=true [Accessed 13 December 2023].

Wilson, J.D., 2015. Multilateral organisations and the limits to international energy cooperation. *New Political Economy* [online], *20*(1), pp. 85–106. Available from: http://www.tandfonline.com/doi/abs/10.1080/13563467.2013.872611 [Accessed 12 February 2017].

Xu, Y. and Ramanathan, V., 2017. Well below 2 C: Mitigation strategies for avoiding dangerous to catastrophic climate changes. *Proceedings of the National Academy of Sciences* [online], *114*(39), pp. 10315–10323. Available from: https://www.pnas.org/content/pnas/114/39/10315.full.pdf [Accessed 7 December 2019].

Yudkowsky, E., 2008. Cognitive biases potentially affecting judgment of global risks. [Kindle reader, Ipad Mini] In: Bostrom, N. and M.M. Cirkovic (eds.), *Global Catastrophic Risks*. Oxford University Press, Oxford, pp. 91–119.

Zhang, J., Lindsay, R., Steele, M. and Schweiger, A., 2008. What drove the dramatic retreat of arctic sea ice during summer 2007? *Geophysical Research Letters* [online], *35*(11). Available from: https://agupubs.onlinelibrary.wiley.com/doi/epdf/10.1029/2008GL034005 [Accessed 7 December 2019].

List of acronyms

AFOLU	Agriculture, Forestry and Other Land Use
AILAC	Independent Association of Latin America and the Caribbean
ASEAN	Association of Southeast Asian Nations
BASIC	Brazil, South Africa, India and China
BAU	Business as Usual
BECCS	Bioenergy with Carbon Capture and Storage
BOE	Bank of England
BRICS	Brazil, India, China and South Africa
C	Celsius
C40	Cities Climate Leadership Group
CACAM	Central Asia, Caucasus, Albania and Moldova Group
CCNIIC	Climate Change, Nation Interests, International Cooperation Model
CCS	Carbon Capture and Storage
CDR	Carbon Dioxide Removals
CfRN	Coalition for Rainforest Nations
CH_4	Methane
CIA	Central Intelligence Agency (United States)
CMA	Conference of the Parties serving as the meeting of the Parties to the Paris Agreement
CMP	Conference of the Parties serving as the meeting of the Parties to the Kyoto Protocol
CO	Carbon monoxide
CO_2	Carbon dioxide
COP	Conference of Parties to the UNFCCC
CSC	Civil Service College (Singapore)
CSF	Centre for Strategic Futures (Singapore)
EIG	Environmental Integrity Group
ENI	Environment and National Interests Model
EPSRC	Engineering and Physical Sciences Research Council
EU	European Union
EVs	Electric Vehicles
FAO	Food and Agriculture Organisation (United Nations)

G7	Group of 7 major developed economies
G-77	Group of 77
GCF	Green Climate Fund
GDP	Gross Domestic Product
GHG	Greenhouse Gas
GOS	Government Office of Science (United Kingdom)
Gt	Gigatonnes
H_2O	Water
IAEA	International Atomic Energy Agency
IAMs	Integrated Assessment Models
IEGL	Institute for Ethics, Governance and Law (Griffith University)
IMF	International Monetary Fund
INDC	Intended Nationally Determined Contribution
IPCC	Intergovernmental Panel on Climate Change
IRENA	International Renewable Energy Agency
LDCs	Least Developed Countries
LUC	Land Use Change
N_2O	Nitrous oxide
NA	Not Applicable
NAMAs	Nationally Appropriate Mitigation Actions
NAPAs	National Adaptation Programmes of Action
NDCs	Nationally Determined Contributions
NPV	Net Present Value
O_3	Ozone
OC	Organic carbon
OECD	Organisation for Economic Cooperation and Development
OPEC	Organization of the Petroleum Exporting Countries
PA	Paris Agreement
RCPs	Representative Concentration Pathways
REDD+	Reducing Emissions from Deforestation and Degradation in Developing Countries
RFCs	Reasons for Concern
RKRs	Representative Key Risks
SBI	Subsidiary Body for Implementation (UNFCCC)
SBSTA	Subsidiary Body for Scientific and Technological Advice (UNFCCC)
SDGs	Sustainable Development Goals (United Nations)
SIDs	Small Island Developing States
SLCP	Short-Lived Climate Pollutants
SLR	Sea Level Rise
SO_2	Sulphur dioxide
SRM	Solar Radiation Management
SSPs	Shared Socio-economic Pathways
STEaPP	Department of Science, Technology, Engineering and Public Policy (UCL)

TCFD	Taskforce on Climate Related Financial Disclosures
ToC	Theory of Change
UCL	University College London
UK	United Kingdom of Great Britain and Northern Ireland
UN	United Nations
UNDP	United Nations Development Programme
UNEP	United Nations Environment Programme
UNFCCC	United Nations Framework Convention on Climate Change
UNGA	United Nations General Assembly
UNSC	United Nations Security Council
UNSG	United Nations Secretary General
UNTC	United Nations Treaty Collection
US	United States
WTO	World Trade Organisation

Glossary

Whereever possible this book uses existing definitions from the IPCC or widely used and reputable dictionaries, but not at the expense of internal consistency. As such, definitions are modified in many cases to ensure internal consistency between terms and the concepts they represent.

Abrupt irreversible change A large-scale change in the climate system that takes place over a few decades or less, persists (or is anticipated to persist) for at least a few decades, and causes substantial disruptions in human and natural systems. *Source: Collins et al. (2013, p. 1114).*

Actor Individuals or groups with the ability to make decisions and influence the system. *Source: Author.* This is consistent with the definition of actor from the Oxford Dictionary: "A participant in an action or process". The Cambridge Dictionary noted in its examples, that actors can include "individuals, communities, and states". For this book, actors include all types of groups including all types of organisations.

Adaptation The process of adjustment to actual or expected climate and its effects. In human systems, adaptation seeks to moderate or avoid harm or exploit beneficial opportunities. In some natural systems, human intervention may facilitate adjustment to expected climate and its effects. *Source: IPCC (2014b).*

Adverse effects of climate change Changes in the physical environment or biota resulting from climate change which have significant deleterious effects on the composition, resilience or productivity of natural and managed ecosystems or on the operation of socio-economic systems or on human health and welfare. *Source: Article 1, UNFCCC (1992).*

Agreement See international agreement and treaty.

Ambition The level of determination, and action required, to fulfil an agreement and achieve its goals, objectives or purpose. *Source: Generalised from Cameron and DeAngelis (2012).*

Apathetic or hopeless response A lack of interest, ambition or concern for climate change and related issues or despair regarding the possibility of addressing climate change. *Source: Author.*

Approach   A way of dealing with a situation or problem. *Source: Oxford Dictionary.*

Architecture   Structure of an international agreement on climate change. *Source: Modified from Stavins et al. (2014, p. 1054).*

At scale   At the required size to solve the problem. At scale typically refers to handling larger volumes. *Source: PCMag Encyclopaedia (2020).*

Baseline scenarios   Scenarios that are based on the assumption that no mitigation policies or measures will be implemented beyond those that are already in force and/or are legislated or planned to be adopted. Baseline scenarios are not intended to be predictions of the future, but rather counterfactual constructions that can serve to highlight the level of emissions that would occur without further policy effort. Typically, baseline scenarios are then compared to mitigation scenarios that are constructed to meet different goals for greenhouse gas (GHG) emissions, atmospheric concentrations or temperature change. The term baseline scenario is used interchangeably with reference scenario and no policy scenario. In much of the literature the term is also synonymous with the term business-as-usual (BAU) scenario, although the term BAU has fallen out of favour because the idea of business as usual in century-long socio-economic projections is hard to fathom. *Modified from: IPCC (2014b).*

Behaviour   Actions, and inaction, by individuals and households. This includes purchasing choices as a consumer. This can also be thought of as the options taken or not taken by individuals and households. *Source: Author.*

Biological system   A group of lifeforms that interact to form a larger entity with its own defining characteristics. *Source: Author.*

Biological systems   Any system in which organisms play a major role. This includes terrestrial ecosystems, wildfire and marine ecosystems. *Source: Author.*

Black swan   An unforeseeable or unforeseen, low probability but high impact event. *Source: Modified from Taleb 2007 and Oxford Dictionary.*

Business and economic activity   Actions, or inaction, by individuals or groups undertaking productive activities, in many cases driven by a profit motive. *Source: Author.*

Business   Actions, or inaction, by individuals or groups undertaking productive activities, in many cases driven by a profit motive. This includes finance and investment and decisions by business leaders and shareholders. *Source: Author.*

Carbon dioxide removal (CDR)   Carbon Dioxide Removal methods refer to a set of techniques that aim to remove CO_2 directly from the atmosphere by either (1) increasing natural sinks for carbon or (2) using chemical engineering to remove the CO_2, with the intent of reducing the atmospheric CO_2 concentration. CDR methods involve the ocean, land and technical systems, including such methods as iron fertilization, large-scale afforestation and direct capture of CO_2 from the atmosphere using engineered chemical means. Some CDR methods fall under the category of geoengineering, though this may not be the case for others, with the distinction being based on the magnitude, scale and impact of the particular CDR activities. The boundary between CDR and mitigation is not clear and there could be some overlap between the two given current definitions. *Source: IPCC (2014b).*

Catastrophe An event with serious impacts on affected actors. This could include endurable but serious or terminal impacts. *Source: Author adapted from Bostrom and Cirkovic (2011).* Also see global catastrophe.

Climate change stressor Manifestations of climate change in specific systems (for example, rainfall variability, droughts, floods, cyclones and tropical storms, glacial melt, sea-level rise, etc.). This could involve extreme weather-related events and more gradual changes. *Source: Adapted from Warner and van der Geest (2013).*

Climate change A change in the state of the climate that can be identified (e.g., by using statistical tests) by changes in the mean and/or the variability of its properties and that persists for an extended period, typically decades or longer. Climate change may be due to natural internal processes or external forcings such as modulations of the solar cycles, volcanic eruptions and persistent anthropogenic changes in the composition of the atmosphere or in land use. Note that the Framework Convention on Climate Change (UNFCCC), in its Article 1, defines climate change as: 'a change of climate which is attributed directly or indirectly to human activity that alters the composition of the global atmosphere and which is in addition to natural climate variability observed over comparable time periods'. The UNFCCC thus makes a distinction between climate change attributable to human activities altering the atmospheric composition and climate variability attributable to natural causes. See also Detection and Attribution. *Source: IPCC (2014b).*

Climate change as a collective action problem Most GHGs accumulate over time and mix globally, and emissions by any actor (e.g., individual, community, company, country) affect other actors. Effective mitigation will not be achieved if individual actors advance their own interests independently. Cooperative responses, including international cooperation, are therefore required to effectively mitigate GHG emissions and address other climate change issues. The effectiveness of adaptation can be enhanced through complementary actions across levels, including international cooperation. The evidence suggests that outcomes seen as equitable can lead to more effective cooperation. *Source: Adapted from (i.e. the word actor has substituted agent) IPCC (2014a, p. 17).*

Climate feedback An interaction in which a perturbation in one climate quantity causes a change in a second and the change in the second quantity ultimately leads to an additional change in the first. A negative feedback is one in which the initial perturbation is weakened by the changes it causes; a positive feedback is one in which the initial perturbation is enhanced. In the Fifth Assessment Report, a somewhat narrower definition is often used in which the climate quantity that is perturbed is the global mean surface temperature, which in turn causes changes in the global radiation budget. In either case, the initial perturbation can either be externally forced or arise as part of internal variability. *Source: IPCC (2014b).*

Climate hazard The potential occurrence of a weather event or trend with physical impacts that may cause loss of life, injury, or other health impacts, as well as damage and loss to property, infrastructure, livelihoods, service provision, ecosystems and environmental resources. *Source: Adapted from IPCC (2014b)*.

Climate projection   The simulated response of the climate system to a scenario of future emission or concentration of greenhouse gases (GHGs) and aerosols, generally derived using climate models. Climate projections are distinguished from climate predictions by their dependence on the emission/concentration/ radiative forcing scenario used, which is in turn based on assumptions concerning, for example, future socio-economic and technological developments that may or may not be realized. *Source: IPCC (2014b)*.

Climate regime   The set of international, national and sub-national institutions and actors involved in addressing climate change. *Source: Moncel et al. (2011)*. Also see regime complex.

Climate scenarios   Plausible representations of future climate conditions (temperature, precipitation and other climatological phenomena). They can be produced using a variety of approaches including: incremental techniques where particular climatic (or related) elements are increased by plausible amounts; spatial and temporal analogues in which recorded climate regimes that may resemble the future climate are used as example future conditions; other techniques, such as extrapolation and expert judgment; and techniques that use a variety of physical climate and Earth system models, including regional climate models. *Source: Moss et al. (2010, p. 749)*.

Climate science   The study of relatively long-term weather conditions, typically spanning decades to centuries but extending to geological timescales. *Source: Nature (2019)*.

Climate stress signal   Information that both indicates the extent to which climate change is a problem and elicits a response. *Source: Author*.

Climate stressor   Manifestations of climate variability and climate change in specific ecosystems (for example, rainfall variability, droughts, floods, cyclones and tropical storms, glacial melt, sea-level rise, etc.). This could involve extreme weather-related events and more gradual changes. *Source: Warner and van der Geest (2013)*.

Climate system   The totality of the atmosphere, hydrosphere, biosphere and geosphere and their interactions. *Source: Article 1, UNFCCC (1992)*.

Co-benefit   The positive effects that a policy or measure aimed at one objective might have on other objectives, thereby increasing the total benefits for society or the environment. *Source: IPCC 2018b*. Note: Potential co-benefits are transition opportunities.

Collective action problems   Issues for which coordinated responses are likely to have better outcomes for all actors, but where actor interests create incentives to act independently of other actors. *Source: Adapted from IPCC (2014a)*. According to Holzinger (2003), collective action can be divided into two main categories of problem, consisting of: problems of coordination and problems involving conflict. Problems of coordination may be addressed politically (e.g. through institutions), while problems involving conflict may addressed by coalitions and coercive means. Ostrom 1990 set out the conditions under which collective action and the formation of institutions can help address collective action problems.

Competitive response   Actors have expectations of some reward, which drives activities for engaging in technologies or practices that make a positive contribution to the global response. *Source: Author.*

Compliance   A matter of whether and to what extent countries do adhere to the provisions of the accord. The concept of compliance includes implementation, but it is generally broader. Compliance focuses not only on whether implementing measures are in effect, but also on whether there is compliance with the implementing actions. Compliance measures the degree to which the actors whose behaviour is targeted by the agreement (whether they be local government units, corporations, organizations, or individuals) conform to the implementing measures and obligations. *Source: Toth et al. (2001).*

Compliance   The degree to which Parties fulfil their obligations under an agreement. *Source: Author.*

Conceptual framework   A logic for organising related ideas. *Source: Author.*

Conceptual model   A representation of a system including related ideas. *Source: Author.*

Condition   Issues actors are willing to live with. *Source: Adapted from Kingdon (1995).*

Contingency   A provision for a possible event or situation. *Source: Oxford Dictionary.*

Contingent strategies   A whole class of planned actions that are contingent on conditions in the world. *Source: Ostrom (1990, p. 36).*

Cooperation response   International cooperation on other issues leads to a realisation that international cooperation could help limit climate change and its impacts. *Source: Author.*

Cooperative response   Actors work together to address climate change and related impacts. *Source: Author.*

Crisis   A time of intense difficulty, trouble, or danger, and a time when a difficult or important decision must be made. *Source: Oxford Dictionary.*

Cynical response   Action based on self-interest that disregards evidence on climate change and related issues that make negative contributions to the global response to climate change. *Source: Author.*

Defence   The ability to protect against harm, or something used to protect against harm. *Source: Modified by author from Cambridge definition for "defense" i.e. defence spelt using the United States spelling.*

Defensive response   Actors attempt to preserve what they have and limit loss and damage. *Source: Author.*

Disaster   A sudden accident or a natural catastrophe that causes great damage or loss of life. *Source: Oxford Dictionary*

Distribution of impacts   Risks/impacts that disproportionately affect particular groups due to uneven distribution of physical climate change hazards, exposure or vulnerability. *Source: IPCC (2018a).* Also see "reasons for concern" (RFCs).

Economic activity   Production and consumption of goods and services by individuals and groups. *Source: Author.*

Effective   Successful in producing a desired or intended result. *Source: Oxford Dictionary.*

Effectiveness   Measures the degree to which international environmental accords lead to changes of behaviour that help to solve environmental problems, that is the extent to which the commitment has actually influenced behaviour in a way that advances the goals that inspired the commitment. *Source: Toth et al. (2001).*

Emergency response   Emergency response is an urgent reaction to climate change and related impacts or risks. *Source: Author.*

Emissions scenarios   A plausible representation of the future development of emissions of substances that are potentially radiatively active (e.g., greenhouse gases (GHGs), aerosols) based on a coherent and internally consistent set of assumptions about driving forces (such as demographic and socio-economic development, technological change, energy and land use) and their key relationships. Concentration scenarios, derived from emission scenarios, are used as input to a climate model to compute climate projections. *Source: IPCC (2014b).*

Emissions   The release of greenhouse gases and/or their precursors into the atmosphere over a specified area and period of time. *Source: Article 1, UNFCCC (1992).*

Emergency   A serious, unexpected, and often dangerous situation requiring immediate action. *Source: Oxford Dictionary.*

Enforcement   The actions taken once violations occur. It is customarily associated with the availability of formal dispute settlement procedures and with penalties, sanctions, or other coercive measures to induce compliance with obligations. Enforcement is part of the compliance process. *Source: Toth et al. (2001).*

Enlightened response   Enlightened responses come about from awareness that creates change. Enlightened responses can be driven by awareness of climate change and related impacts and risks as well as awareness of other things not directly related to climate change, for example the value of international cooperation. *Source: Author.*

Environment   Physical surroundings. *Source: UN (2012, p. 16).*

Environmental scenarios   Analysis of the potential impact of a particular climate scenario requires environmental scenarios of ecological and physical conditions at greater detail than is included in climate models. These scenarios focus on changes in environmental conditions other than climate that may occur regardless of climate change. Such factors include water availability and quality at basin levels (including human uses), sea level rise incorporating geological and climate factors, characteristics of land cover and use, and local atmospheric and other conditions affecting air quality. *Source: Moss et al. (2010, p. 749).*

Explore   Travel through (an unfamiliar area) in order to learn about it. *Source: Oxford Dictionary.*

Exposure   The presence of people, livelihoods, species or ecosystems, environmental functions, services, and resources, infrastructure, or economic, social, or cultural assets in places and settings that could be adversely affected. *Source: IPCC (2014b).*

Extreme weather events   Risks/impacts to human health, livelihoods, assets and ecosystems from extreme weather events such as heat waves, heavy rain, drought and associated wildfires, and coastal flooding. *Source: IPCC (2018a).* Also see "reasons for concern" (RFCs).

Failure scenario   In the context of this book, a failure scenario is: a possible set of conditions or sequence of events that either result in violation of the UNFCCC objective or contribute towards violation of the UNFCCC objective. *Source: Author.*

Failure   In the context of this book failure: is a violation of the UNFCCC objective. This means not stabilising atmospheric greenhouse gas concentrations; or ecosystems unable to adapt naturally to anthropogenic climate change; or food production threatened by anthropogenic climate change; or economic development not able to proceed sustainably due to anthropogenic climate change. *Source: Author.*

Finance   A supply of money. *Source: Cambridge Dictionary.*

Flexibility   Adjustable institutional arrangements. *Adapted from Stavins et al. (2014, p. 1010).*

Food security   A state that prevails when people have secure access to sufficient amounts of safe and nutritious food for normal growth, development, and an active and healthy life. *Source: IPCC (2014b).*

Foresight   Methodologies and approaches that explore the future and alternative actions taking into account complexity and uncertainty. *Source: Modified from UNDP (2014, p. 5).*

Framework convention   A legally binding treaty (Treaties) of international law that establishes broad commitments for its parties and a general system of governance, while leaving more detailed rules and the setting of specific targets either to subsequent agreements between the parties, usually referred to as protocols, or to national legislation. *Source: OPIL (2011).* Generally speaking: a framework convention serves as an umbrella document which lays down the principles, objectives and the rules of governance of the treaty regime. *Source: Adapted from UNECE (2011).*

Geoengineering   Geoengineering refers to a broad set of methods and technologies that aim to deliberately alter the climate system in order to alleviate the impacts of climate change. Most, but not all, methods seek to either (1) reduce the amount of absorbed solar energy in the climate system (Solar Radiation Management) or (2) increase net carbon sinks from the atmosphere at a scale sufficiently large to alter climate (Carbon Dioxide Removal). Scale and intent are of central importance. Two key characteristics of geoengineering methods of particular concern are that they use or affect the climate system (e.g., atmosphere, land or ocean) globally or regionally and/or could have substantive unintended effects that cross national boundaries. Geoengineering is different from weather modification and ecological engineering, but the boundary can be fuzzy. *Source: IPCC (2014b).*

Global aggregate impacts   Global monetary damage, global-scale degradation and loss of ecosystems and biodiversity. *Source: IPCC (2018a)*. Also see "reasons for concern" (RFCs).

Global catastrophe   An event with serious impacts on actors around the world. *Source: Modified from Bostrom and Cirkovic (2008)*.

Global governance   The totality of institutions, policies, norms, procedures and initiatives through which States and their citizens try to bring more predictability, stability and order to their responses to transnational challenges. *Source: CDP (2014, p. vi)*.

Global problems   Issues that are either, global in scale, due to interconnections around the world, or are localised but common to people in many countries. *Source: Adapted from Tanter (2008)*.

Global response to climate change   Working definition: All human actions and inaction, influencing achievement of the UNFCCC objective. This working definition reflects how the global response to climate change is measured. Note: it would be semantically correct if the global response was defined as: All reactions to climate change; or, all human actions and inaction, where climate change was a factor in the decision to act or not. However, these definitions do not reflect how the global response to climate change is measured in practice, for example emissions inventories or the emissions gap includes all GHG emissions regardless of whether decisions that led to the emissions considered climate change or not. *Source: Author.*

Global response problem   An issue that requires coherent actions by actors around the world for the issue to be resolved with minimal negative impacts. *Source: Author.*

Global warming   Global warming refers to the gradual increase, observed or projected, in global surface temperature, as one of the consequences of radiative forcing caused by anthropogenic emissions. *Source: IPCC (2014b)*.

Global   Relating to the whole world; worldwide. *Source: Oxford Dictionary (2019)*.

Governance   The process of decision making on problems affecting groups of people. These processes may include formal or informal processes, formal or informal institutions, government, civil society, business, communities or individuals. *Source: Author.*

Government   Political body within a state which holds executive power to establish the general policy of the state and enforce the law. *Source: UNTERM (2019a)*.

Greenhouse gas emissions   See Emissions.

Greenhouse gas removals   The extraction of chemicals that drive global warming and climate change, from the atmosphere. *Source: Author.*

Greenhouse gases   Those gaseous constituents of the atmosphere, both natural and anthropogenic, that absorb and re-emit infrared radiation. *Source: UNFCCC Article 1.*

Greenhouse gases   Those gaseous constituents of the atmosphere, both natural and anthropogenic, that absorb and emit radiation at specific wavelengths within the spectrum of terrestrial radiation emitted by the Earth's surface, the atmosphere itself and by clouds. This property causes the greenhouse effect. Water vapour (H_2O), carbon dioxide (CO_2), nitrous oxide (N_2O), methane (CH_4) and ozone (O_3) are the primary GHGs in the Earth's atmosphere. Moreover, there are a number of entirely human-made GHGs in the atmosphere, such as the halocarbons and other chlorine- and bromine-containing substances, dealt with under the Montreal Protocol. *Source: IPCC (2018b)*.

Hazard   The potential occurrence of a natural or human-induced physical event or trend or physical impact that may cause loss of life, injury, or other health impacts, as well as damage and loss to property, infrastructure, livelihoods, service provision, ecosystems, and environmental resources. In this report, the term hazard usually refers to climate-related physical events or trends or their physical impacts. *Source: Oppenheimer et al. (2014)*.

Human security   A condition that is met when the vital core of human lives is protected, and when people have the freedom and capacity to live with dignity. In the context of climate change, the vital core of human lives includes the universal and culturally specific, material and nonmaterial elements necessary for people to act on behalf of their interests and to live with dignity. *Source: IPCC (2018b)*.

Human systems   Any system in which human organizations and institutions play a major role. *Source: IPCC (2018b)*. This includes food production, livelihoods, health and economics.

Hypocritical response   Human actions or inaction that affect the climate, by an actor that has expressed concern about the very same actions or inaction. *Source: Author*.

Impact response   Impact response is a reaction to climate change and related effects on physical, biological or human systems. An impact response can also be defined as actions, or inaction, where climate change and related effects on physical, biological or human systems are a factor in the decision to act or not. *Source: Author*.

Impacts   Effects on natural and human systems. Source: Author. The IPCC defines Impacts as consequences or outcomes, specifically: The consequences of realized risks on natural and human systems, where risks result from the interactions of climate-related hazards (including extreme weather and climate events), exposure, and vulnerability. Impacts generally refer to effects on lives; livelihoods; health and well-being; ecosystems and species; economic, social and cultural assets; services (including ecosystem services); and infrastructure. Impacts may be referred to as consequences or outcomes and can be adverse or beneficial. *Source: IPCC (2018b)*.

Implementation   The actions (legislation or regulations, judicial decrees, or other actions) that governments take to translate international accords into domestic law and policy. It includes those events and activities that occur after authoritative public policy directives have been issued, such as the effort to administer the substantive impacts on people and events. It is important to distinguish between the legal implementation of international commitments (in national law) and the effective implementation (measures that induce changes in the behaviour of target groups). *Source: Toth et al. (2001).*

Institutions   Systems of established and embedded social rules that structure social interactions. *Source: Modified from Hodgson (2006), Duverger (1972) in Archer (2015).* These include formal and informal rules and structures. Ostrom defines institutions as the sets of working rules that are used to determine who is eligible to make decisions in some arena, what actions are allowed or constrained, what aggregation rules will be used, what procedures must be followed, what information must or must not be provided, and what payoffs will be assigned to individuals dependent on their actions. *Source: Ostrom (1990, p. 51).*

Interests   The things that an actor cares about. *Source: Author.*

Intergovernmental   Something between two or more sovereign states and their representatives. Synonymous with interstate. *Source: Modified from Archer (2015, p. 1).*

International agreement   An agreement between actors from more than one state or between an international organisation and other actors. In many cases international agreements refer to agreements between states or states and international organisations (see Treaty), but this book also includes agreements between non-state actors from different states or between non-state actors and international organisations. *Source: Author.*

International cooperation on climate change   Actors from different states acting together to address climate change. This includes state actors and non-state actors. *Source: Author.*

International cooperation   Actions or resources exchanged between actors from different countries, voluntarily and according to their own interests and strategies. *Source: Modified from Vázquez (2011).*

International institutions   Systems of established and embedded social rules that structure social interactions between actors from different sovereign states. This includes international organisations and international conventions. *Source: Adapted from the definition of "institutions".*

International organisation   Any entity established by actors from different states. This term commonly refers to intergovernmental organisations established by state actors, but in this book, it also refers to international non-government organisations and international corporations (i.e. multinational corporations) reflecting the fact that these entities are actors that can each engage in, and influence, international cooperation. *Source: Author.*

International regime   Sets of implicit or explicit principles, norms, rules, and decision-making procedures around which actors' expectations converge in a given area of international relations. *Source: Krasner (1982, p. 2) in Archer (2015, p. 2).*

International regime   The rules and norms that guide interactions between actors from different states. This includes state actors and non-state actors. *Source: Author.*

International   Something involving more than one sovereign state including their governments, people, economy or environments, or something beyond the jurisdiction of sovereign states. *Source: Modified from Archer (2015).* Also see intergovernmental, interstate and transnational.

Interstate Between governments. Synonymous with intergovernmental. *Source: Modified from Archer (2015).*

Intervention The act or fact of becoming involved intentionally in a difficult situation. *Source: Cambridge Dictionary.*

Irreversibility A perturbed state of a dynamical system is defined as irreversible on a given timescale, if the recovery timescale from this state due to natural processes is substantially longer than the time it takes for the system to reach this perturbed state. In the context of this report, the time scale of interest is centennial to millennial. See also Tipping point. *Source: IPCC (2014b).*

Issue A vital concern or unsettled problem. *Source: adapted from Merriam-Webster Dictionary.*

Key factors Characteristics, groupings and entities that best describe the situation and have a meaningful influence on events. *Source: Author.* Also see other factors.

Key risks have potentially severe adverse consequences for humans and social-ecological systems resulting from the interaction of climate related hazards with vulnerabilities of societies and systems exposed. *Source: IPCC Online Glossary.*

Land use The total of arrangements, activities and inputs undertaken in a certain land cover type (a set of human actions). The term land use is also used in the sense of the social and economic purposes for which land is managed (e.g., grazing, timber extraction, conservation and city dwelling). In national greenhouse gas inventories, land use is classified according to the IPCC land use categories of forest land, cropland, grassland, wetland, settlements, other. *Source: IPCC (2018b).* See also Land-use change (LUC).

Land-use change (LUC) Land-use change involves a change from one land use category to another. *Source: IPCC (2018b).*

Large-scale singular events Relatively large, abrupt and sometimes irreversible changes in systems that are caused by global warming. Examples include disintegration of the Greenland and Antarctic ice sheets. *Source: IPCC (2018a).* Also see "reasons for concern" (RFCs).

Legitimacy The belief by sovereign states and others, that the institution is credible with some degree of authority. *Source: Adapted from Stavins et al. (2014).*

Liability risk The potential for legal compensation claims based on attributable climate change impacts. *Source: Adapted from BOE (2015).*

Liability risk The potential for legal compensation claims related to climate change. *Source: Adapted from Carney (2015).*

Livelihood The resources used and the activities undertaken in order to live. Livelihoods are usually determined by the entitlements and assets to which people have access. Such assets can be categorised as human, social, natural, physical or financial. *Source: IPCC (2018b).*

Loss and damage Negative effects of climate variability and climate change that people have not been able to cope with or adapt to. *Source: Warner et al. (2012, p. 20).*

Measurable, reportable and verifiable (MRV) A process/concept that potentially supports greater transparency in the climate change regime. *Source: UNFCCC Glossary.*

Mitigation scenarios Scenarios that are constructed to meet different goals for greenhouse gas (GHG) emissions, atmospheric concentrations or temperature change. *Modified from: IPCC (2014b).*

Mitigation A human intervention to reduce the sources or enhance the sinks of greenhouse gases (GHGs). This report also assesses human interventions to reduce the sources of other substances which may contribute directly or indirectly to limiting climate change, including, for example, the reduction of particulate matter emissions that can directly alter the radiation balance (e.g., black carbon) or measures that control emissions of carbon monoxide, nitrogen oxides, Volatile Organic Compounds and other pollutants that can alter the concentration of tropospheric ozone which has an indirect effect on the climate. *Source: IPCC (2014b).*

National interest Things that the State, including domestic coalitions of actors, care about. *Source: Adapted from definition of interests based on the work of Putnam (1988).*

Non-response with negative incidental contributions Human actions, and inaction, that affect the climate, where decisions to act were made for reasons other than climate change or related impacts. In this case the incidental contributions to the global response to climate change are unhelpful, for example reducing climate resilience and increasing atmospheric concentrations of greenhouse gases. *Source: Author.*

Non-response with positive incidental contributions Human actions, and inaction, that affect the climate, where decisions to act were made for reasons other than climate change or related impacts. In this case the incidental contributions to the global response to climate change happen to be helpful, for example increasing climate resilience and limiting atmospheric concentrations of greenhouse gases. *Source: Author.*

Option Some action or inaction that can be decided upon by an actor or set of actors. *Source: Author.*

Organisation A unique framework of authority within which a person or persons act, or are designated to act, towards some purpose. *Source: OECD Glossary of Statistical Terms.*

Other factors Other things that can influence the global response to climate change. *Source: Author.*

Other scenario In the context of this book, an "other scenario" is: a possible set of conditions or sequence of events that highlight some issue or dynamic related to the global response to climate change without necessarily contributing to fulfilling or violating the UNFCCC objective; or, regards an extreme set of conditions or events that affect the climate or the global response to climate change. *Source: Author.*

Participation (official) Being a Party to an agreement. *Source: Author.* Furthermore: Participation in an international climate agreement might refer to the number of parties, geographical coverage, or the share of global GHG emissions covered. *Source: Stavins et al. (2014, p. 1010).*

Participation (unofficial) Conforming to the objective or purpose of an agreement between sovereign states, without being a party to the agreement. This might include sovereign states that are not parties or non-state actors that cannot be parties to the agreement. *Source: Author.*

Physical risk Potential physical impact of climate change. *Source: Adapted from Carney (2015).*

Physical stress test An assessment with the purpose of understanding how something responds to the physical impact of climate change, informing risk mitigation and decision making. *Source: Author.*

Physical systems Any system in which physical processes play a major role. This includes glaciers snow, ice, permafrost, rivers lakes, floods, drought, coastal erosion, and sea level effects. *Source: Author.*

Plain English Clear and unambiguous language, without the use of unnecessary technical terms. *Source: Oxford Dictionary.*

Policy A course or principle of action adopted or proposed by an organization or individual. *Source: Oxford Dictionary.*

Policy Positions and interventions. *Source: Author.*

Political will and policy The ambition level of government leaders, and others in government when it comes to positions on climate change, and the interventions they make that influence fulfilment of the UNFCCC objective. *Source: Author.*

Political will The ambition level of government leaders, and others in government, to act on climate change and related options. *Source: Author.*

Power The capacity or ability to direct or influence the behaviour of others, including institutions. *Source: Modified from Oxford Dictionary.*

Practice response Actors expect there are human, ecological or environmental system knowledge and processes that can be used to solve the problem. *Source: Author.*

Practices Methods used when attempting to achieve something, that rely on properties of human, ecological or environmental systems. *Source: Author.*

Precondition A state that must be fulfilled before other things can happen or be done. *Source: Modified from the Oxford Dictionary.*

Preconditions for an effective global response to climate change The system characteristics necessary for achieving the UNFCCC objective under a given set of assumptions. *Source: Author.*

Problem Something that actors find undesirable and want changed. *Source: Adapted from Kingdon (1995).*

Protocol An additional legal instrument that complements and add to a treaty. A protocol is 'optional' because it is not automatically binding on States that have already ratified the original treaty; States must independently ratify or accede to a protocol. *Source: UNICEF (2017, p. 2).*

Public policy A course or principle of action adopted or proposed by an organization or individual that has a bearing on wider societal wellbeing. *Source: Adapted from definitions of "policy" and "public policy" from the Oxford Dictionary.*

Rapid onset event   A single, discrete hazards that occurs in a matter of days or even hours. *Source: Adapted from UNFCCC (2012) and Siegele (2012).*

Ratification (of a treaty)   the international act whereby a state indicates its consent to be bound to a treaty if the parties intended to show their consent by such an act. In the case of bilateral treaties, ratification is usually accomplished by exchanging the requisite instruments, while in the case of multilateral treaties the usual procedure is for the depositary to collect the ratifications of all states, keeping all parties informed of the situation. The institution of ratification grants states the necessary time-frame to seek the required approval for the treaty on the domestic level and to enact the necessary legislation to give domestic effect to that treaty. *Source: UNTC (2019).*

Reasons for concern   Unique and threatened systems; extreme weather events; distribution of impacts; global aggregate impacts; and, large-scale singular events. *Source: IPCC (2018a).*

Regime complex   A loosely coupled set of specific regimes. *Source: Keohane and Victor (2011, p. 7).* Regime complexes are marked by connections between the specific and relatively narrow regimes but the absence of an overall architecture or hierarchy that structures the whole set. *Source: Keohane and Victor (2011, p. 8).* Also see regime and climate regime.

Regime   A system or way of doing things, including interacting institutions at international, national and subnational levels. *Source: Adapted from Oxford Dictionary and definition of climate regime.*

Representative Concentration Pathways (RCPs)   Scenarios that include time series of emissions and concentrations of the full suite of greenhouse gases (GHGs) and aerosols and chemically active gases, as well as land use/land cover. *Source: Moss et al. (2008).* The word representative signifies that each RCP provides only one of many possible scenarios that would lead to the specific radiative forcing characteristics. The term pathway emphasizes that not only the long-term concentration levels are of interest, but also the trajectory taken over time to reach that outcome. *Source: Moss et al. (2010).* Representative Concentration Pathways (RCPs) Scenarios that include time series of emissions and concentrations of the full suite of greenhouse gases (GHGs) and aerosols and chemically active gases, as well as land use/land cover (Moss et al., 2008). The word representative signifies that each RCP provides only one of many possible scenarios that would lead to the specific radiative forcing characteristics. The term pathway emphasizes that not only the long-term concentration levels are of interest, but also the trajectory taken over time to reach that outcome. *Source: Moss et al. (2010).*

Representative Key Risks (RKRs)   Representative, thematic clusters of key risks. *Source: IPCC Online Glossary.* Also see "key risks".

Renewable resources   Stocks that are capable, under favourable conditions, of producing a maximum quantity of a flow variable without harming the stock or the resource system itself. *Source: modified from Ostrom (1990, p. 30).*

Reservoir   A component or components of the climate system where a greenhouse gas or a precursor of a greenhouse gas is stored. *Source: Article 1, UNFCCC (1992).*

Response attitudes and inclinations Ways of thinking about climate change and related issues, including predispositions influencing options considered and actions taken. *Source: Author.*

Response to climate change Any action and inaction affecting the achievement of the UNFCCC objective. *Source: Author.* Note this definition reflects the way the global response to climate change is assessed. Also see "global response to climate change".

Responsiveness The extent to which human actions, or inactions, are timely and sufficiently scaled to fulfil the UNFCCC objective. *Source: Author.*

Risk management The plans, actions or policies to reduce the likelihood and/or consequences of risks or to respond to consequences. *Source: IPCC (2014b).*

Risk response Risk response is a reaction to the potential for climate change and related effects on physical, biological or human systems. A risk response can also be defined as actions, or inaction, where the potential for climate change and related effects on physical, biological or human systems are a factor in the decision to act or not. *Source: Author.*

Risk to low-lying coastal socio-ecological systems (RKR-A) Risks to ecosystem services, people, livelihoods and key infrastructure in low-lying coastal areas, and associated with a wide range of hazards, including sea level changes, ocean warming and acidification, weather extremes (storms, cyclones), sea ice loss, etc. *Source: O'Neill et al. (2022).*

Risk to terrestrial and ocean ecosystems (RKR-B) Transformation of terrestrial and ocean/coastal ecosystems, including change in structure and/or functioning, and/or loss of biodiversity. *Source: O'Neill et al. (2022).*

Risk to living standards (RKR-D) Economic impacts across scales, including impacts on gross domestic product (GDP), poverty and livelihoods, as well as the exacerbating effects of impacts on socioeconomic inequality between and within countries. *Source: O'Neill et al. (2022).*

Risk to human health (RKR-E) Human mortality and morbidity, including heat-related impacts and vector-borne and waterborne diseases. *Source: O'Neill et al. (2022).*

Risk to food security (RKR-F) Food insecurity and the breakdown of food systems due to climate change effects on land or ocean resources. *Source: O'Neill et al. (2022).*

Risk to water security (RKR-G) Risk from water-related hazards (floods and droughts) and water quality deterioration. Focus on water scarcity, water-related disasters and risk to indigenous and traditional cultures and ways of life. *Source: O'Neill et al. (2022).*

Risks associated with critical physical infrastructure, networks and services (RKR-C) Systemic risks due to extreme events leading to the breakdown of physical infrastructure and networks providing critical goods and services. *Source: O'Neill et al. (2022).*

Risks to peace and to human mobility (RKR-H) Risks to peace within and among societies from armed conflict as well as risks to low-agency human mobility within and across state borders, including the potential for involuntarily immobile populations. *Source: O'Neill et al. (2022).*

Risk transfer   The practice of formally or informally shifting the risk of financial consequences for particular negative events from one party to another. *Source: IPCC (2014d).*

Risk   The potential for consequences where something of value is at stake and where the outcome is uncertain, recognizing the diversity of values. Risk is often represented as probability or likelihood of occurrence of hazardous events or trends multiplied by the impacts if these events or trends occur. In this report, the term risk is often used to refer to the potential, when the outcome is uncertain, for adverse consequences on lives, livelihoods, health, ecosystems and species, economic, social and cultural assets, services (including environmental services) and infrastructure. *Source: IPCC (2014b).*

Rule of law   Formal laws and working rules are closely aligned and that enforcers are held accountable to the rules as well as others. *Source: Ostrom (1990, p. 51).*

Scale   The relative size or extent of something. *Source: Oxford Dictionary.* Also see "at scale".

Scenario   A possible set of conditions, situations or sequence of events. *Source: Modified from Oxford Dictionary and The Free Dictionary.* Also see baseline scenario, climate scenarios, emissions scenario, environmental scenario, mitigation scenario, and vulnerability scenario.

Science   The observation, knowledge and description of consistency. *Source: The author.*

Securitisation   The process by which an issue comes to be represented as not only a political problem but as an existential threat to a valued reference object. *Source: Moncel and van Asselt (2012, p. 168).*

Serendipity   The occurrence and development of events by chance in a happy or beneficial way. *Source: Oxford Dictionary.* In the context of climate change, serendipity refers to the occurrence and development of events that by chance, rather than intent, contribute towards fulfilling the UNFCCC objective.

Shared Socio-Economic Pathways (SSPs)   Reference pathways describing plausible alternative trends in the evolution of society and ecosystems over a century timescale, in the absence of climate change or climate policies. *Source: O'Neill et al. (2014).*

Shock   A sudden stressful event. *Source: Author.*

Signal   Information about a situation, system conditions, or phenomena that influences decisions. *Source: Author.*

Sink   Any process, activity or mechanism which removes a greenhouse gas, an aerosol or a precursor of a greenhouse gas from the atmosphere. *Source: Article 1, UNFCCC (1992).*

Slow onset event   Hazardous conditions that evolve gradually from incremental changes occurring over many years or from an increased frequency or intensity of recurring events. Slow onset events include sea level rise, increasing temperatures, ocean acidification, glacial retreat and related impacts, salinization, land and forest degradation, loss of biodiversity and desertification. *Source: Adapted from UNFCCC (2012), Siegele (2012) and IPCC (2012).*

Social change and behaviour   The interests of people, as individuals, households and communities, and the actions, or inactions, these people might individually or collectively take. *Source: Author.*

Social change   Social change refers to a shift in common values, norms and expectations that influence the behaviours of individuals, households, groups and communities, businesses and governments within a society. In short, it's a shift in the way individuals and groups live together and behave. *Source: Author.*

Social contract   An implicit agreement among the members of a society to cooperate for social benefits, for example by sacrificing some individual freedom for state protection. *Source: Oxford Dictionary.*

Solar radiation management   See solar radiation modification.

Solar radiation modification (SRM)   The intentional modification of the Earth's shortwave radiative budget with the aim of reducing warming. Artificial injection of stratospheric aerosols, marine cloud brightening and land surface albedo modification are examples of proposed SRM methods. SRM does not fall within the definitions of mitigation and adaptation. Note that in the literature SRM is also referred to as solar radiation management or albedo enhancement. *Source: IPCC (2018b).*

Source   Any process or activity which releases a greenhouse gas, an aerosol or a precursor of a greenhouse gas into the atmosphere. *Source: Article 1, UNFCCC (1992).*

Strategy   A plan designed to achieve a specific goal or set of objectives under conditions of uncertainty. *Modified from Oxford Dictionary and Cambridge Dictionary.*

Stress test   An assessment with the purpose of understanding how something responds under difficult conditions, informing risk management and decision making. *Source: Author.*

Stress   Pressure or tension experienced by something. *Source: Author.*

Stressors   Events and trends, often not climate-related, that have an important effect on the system exposed and can increase vulnerability to climate-related risk. *Source: Oppenheimer et al. (2014, p. 1048).*

Stringency   The formal tightness or substantive ambition of an agreement. Formal tightness means the agreement is legally binding, highly precise, with strong compliance mechanisms. Substantive ambition means the agreement is comprehensive in scope and has obligations sufficient to fulfil the goals, objectives, or purpose of the agreement. Note: Agreements with formal tightness or substantive ambition require participation and compliance to be effective. *Source: Modified from Hofmann (2019).*

Success scenario   A possible set of conditions or sequence of events that either result in fulfilling the UNFCCC objective or contribute towards fulfilling the UNFCCC objective. *Source: Author.*

Success   Fulfilling the UNFCCC objective which consists of limiting concentrations of greenhouse gases in the atmosphere to levels that allow ecosystem to adapt naturally, food production not to be threatened and economic development to proceed sustainably. *Source: Author.*

Technological response Actors expect there are physical, chemical, or biological properties, related objects, knowledge and processes that can be used to solve the problem. *Source: Author.*

Technology Methods used when attempting to achieve something, that rely on physical, chemical, or biological properties. This includes objects used for a particular purpose, the processes used, and information related to these things. *Source: Author.* Note: This is consistent with the Oxford Dictionary which defines technology as "the application of scientific knowledge for practical purposes". Also see practices.

Technology and practices Methods including objects, activities and rules, used when attempting to achieve something. *Source: Author.*

Theory of change The description of a sequence of events that is expected to lead to a particular outcome. *Source: Davies (2012) in Vogel (2012).*

Tipping point A level of change in system properties beyond which a system reorganizes, often abruptly, and does not return to the initial state even if the drivers of the change are abated. For the climate system, it refers to a critical threshold when global or regional climate changes from one stable state to another stable state. *Source: IPCC (2018b).*

Transformation A change in the fundamental attributes of natural and human systems. *Source: IPCC (2018b).*

Transformation risk Potential disruption due timely implementation of climate resilient low greenhouse gas emissions technologies and practices at a scale sufficient to limit climate change and related impacts. *Source: Author.*

Transition opportunity See co-benefits.

Transition risk Potential disruption due to adjustment towards a climate resilient low greenhouse gas emissions economy. *Source: Author adapted from Carney (2015).* Also see adverse side effects.

Transnational Something between sovereign states not involving governments. *Source: Modified from Archer (2015).*

Treaty A generic term embracing all instruments binding under international law, regardless of their formal designation, concluded between two or more international juridical persons. Thus, treaties may be concluded between: States; international organizations with treaty-making capacity and States; or international organizations with treaty-making capacity. *Source: UNTERM (2019b).*

Unique and threatened systems Ecological and human systems that have restricted geographic ranges constrained by climate-related conditions and have high endemism or other distinctive properties. Examples include coral reefs, the Arctic and its indigenous people, mountain glaciers and biodiversity hotspots. *Source: IPCC (2018a).* Also see "reasons for concern".

Viable Ability to work successfully. *Source: Oxford Dictionary.*

Vulnerability scenarios Scenarios of factors affecting vulnerability, such as demographic, economic, policy, cultural and institutional characteristics are needed for different types of impact modelling and research. This information is crucial for evaluating the potential of humankind to be affected by changes in climate, as well as for examining how different types of economic growth and social change affect vulnerability and the capacity to adapt to potential impacts. Although some of these factors can be modelled and applied at regional or national scales, for the most part data at finer spatial resolution are required. *Source: Moss et al. (2010, p. 749).*

Vulnerability   The propensity or predisposition to be adversely affected. Vulnerability encompasses a variety of concepts and elements including sensitivity or susceptibility to harm and lack of capacity to cope and adapt. *Source: IPCC (2018b).*

Water security   The capacity of a population to safeguard sustainable access to adequate quantities of acceptable quality water for sustaining livelihoods, human well-being, and socio-economic development, for ensuring protection against water-borne pollution and water-related disasters, and for preserving ecosystems in a climate of peace and political stability. *Source: UNWater (2019).*

Appendix A

Physical signals from IPCC assessments

Table A.1 Physical signals from IPCC assessments based on representative quotes

Year	Climate science	Risks	Impacts
1992	"We are certain of the following: there is a natural greenhouse effect which already keeps the Earth warmer than it would otherwise be; Emissions resulting from human activities are substantially increasing the atmospheric concentrations of the greenhouse gases... These increases will enhance the greenhouse effect, resulting on average in an additional warming of the Earth's surface. The main greenhouse gas, water vapour, will increase in response to global warming and further enhance it." (IPCC 1990, p. 52)	Potential future impacts identified in the areas of: (1) Agriculture and forestry; (2) Natural terrestrial ecosystems; (3) Hydrology and water resources; (4) Human settlements, energy, transport, and industrial sectors, human health and air quality; (5) Oceans and coastal zones; and (6) Seasonal snow cover, ice and permafrost.	No impacts identified
1995	"Global mean surface temperature has increased by between about 0.3 and 0.6°C since the late 19th century, a change that is unlikely to be entirely natural in origin. The balance of evidence... ... suggests a discernible human influence on global climate." (IPCC 1995, p. 5).	"Whereas many regions are likely to experience the adverse effects of climate change — some of which are potentially irreversible — some effects of climate change are likely to be beneficial. Hence, different segments of society can expect to confront a variety of changes and the need to adapt to them." (IPCC 1995, p. 6).	The Synthesis report does not identify any attributable impacts bu stated "Unambiguous detection of climate-induced chang in most ecological and social systems will pro extremely difficult in t coming decades." (IPC 1995, p. 6).

01	"The Earth's climate system has demonstrably changed on both global and regional scales since the pre-industrial era, with some of these changes attributable to human activities." (IPCC 2001, p. 4).	Five reasons for concern were identified. "Projected climate change will have beneficial and adverse effects on both environmental and socioeconomic systems, but the larger the changes and the rate of change in climate, the more the adverse effects predominate" (IPCC 2001, p. 9)	"Observed changes in regional climate have affected many physical and biological systems, and there are preliminary indications that social and economic systems have been affected." (IPCC 2001, p. 6)
07	"Warming of the climate system is unequivocal" (IPCC 2007, p. 2) "Most of the observed increase in global average temperatures since the mid-20th century is very likely due to the observed increase in anthropogenic GHG concentrations." (IPCC 2007, p. 5)	Five reasons for concern were updated. It was also noted that "Anthropogenic warming could lead to some impacts that are abrupt or irreversible, depending upon the rate and magnitude of the climate change." (IPCC 2007, p. 13).	"There is medium confidence that other effects of regional climate change on natural and human environments are emerging, although many are difficult to discern due to adaptation and non-climatic drivers." (IPCC 2007, p. 3)
14	"Warming of the climate system is unequivocal" (IPCC 2014a, p. 2) Anthropogenic greenhouse gas emissions and other anthropogenic drivers "are extremely likely to have been the dominant cause of the observed warming since the mid-20th century." (IPCC 2014a, p. 4)	Five reasons for concern updated Risks to physical systems, biological systems and human systems identified Eight key risks identified "Climate change will amplify existing risks and create new risks for natural and human systems." (IPCC 2014a, p. 13)	"Evidence of observed climate change impacts is strongest and most comprehensive for natural systems." and "Some impacts on human systems have also been attributed to climate change, with a major or minor contribution of climate change distinguishable from other influences." (IPCC 2014a, p. 6).
23	"Human activities, principally through emissions of greenhouse gases, have unequivocally caused global warming, with global surface temperature reaching 1.1°C above 1850–1900 in 2011–2020." (IPCC 2023, p. 4)	"…projected long-term impacts are up to multiple times higher than currently observed (high confidence)… Climatic and non-climatic risks will increasingly interact, creating compound and cascading risks that are more complex and difficult to manage (high confidence)." (IPCC 2023 p. 14)	"…widespread adverse impacts and related losses and damages to nature and people (high confidence). Vulnerable communities …disproportionately affected (high confidence)." (IPCC 2023, p. 5)

rce: Compiled by the author based on an assessment of physical signals in IPCC assessment report summaries policy makers.

Appendix B

Environment and national interests model

The CCNIIC model presented in Section 3.4.2 can be adapted to include more detail on climate, for example including forcing agents and feedbacks as discussed in Section 2.2. Socio-economic processes and physical risks can be framed in the context of national interests and the interests of various actors such as economic units, and other legal entities including government actors, forming the Environment and National Interests (ENI) model (Figure B.1).

Within the environment part of the ENI model, forcing agents, climate feedbacks, climate and other physical changes are highlighted (Figure B.1). Forcing agents include greenhouse gases as well as other forcing agents. Forcing agents are shown to influence climate directly as well as indirectly through feedbacks

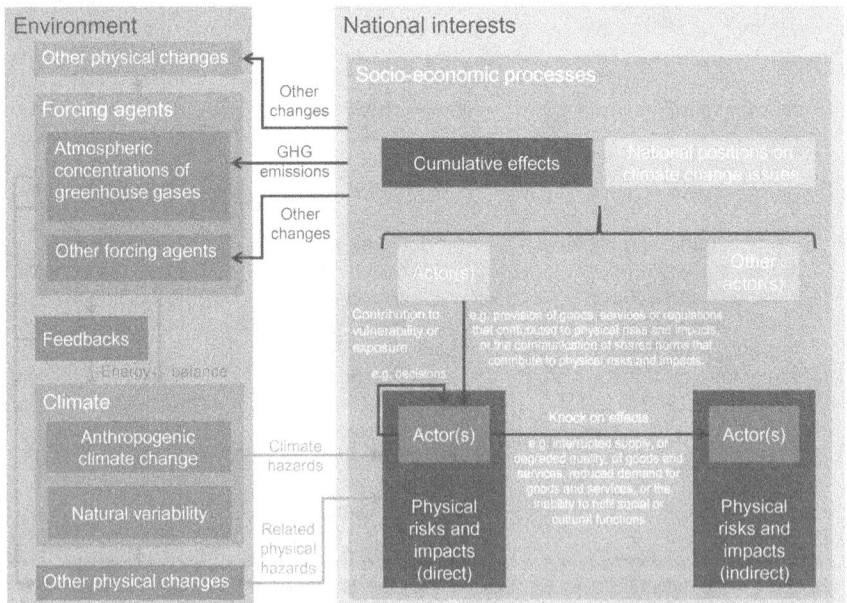

Figure B.1 Environment National Interests (ENI) Model showing the interactions linking climate-related physical risks, national interests, and greenhouse gas emissions.

which as noted in Section 2.2.3 influence climate for long periods of time before reaching equilibrium. Climate includes natural variation as well as climate change. Linking climate to national interests are hazards that, coupled with vulnerability and exposure, create risks to actors and their interests. Climate hazards are linked to climate rather than climate change, reflecting the difficulty of attributing climate events and impacts to climate change (Allen 2003, Bindoff et al. 2013, National Academies of Sciences, Engineering, and Medicine 2016).

In addition to climate hazards are other related physical hazards, such as sea level rise and ocean acidification, which are not climatic but are either related to climate change (e.g., sea level rise due to absorption of heat by seas) or are related to the accumulation of greenhouse gases (e.g., ocean acidification due to increased carbon dioxide absorption by the sea). Other related physical hazards may arise from other physical changes such as the impacts of climate change on physical systems and biological systems. An example might include changes in ecosystem services to remote communities (e.g., provisioning of fish) resulting from the impacts of climate change on marine ecosystems (Oppenheimer et al. 2014).

The national interests part of the ENI model focuses on actors. Contributions to vulnerability and exposure of actor(s) include actor(s) decisions as well as regulations including zoning, plans, and standards, the advice of other actors for example developers or designers, or the quality of goods such as building materials or heating and cooling systems (Table B.1). In addition to these things, there is the possibility of shared norms and practices to contributing to physical risks and impacts. Indirect physical risks and impacts are differentiated from direct physical risks and include interrupted supply, or degraded quality, of goods and services as well as the potential inability of impacted actors to provide social or cultural functions.

The combined actions of actors create cumulative effects including GHG emissions as well as "other changes" to forcing agents and other physical systems. Furthermore, the interests of actors influence national positions on climate change issues (Marchiori et al. 2017).

Table B.1 Relationship between vulnerability and exposure and the provision of goods, services and regulations

Contribution to:	Vulnerability	Exposure
Goods	Quality of materials used in construction Quality of heating and cooling systems	
Services	Design of buildings, infrastructure	Planning and location of buildings and infrastructure
Regulations	Building and infrastructure standards	Zoning and permitting

Note: The ENI model could be generalised to include possible benefits if hazards were substituted for "influences" good and bad.

Appendix C
Jellyfish model of the climate regime complex

Griffith University's Institute for Ethics, Governance and Law (IEGL) compiled a database cataloguing organisations formally linked to the UNFCCC in some way. Together with the Dutch IT company, LUST, this information was mapped graphically and organised into nine interactive and hierarchical layers starting with the Convention at the top. Table C.1 highlights the layers of the climate regime that are within the institution of the UNFCCC and the parts of layers of the regime outside the institution of the UNFCCC but supporting its objective.

The climate regime mapped by IEGL (2017) only consisted of organisations formally linked to the UNFCCC and supporting the achievement of the UNFCCC objective. IEGL (2017) mapped over 300 such organisations. However, there are many more climate-related organisations or initiatives that aren't included in the climate regime mapped by IEGL, for example there are thousands of organisations registered as observers to each COP (UNFCCC 2018b).

IPCC Working Group III showed cooperation in support of the UNFCCC objective can happen between international, national, and sub-national actors (Stavins et al. 2014). Table C.2 lists the types of institutions related to climate change with relevant examples. This includes cooperation between international organisations, between state actors for example through environmental treaties and bilateral arrangements, as well as non-state actors for example city networks including transnational city networks.

Much of the IPCC's chapter on international cooperation regarded the climate regime although there are important sections on the overlap between the climate regime and trade, technology, intellectual property, investments and finance for example. Keohane and Victor's (2011) paper included an indicative map of institutions and initiatives inside and outside of the climate regime, that have a bearing on the global response to climate change. This IEGL map and related information (Table C.2) indicated there are links between the UNFCCC and other institutions but only indicated to a limited extent, how these institutions interact within the climate regime complex.

Drawing inspiration from IEGL's work representing the climate regime and the fact that the UNFCCC has limited direct influence on the global response to

Table C.1 Levels used by IEGL to organise and map the climate regime

Regime	Level	Description
Within UNFCCC	The Convention	The UNFCCC entered into force on 21 March 1994. Today it has near-universal membership. The 195 countries that have ratified the Convention are called Parties to the Convention. Preventing "dangerous" human interference with the climate system is the ultimate aim of the UNFCCC.
	Conference	The supreme body of the Convention. It currently meets once a year to review the Convention's progress.
	Protocol/other instrument	An international agreement linked to an existing convention, but as a separate and additional agreement which must be signed and ratified by Parties to the convention concerned. Protocols typically strengthen a convention by adding new, more detailed commitments.
	Permanent subsidiary bodies	Subsidiary body: A committee that assists the Conference of Parties. Two permanent subsidiary bodies are created by the Convention: the Subsidiary Body for Implementation (SBI) and the Subsidiary Body for Scientific and Technological Advice (SBSTA).
	Secretariat	The office staffed by international civil servants responsible for "servicing" the UNFCCC Convention and ensuring its smooth operation. The secretariat make arrangements for meeting, compiles and prepares reports and coordinates with other relevant international bodies and is based in Bonn, Germany.
Outside UNFCCC	Finance	Climate finance refers to local, national or transnational financing, which may be drawn from public, private and alternative sources of financing. Climate finance is critical to addressing climate change as significant financial resources are needed to allow countries to adapt to the adverse effects and mitigate the impacts of climate change.
	Implementation agencies	Agencies responsible for putting commitments into practice through programmes and projects. Usually used for UN agencies and bi-lateral and multilateral banks, but may refer to UNEP, non-governmental and governmental organisations.
	National level activities	Actions undertaken at the national level to address climate change usually through mitigation and adaptation activities.
	Reporting	Parties to the Convention must submit national reports on the implementation of the Convention to the Conference of Parties (COP). The required contents of national reports and the timetable for their submission are different for Annex I and non-Annex I parties in accordance with the principle of "common but differentiated responsibilities."

Source: Compiled from EIGL 2017.

Table C.2 Types of institutions related to climate change

Types of institutions related to climate change	Examples
UNFCCC	Kyoto Protocol, Clean Development Mechanism, International Emissions Trading
Other UN intergovernmental organizations	Intergovernmental Panel on Climate Change, UN Development Programme, UN Environment Programme, UN Global Compact, International Civil Aviation Organization, International Maritime Organization, UN Fund for International Partnerships
Non-UN international organisations	World Bank, World Trade Organization
Other environmental treaties	Montreal Protocol, UN Conference on the Law of the Sea, Environmental Modification Treaty, Convention on Biological Diversity
Other multilateral 'clubs'	Major Economies Forum on Energy and Climate, G20, REDD+ Partnerships
Bilateral arrangements	US-India, Norway-Indonesia
Partnerships	Global Methane Initiative, Renewable Energy and Energy Efficiency Partnership, Climate Group
Offset certification systems	Gold Standard, Voluntary Carbon Standard
Investor governance initiatives	Carbon Disclosure Project, Investor Network on Climate Risk
Regional governance	EU climate change policy
Subnational regional initiatives	Regional Greenhouse Gas Initiative, California emissions-trading system
City networks	US Mayors' Agreement, Transition Towns
Transnational city networks	C40, Cities for Climate Protection, Climate Alliance, Asian Cities Climate Change Resilience Network
NAMAs, NAPAs	Nationally Appropriate Mitigation Actions (NAMAs) of developing countries; National Adaptation Programmes of Action (NAPAs)

Source: Stavins et al. (2014).

climate change, the UNFCCC and climate regime can be represented as a jellyfish (Figure C.1). In this case the hood (also known as the bell) of the jellyfish covers the most coherent and easily identifiable parts of the UNFCCC, specifically the Convention, other agreements (e.g., the Paris Agreement) and instruments under the UNFCCC, the COP, permanent subsidiary bodies, and the UNFCCC Secretariat. Below the hood are tentacles representing the most difficult to map elements of the climate regime, that reach out from the international level, down to the national and sub-national levels where much of the global response to climate change happens. These tentacles consist of implementing agencies, finance, activities and actions that have a bearing on the global response to climate change for example by businesses, communities or households, and the reporting of progress which feeds back to the UNFCCC, for example to the Secretariat and the COP. The tentacles

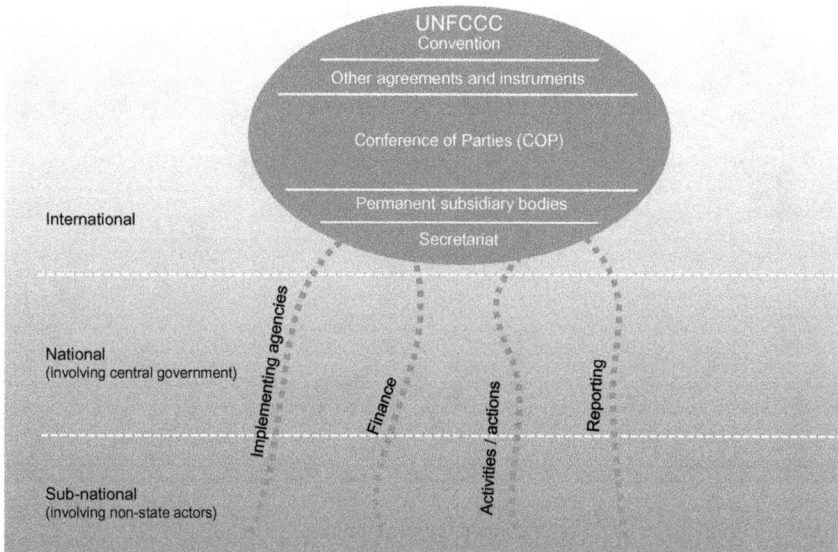

Figure C.1 Jellyfish model of the UNFCCC and global response to climate change.

of the UNFCCC attempt to influence the national actors and non-state actors at the sub-national level.

The climate regime complex can also be represented using the jellyfish model expanded to include other international institutions and initiatives (Figure C.2). The UNFCCC (and climate regime) is represented as jellyfish with tentacles stretching down from the international level to the national and sub-national levels. However, there are other international institutions that include a hood consisting of international agreements, meetings of parties or members, a secretariat and other subsidiary bodies. These other international institutions also have tentacles that reach down from the international level to national governments and sub-national actors influencing their activities and actions. These influences can either help or hinder the global response to climate change.

In addition to international institutions, which are established by sovereign states, there are international initiatives by non-state actors for example between cities, civil society and businesses from different countries. These international initiatives typically include some sort of agreement and may involve a secretariat for example. These initiatives also have mechanisms reaching down to the national and sub-national levels. The net result of applying the jellyfish model to the climate regime complex is that there are many competing influences affecting activities and actions at national and subnational levels and the global response to climate change as a whole.

IEGL (2017), Stavins et al. (2014), Keohane and Victor (2011) and the Jellyfish model all focus on formal institutions. Formal institutions, in particular the

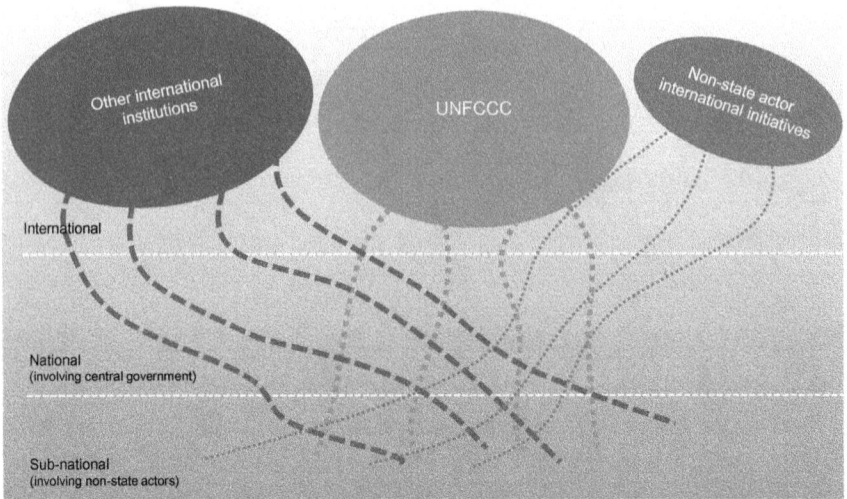

Figure C.2 Jellyfish model of the climate regime complex.

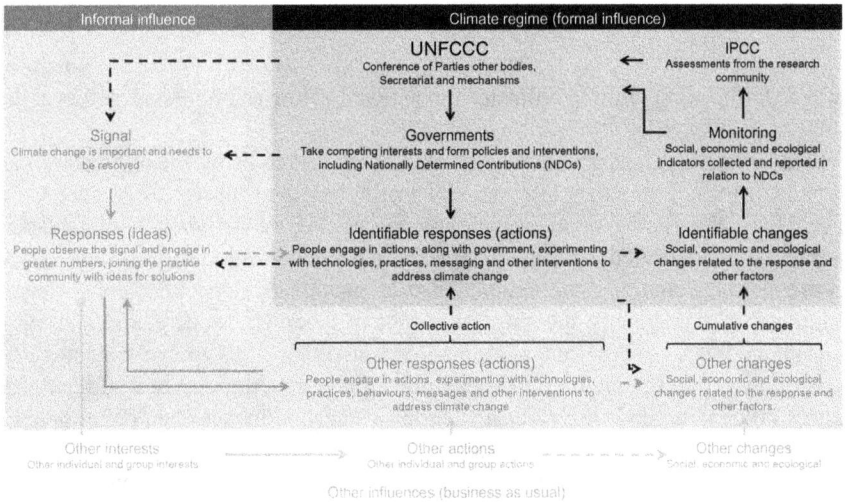

Figure C.3 The climate regime is coupled with informal influences.

spectacle of the COP with world leaders in attendance every year, generate signals indicating climate change is a problem that should be taken seriously. Figure C.3 indicates greater numbers of people are coming up with ideas to address the climate change problem because of this signal. This "informal influence" of the UNFCCC generates research and development activities that involve risk and cost to the individuals involved, akin to Hekkert et al.'s (2007) "entrepreneurial activities." The people engaging in the generation of ideas and related activities are leaders and innovators, creating options for addressing climate change i.e. potential solutions. In some cases, other actors may engage in these ideas and activities for reasons other than climate change, for example when options appear commercially viable or politically attractive. While all activities contribute "collectively" to the global response and "cumulatively" to the effectiveness of the response, many activities might never be formally recognised or integrated into the climate regime. These actors and activities exist in the wide spaces between the tentacles of the Jellyfish Model (Figure C.2).

Note: There is scope for further investigation into Hekkert et al.'s (2007) functions of innovation systems and how these functions might influence the global response to climate change.

Appendix D
Methods used to explore the future

D.1 Introduction

From Chapter 2, climate change is driven by greenhouse gas (GHG) emissions, and there will be no new normal until GHG emissions stabilise in the atmosphere at safe levels. Even then there will be further sea level rise and other long-term changes locked into the system. How we are impacted and will respond (socially, technologically, and economically) to these physical changes is unclear.

To address the questions from Section 1.5, it would be ideal to have a representative sample of possible futures, and be able to analyse these futures, including the actors involved and the conditions that support effective global responses to climate change. Of course, the dilemma of how to collect information on possible futures is not unique to this book. Scenarios have been widely used where there is uncertainty regarding future conditions or how these conditions might influence actors and their interests. At the same time, scenario methods to date have included limitations, for example often focusing on a small set of scenarios for practical reasons, even though there are many possible scenarios.

The purpose of the data collection in this book was to get information on possible futures (i.e., a sample of possible futures) so that the questions identified in Chapter 1 could be addressed. The structure of this chapter follows the process illustrated in Figure D.1. Data collection is addressed in Section D.2, and focuses on how the research presented in this book collected information on possible futures including success, failure, and other scenarios. Data processing is addressed in Section D.3 and focuses on how a sample of possible futures was compiled. Data analysis is addressed in Section D.4 including the methods used to analyse themes in the sample of possible futures and preconditions for effective global responses to climate change.

D.2 Data collection

Data collection had five key steps, consisting of deciding the data collection purpose and approach (see Section 4.3.2 in the main part of the book), followed by semi-structured survey question design (Section D.2.1), sample design (Section D.2.2), conducting interviews (Section D.2.3), and transcription of interviews (Section D.2.4).

A. Data collection of information about possible futures, including effective (i.e. successful) global responses to climate change, failure and other possibilities	1. Decide on data collection purpose and approach
	2. Design semi-structured survey (reverse stress test / backcasting approach)
	3. Design sample and contact potential respondents
	4. Conduct semi-structured survey interviews
	5. Transcribe interviews

B. Data processing forming a thematically coded set of scenarios, i.e. a searchable sample of possible futures	1. Compile scenarios (theory of change analyses of interview transcripts)
	2. Thematic analysis and coding of scenarios (creating searchable sample of possible futures)
	3. Identification of important themes for success, failure and other scenarios

C. Data analysis explore the sample of possible futures and themes contained within, identifying preconditions for effective global responses to climate change	1. Thematic chain analyses of scenarios (from the searchable sample of possible futures)
	2. Climate change impacts, risks and response scenarios and themes
	3. Actors and interest scenarios and themes
	4. International cooperation scenarios and themes
	5. Other scenarios and themes (influencing climate or the global response)
	6. Definition of preconditions for effective global responses to climate change

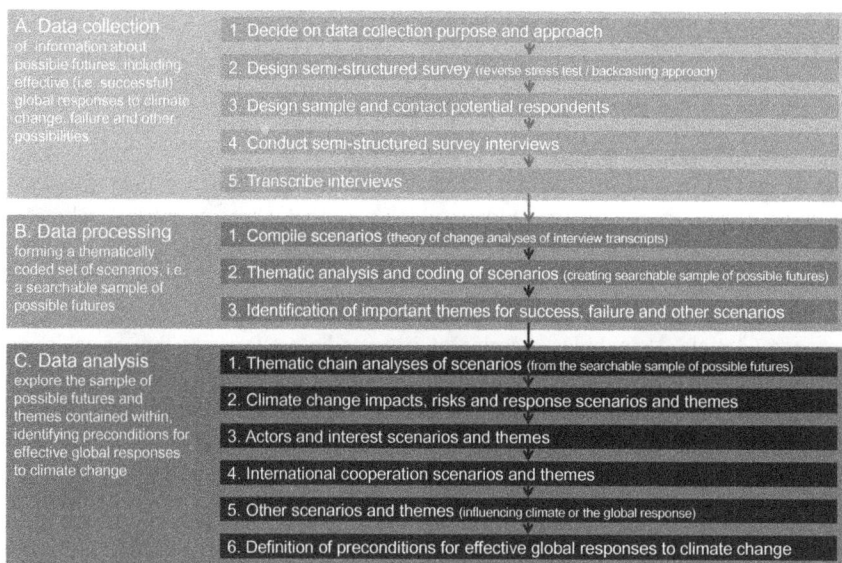

Figure D.1 Summary of methods used to collect and compile information about possible futures and analyse preconditions for effective global responses to climate change.

D.2.1   Semi-structured survey question design

With regards to the mode of data collection, it was decided to use semi-structured interviews. Semi-structured interviews are flexible, can be conducted in 30 minutes or an hour, and whenever a respondent is available. Workshops were considered impractical given the difficulty of organising participants from around the world many of whom have very limited time available.

The first step in designing the survey was defining success and failure criteria (Section D.2.1.1). The survey also included background questions (Section D.2.1.2), backcasting and reverse stress test-related questions (Section D.2.1.3), and reflective questions (Section D.2.1.4).

As discussed in Section 4.2, backcasting and reverse stress testing are methods for eliciting success and failure scenarios respectively. Following methods outlined by Füser et al. (2012), questions addressed current conditions, and possible success and failure scenarios. During the early stages of the survey data collection process, respondents were asked about other people that should be interviewed, applying the snowballing method of finding respondents.

D.2.1.1   Success and failure criteria

An important part of backcasting and reverse stress testing is defining success and failure conditions. For this book it was important to define what an effective global response (i.e., success) would look like. Given that there are 197 Parties to the

UNFCCC accounting for the vast majority of global GHG emissions, the UNF-CCC objective serves as an important reference, as does the Paris Agreement.

The UNFCCC objective (Section 3.3.1) and Paris Agreement purpose (Section 3.3.2) are interrelated (Table D.1). The Paris Agreement is "In pursuit of the objective of the Convention" and is "guided by its principles" (Article 1, Paris Agreement) (UNFCCC 2015). The Paris Agreement aims to strengthen the global response to climate change and puts in place a system where countries make contributions to the global response (i.e., NDCs) and progress is measured (i.e., global stocktakes) (UNFCCC 2015). As such, the Paris Agreement is an important step towards operationalising the global response to climate change. The Paris Agreement purpose (Article 2) includes a series of intermediate outcomes that if achieved, would help fulfil the ultimate outcomes of holding the increase in the global average temperature to well below 2°C above pre-industrial levels (Article 2, Paris Agreement), stabilising GHG concentrations, allowing ecosystems to adapt naturally, ensuring food production is not threatened, and economic development can proceed in a sustainable manner (Article 2, UNFCCC) (Section 3.3.2).

Table D.1 Ultimate outcomes and intermediate outcomes from the UNFCCC objective and Paris Agreement purpose

Context	Source	Text from Article 2
Ultimate outcomes	UNFCCC	Stabilization of greenhouse gas concentrations in the atmosphere at a level that would prevent dangerous anthropogenic interference with the climate system
	PA	Holding the increase in the global average temperature to well below 2°C above pre-industrial levels
	UNFCCC	Allow ecosystems to adapt naturally to climate change
	UNFCCC	Ensure that food production is not threatened
	UNFCCC	Enable economic development to proceed in a sustainable manner
Intermediate outcomes	PA	Strengthen the global response to the threat of climate change[a]
	PA	Pursue efforts to limit the temperature increase to 1.5°C above pre-industrial levels
	PA	Increasing the ability to adapt to the adverse impacts of climate change
	PA	Making finance flows consistent with a pathway towards low greenhouse gas emissions and climate-resilient development
	PA	Climate resilience and low greenhouse gas emissions development
	PA	Response does not threaten food production
	PA	Equity in the light of different national circumstances[b]

Source: Compiled by the Author.

PA = Paris Agreement purpose.

UNFCCC = United Nations Framework Convention on Climate Change objective

[a] …in the context of sustainable development and efforts to eradicate poverty.

[b] This Agreement will be implemented to reflect equity and the principle of common but differentiated responsibilities and respective capabilities, in the light of different national circumstances.

The ultimate outcomes identified in Section 3.3.2, largely come from the UN-FCCC objective, with the only ultimate outcome from the Paris Agreement being "Holding the increase in the global average temperature to well below 2°C above pre-industrial levels" (Article 2, Paris Agreement) which essentially provides a goal and metric for assessing progress towards "Stabilization of greenhouse gas concentrations in the atmosphere at a level that would prevent dangerous anthropogenic interference with the climate system" (Article 2, UNFCCC) (Randalls 2010). As such, the UNFCCC objective provides success and failure conditions for the global response to climate change. Success is considered to be stabilising atmospheric concentrations of GHGs at safe levels, where safe levels mean ecosystems are able to adapt naturally, food production is not threatened and economic development is able to proceed in a sustainable manner. Failure is simply failure to achieve one, or more, of the conditions specified in the UNFCCC objective.

Another approach for assessing the effectiveness of the global response to climate change is to avoid binary success failure determinations and instead analyse the nature of success or failure along a spectrum of possible conditions. This could include achieving the UNFCCC objective or being on path towards achieving the UNFCCC objective, through to being on path towards failure, actual failure to achieve the UNFCCC objective and having a plan B, or failure without any alternative plans (see column headings in Table D.2).

International cooperation is widely regarded as being essential for effective global responses to climate change (Stavins et al. 2014). As discussed in Section

Table D.2 Degrees of success (white) and failure (dark grey) relative to the UNFCCC objective and UNFCCC legitimacy (i.e., universality of membership)

Membership of the UNFCCC as an indicator of legitimacy	UNFCCC objective				
	Achieved	On path towards being achieved	On path towards failure	Failed (i.e., either emissions have not been stabilised or ecosystems unable to adapt naturally, food production threatened or economic development unable to proceed in a sustainable manner)	
				There is plan B	The is no plan B
Universal membership					
Almost universal membership					
Majority membership					
Around 50% membership					
Minority membership					
No members					

(Legitimacy)

3.4.3, participation in international agreements and international institutions is essential for these agreements and institutions to be effective. Participation also serves as an important indicator of institutional legitimacy (Stavins et al. 2014). As such, the level of membership (ranging from universal to no members) is an important indicator (see the row headings in Table D.2).

These success and failure criteria informed the design of backcasting and reverse stress test questions (Section D.2.1.3).

D.2.1.2   Background questions

The semi-structured survey started with background questions designed to get a sense of general optimism regarding the future and faith in the institution of the UNFCCC. Each respondent was asked to fill in a form with background questions, either online or on hardcopy. These background questions asked respondents to indicate whether they strongly disagreed, disagreed, neither disagreed or agreed, agreed, or strongly agreed with the following statements:

- The world is worse place to live today than it was in the past
- The world will be a better place to live in the future
- The UNFCCC should be abandoned
- The UNFCCC is essential for an effective global response to climate change.

Respondents were also asked if they had attended the UNFCCC conference of parties.

From these background questions, the aim was to verify that the sample of respondents included participants across the spectrum of optimism to pessimism and levels of faith in the UNFCCC. For more information on survey sample design see Section D.2.2.

D.2.1.3   Backcasting and reverse stress test questions

Backcasting and reverse stress test questions are the most important questions in the survey. These are the main questions used to elicit possible futures from respondents. Backcasting and reverse stress test questions were supplemented, if necessary, by follow-up questions to ensure responses were sufficiently detailed and coherent to allow analysis and comparison with other responses. Backcasting and reverse stress test questions addressed the current situation, failure scenarios, success scenarios, and other scenarios (Figure D.2).

To get information on each respondent's perceptions of current climate change and development system conditions, they were asked:

How would you describe the current climate change situation?

Responses to this question allowed comparison of the respondent's perceptions with IPCC assessments and peer-reviewed literature on current climate change and development conditions summarised in Chapters 2 and 3. General follow-up questions were asked if required, to get more information (Section D.2.1.4). The

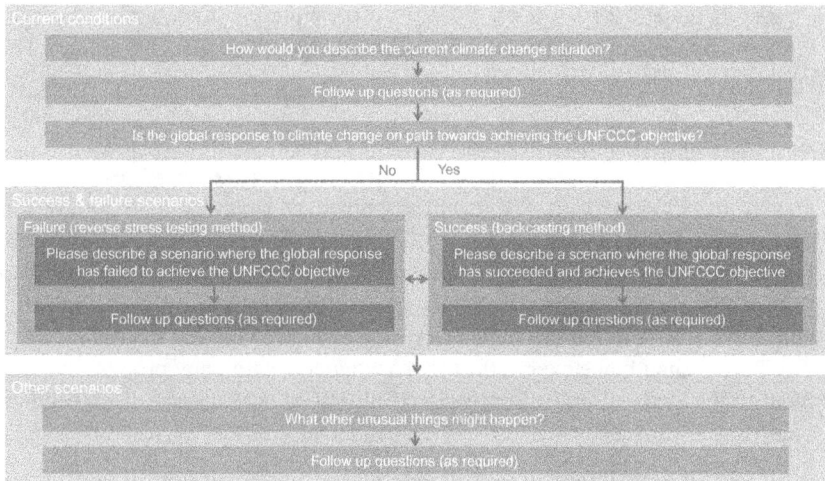

Figure D.2 Reverse stress test question order in the semi-structured interview.

respondent's description of current conditions also helped provide a starting point for the compilation of scenarios (Section D.3.1).

The last follow-up question, regarding current conditions, asked whether the respondent thought the global response was on path towards achieving the UNFCCC objective. In all cases respondents indicated that the global response was not on path towards achieving the UNFCCC objective.[1] Hence, the next question regarded failure scenarios, with respondents asked to:

Please describe a scenario where the global response has failed to achieve the UNFCCC objective.

Some respondents indicated they knew Article 2 of the UNFCCC and as such did not need to have it described to them. However, in most cases failure to fulfil Article 2 was explained to the respondent for clarity's sake, as atmospheric greenhouse concentrations have not been stabilised or ecosystems are unable to adapt naturally, food production is threatened or economic development is not able to proceed sustainable manner.

General follow-up questions were asked as required (Section D.2.1.5). In addition, two specific follow-up questions related to ineffective global responses were asked if the respondent had not already addressed the issues. These specific follow-up questions consisted of:

- **Please describe a scenario where the UNFCCC objective (Article 2) is amended by the Conference of Parties to the UNFCCC.**
- **Please describe a scenario where Parties withdraw from the UNFCCC.**

The question on amending the UNFCCC objective helped address the possibility of failing to achieve the UNFCCC objective and having a so-called "plan B" as illustrated in Table D.2 from Section D.2.1.1. It is important to note that Article 15 of the Convention set the conditions for amending the UNFCCC (see below). This information was shared with respondents.

1. Any Party may propose amendments to the Convention.
2. Amendments to the Convention shall be adopted at an ordinary session of the Conference of the Parties. The text of any proposed amendment to the Convention shall be communicated to the Parties by the secretariat at least six months before the meeting at which it is proposed for adoption. The secretariat shall also communicate proposed amendments to the signatories to the Convention and, for information, to the Depositary.
3. The Parties shall make every effort to reach agreement on any proposed amendment to the Convention by consensus. If all efforts at consensus have been exhausted, and no agreement reached, the amendment shall as a last resort be adopted by a three-fourths majority vote of the Parties present and voting at the meeting...

(Article 15, UNFCCC 1992)

The legitimacy of the UNFCCC is widely regarded to be a function of its near universal membership (Sections 3.4.3 and D.2.1.1). The question of UNFCCC Parties withdrawing from the UNFCCC could be a situation where only a few parties withdraw or a situation where many parties withdraw from the UNFCCC. As such, responses could fall across the vertical axis of Table D.2 from Section D.2.1.1.

To get information on effective global responses to climate change, respondents were asked to:

Please describe a scenario where the global response to climate change succeeds in achieving the UNFCCC objective.

Those respondents that indicated they knew Article 2 of the UNFCCC did not need to have it described. However, for the sake of clarity achieving the UNFCCC objective was described to most respondents, as being atmospheric concentrations of greenhouse being stabilised at levels that allow ecosystems to adapt naturally, food production not being threatened and economic development being able to proceed in a sustainable manner.

General follow-up questions were asked as required (Section D.2.1.5). In addition, two specific follow up questions related to effective global responses were asked if the respondent had not already addressed the issues in their responses. These specific follow up questions consisted of:

- **Please describe a scenario where a stringent international climate change agreement is enforced.**
- **Please describe a scenario where greenhouse gas removals successfully stabilise atmospheric concentrations of greenhouse gases at safe levels.**

These two questions focus on dilemmas around how the UNFCCC objective might be fulfilled. The question on a stringent enforced agreement helps elicit responses related to possible institutional arrangements including the possibility of a protocol (i.e., a legally enforced agreement under a pre-existing agreement). The question on GHGs removals directly addressed a key question from Section 1.5 regarding preconditions for effective GHG removals.

Based on a review of the stress testing literature which highlighted the need to address all possible risks, questions were included on other scenarios that might affect climate change or the global response. These other scenarios were elicited with two questions:

What other unusual things might happen?

What other scenarios can you imagine for the UNFCCC?

As experience developed in the delivery of the survey, the interviewer would clarify these questions, asking: what other things might influence climate or the global response to climate change? The aim of these questions was to extend the range of scenarios collected, while also addressing all possible risks including low probability high impact scenarios (i.e., so-called black swans), or more mundane extreme scenarios for example a year of drizzle in the United Kingdom (pers. comm. Adam Cooper). Other UNFCCC-related scenarios were also explored.

D.2.1.4 *Reflective questions*

At the end of the survey reflective questions were asked, to help understand the respondent's role in relation to current climate change conditions and possible scenarios. The main question consisted of:

What role do you and the organisation you are with have in relation to climate change?

Specific follow-up questions consisted of:

• **How would you describe your organisation's contribution towards the global response to climate change?**
• **Which scenario would you say your organisation contributing towards?**

As experience developed in the delivery of the survey, respondents were asked: With regards the scenarios described, which scenario is your organisation contributing towards? And, which scenario are you personally contributing towards?

D.2.1.5 *General follow-up questions*

General follow-up questions were asked if a response was: unclear; did not fully address the question being asked; or, if there was some interesting element to be further explored. Follow-up questions were designed to ensure the responses to each question included enough information for a robust analysis and that all elements of the UNFCCC objective were addressed. Follow-up questions also addressed actors and the reasons for their actions in each scenario.

General follow-up questions included:

• How would you describe the situation?

For example, in relation to:

- the ability of ecosystems to adapt naturally
- food production
- economic development
- society, politics and other things

- Who are the main actors?

  - Countries
  - Groups of countries
  - Businesses and business groups
  - Civil society and civil society groups
  - The UNFCCC and other international organisations
  - Individuals and households

- How do they influence the situation and for what reasons?
- How would you describe the UNFCCC's role?
- Outside of climate change processes, what are the key things driving the situation?

D.2.2   Semi-structured survey sample design

It is important to note, **the aim of the survey was to get a representative sample of possible futures rather than responses representative of people in the sample**. However, to collect information on possible futures it was important to have a sample of experts from which to collect information on possible futures. As such, a survey sample was designed (Section D.2.2.1) and respondents were recruited (Section D.2.2.2).

D.2.2.1   Survey sample

The survey sample was designed to get a broad sample of possible futures (i.e., scenarios) by eliciting global response scenarios from a diverse group of people with different backgrounds, experiences and views on climate change and the global response.

Sample selection criteria are presented in Table D.3 and include nationality, gender, level of engagement with the UNFCCC, and institutional affiliations. In addition to these criteria, climate change sceptics (i.e., people sceptical of the scientific basis for anthropogenic climate change) were excluded from the sample because this book is focused on collecting evidence based global responses to climate change rather than interrogating climate science, which is accepted as being unequivocal (IPCC 2014a).

The level of engagement respondents had with the UNFCCC was considered important because it was essential to have respondents with a knowledge of the UNFCCC, its institutional arrangements and how this affects the global response

Table D.3 Sample selection criteria

Development and geographic distribution	Gender	Engagement with UNFCCC	Current institutional affiliation
Nationality(s) including developed and developing country nationalities	Female Male	Secretariat staff member Representative on a UNFCCC body Negotiator and delegate to COP Other delegate to COP Observer to COP IPCC Distant observer (never attended COP)	**UNFCCC:** UNFCCC IPCC **Other international organisations:** Multilateral Development Bank Other international organisations **Sovereign States:** National government policy National security **Non-state actors:** Local Government Business Civil Society Research Media Religious

to climate change. At the same time, only selecting people deeply engaged with the UNFCCC could bias the sample towards those who see value in the institution. As such, some people who have not been deeply engaged were wanted in the sample, for example people involved in climate-related interventions in the field. Furthermore, the study presented in this book deliberately sought people sceptical of the global response to climate change and the UNFCCC for inclusion in the survey sample. Background questions regarding faith in the UNFCCC (Section D.2.1.2) were included in the survey, making it possible to verify the survey sample included global response sceptics.

A key issue of concern in the Fifth IPCC Assessment Report chapter on international cooperation was competing international institutions, the possibility of forum shopping and whether this enhances or weakens the global response to climate change (Stavins et al. 2014). As such, it was important to have people in the sample with a range of institutional affiliations, including people from potentially competing and complementary institutions and organisations. The right-hand column of Table D.3 has a list of the institutional affiliations of interest.

D.2.2.2   *Recruitment of respondents*

A list of potential respondents was compiled based on the sample selection criteria discussed above. However, once interviews began, snowballing was also used to recruit new participants as well, although this only resulted in a couple of actual interviews.

Prior to being interviewed, each respondent was sent an email with an invitation letter and a participant information sheet. They were also asked if there is a particular time or place that would be best for the interview to take place. In some cases, the invitation was for an in-person interview and in other cases it was for an interview by telephone, via skype or some other online communication provider.

Data collection and recruitment stopped once theoretical saturation was achieved i.e., when interviews stopped yielding new information.

D.2.3 Semi-structured survey interviews – sample and response rate

Invitations to participate in the survey were sent to 44 people, of which 18 were women and 26 were men (Table D.4). Of these, 27 people responded and participated in the survey, of whom 8 were women and 19 were men. Of the 27 interviews conducted, 19 were conducted in person and 8 were conducted by telephone or over the internet. Before starting the interview, each respondent was asked to go through the participant information sheet and consent to the interview. They were also provided with a copy of the participant information sheet. Interviews took between 30 and 85 minutes and were recorded for transcription purposes.

With regards to primary national and regional affiliations (Table D.5), 3 respondents were from Africa, 2 from Asia, 11 from Europe, 3 from North America, 7 from Oceania, and 1 from South America. As such there is a developed country bias in the sample, however, responses were not markedly different from developed or developing country respondents, for example, global response scenarios focused on powerful actors regardless of whether the respondent was from developed or developing countries. Many of the developed country respondents had spent time working in or with developing countries.

With regards to primary institutional affiliations, 1 was primarily affiliated with a United Nations-related climate change institution, 4 were affiliated with other international organisations, 5 with sovereign states and their governments, and 17 were primarily affiliated with non-state actors such as research organisations or civil society (Table D.6). As such, there is a bias in the types of respondents (i.e., survey sample), with high numbers of researchers and civil society respondents and no business, local government or religious respondents. The extent to which "survey sample" bias is a problem depends on the extent to which this biases the "sample of possible futures" collected. **As noted in Section 4.3.2, the survey attempts to collect a representative sample of possible futures i.e., a set of scenarios broadly representing the range of possibilities**. However, the sample of

Table D.4 The number of female and male invitees and respondents along with response rates

	Invitees	Respondents	Response rate (%)
Female	18	8	44
Male	26	19	73
Total	44	27	61

Table D.5 Primary national affiliations of invitees and respondents along with response rates

	Invitees	Respondents	Response rate (%)
Africa:	**4**	**3**	**75**
Ethiopia	1	1	100
Mali	1	1	100
South Africa	1	0	0
Sudan	1	1	100
Asia:	**3**	**2**	**67**
India	1	0	0
Japan	1	1	100
Philippines	1	1	100
Europe:	**19**	**11**	**58**
Austria	1	0	0
Denmark	1	1	100
France	1	0	0
Germany	2	1	50
Hungary	1	1	100
Ireland	1	0	0
Italy	1	1	100
Sweden	1	0	0
United Kingdom	10	7	70
North America:	**7**	**3**	**43**
United States of America	7	3	43
Oceania:	**9**	**7**	**78**
Australia	1	1	100
Fiji	1	0	0
New Zealand	7	6	86
South America:	**2**	**1**	**50**
Argentina	1	1	100
Costa Rica	1	0	0

possible futures might lack global response scenarios involving business, local government or religion as key drivers for change.

Individually, the respondents have a wealth of experience and in many instances have had multiple roles in life, for example in the United Nations system, for their government or at the community level (Table D.7).

D.2.4 Transcription of interviews

The 27 interviews were transcribed verbatim including time tags making it possible to go back to relevant sections of audio files. Transcripts were imported into excel where each paragraph became a row. The 27 imported transcripts consist of over 13,500 rows and just over 238,000 words. Each transcript was checked for quality, against the recorded response and any errors in transcription were corrected. A copy of each response was made and edited removing stutters, repeated words, and words such as "um" or "uh." These clean transcripts of responses were then used as a basis for scenario identification. The data set was read in its entirety and notes were made.

Table D.6 Primary institutional affiliations of invitees and respondents along with response rates

Primary institutional affiliation	Invitees	Respondents	Response rate (%)
United Nations Climate Change Institutions:	**2**	**1**	**50**
UNFCCC	1	0	0
IPCC	1	1	100
Other international organisations:	**6**	**4**	**67**
Multilateral Development Bank	2	1	50
Other UN or international organisation	4	3	75
Sovereign States:	**9**	**5**	**56**
National government policy	8	4	50
National security	1	1	100
Non-state actors:	**27**	**17**	**63**
Local Government	0	0	NA
Business	0	0	NA
Civil Society	13	7	54
Research	13	9	69
Media	1	1	100
Religious	0	0	NA

Table D.7 Summary of respondents including respondent numbers

Respondent number	Information regarding the respondent
1	The respondent has a background in history and has published on issues of climate change.
2	The respondent comes from a security background and has been involved in climate change issues internationally.
3	The respondent has worked in international development, climate change and rural development both for their government as well as in civil society.
4	The respondent is a recently retired senior UN official with international development experience in Asia, Africa and Europe including dealing with international crises.
5	The respondent has been involved in the IPCC at a high level as well as energy research and civil society.
6	The respondent has worked on geopolitical and security-related climate change scenarios with governments from around the world.
7	The respondent has held science diplomacy roles with a developed country government interacting with large developing countries and has also been involved in climate change research.
8	The respondent has been an IPCC author and researches issues related to land and agriculture.
9	The respondent works for an international civil society organisation engaged in issues of climate change in developing countries.
10	The respondent works in a multilateral development bank, has extensive experience supporting developing countries, and has supported climate negotiators in the past.
11	The respondent works in a faith-based non-profit organisation working in developing countries on issues related to ecosystems and agriculture.
12	The respondent has worked on issues of disaster response in a developing country for government and civil society.

13	The respondent is a former politician and diplomat who has been active in climate change, foreign affairs and the United Nations.
14	The respondent has been involved in environmental consulting in developed and developing countries, environmental policy and national politics in a developed country.
15	The respondent is a prominent researcher on economics, climate change and other global issues.
16	The respondent has worked in civil society and international organisations supporting governments in Africa on issues related to climate change, development and the environment.
17	The respondent is involved in energy modelling and planning for an international energy system including stakeholder engagement.
18	The respondent is employed by an organisation that works on renewable energy, and the respondent has been a part of international initiatives related to climate change and sustainable development.
19	The respondent is a senior figure involved in energy policy in a developed country.
20	The respondent has a background in paleoclimate and hazards-related research.
21	The respondent is an academic involved in modelling energy systems and related greenhouse gas emissions as well as GHG balances.
22	The respondent has a background in media and communications including in government (agriculture related) and consulting roles.
23	The respondent works for an international humanitarian organisation and has been involved in issues of climate change and development.
24	The respondent has advised senior figures in the United Nations and has a leadership role in civil society and research on issues related to climate change.
25	The respondent has been involved in civil society, the promotion of renewable energy and government delegations to climate negotiations.
26	The respondent is a researcher with a legal background, teaching climate change law.
27	The respondent has advised governments and cities on issues related to climate change and development.

D.3   Data processing

Data processing had two key steps, consisting of scenario compilation using theory of change (Section D.3.1), and thematic analysis including the coding of scenarios (Section D.3.2). It should be noted that the division between data processing and data analysis is somewhat arbitrary, as thematic coding is a form of analysis and contributes to the definition of themes that are further analysed. However, for the purposes of presenting the methods used, data processing stops at the point where the sample of possible futures is compiled and coded, ready for "thematic chain analysis" and the exploration of preconditions.

D.3.1   Scenario compilation using theory of change analysis

While the current situation is not the main focus, the survey started with the current situation to get a sense of what each respondent thinks about the climate change as an issue, and also to get a sense of the extent to which responses and scenarios provided were consistent with published literature. In many interviews respondents deferred to, and repeated, the IPCC and related research on climate change.

For each respondent, the clean transcript of their response was analysed using theory of change. The theory of change analysis included reading transcripts and looking for statements and quotes that had a logical flow from one set of conditions to another set of conditions. Statements that indicated "if... then" were identified and highlighted. These quotes were then taken from the transcript and plotted as "scenario elements" on a two-dimensional plot, with time running left to right. With regards to the vertical axis, localised issues were plotted lower while overarching or international issues were plotted higher. In most cases scenario elements were based on quotes from a respondent (in quotation marks) but in some cases scenario elements consisted of short summaries based on an analysis of the respondent's responses (presented without quotation marks).

For each scenario, distinctions were made between essential and other information. Dark grey boxes were used to plot quotes deemed essential to the scenario, grey boxes presented quotes related to, or of interest, but not essential to the scenario, and light grey boxes were used to present related issues and assumptions (Table D.8). However, it should be noted that the distinction between essential and non-essential elements was not not useful during the analysis of scenarios, and as such, dark grey and grey elements were analysed in the same way.

When plotting scenarios elements, boxes were aligned with each other to indicate the extent to which they may be related, for example boxes that are parallel along the x-axis indicate these scenario elements (i.e., boxes) are relevant over the same period of time. Boxes that are parallel along the y-axis indicate that these scenario elements (i.e., boxes) are related in terms of theme (see Figure F.1 in Appendix F). It should be noted that alignment was indicative and imperfect. In some cases, for example where there are branching follow-on elements, then it may be arbitrary which element is above or below. But to the extent possible, the criteria above were used to plot scenarios.

When asked to describe the current situation, failure, success or other scenarios, in many cases respondents provided a series of statements, sometimes switching between the current situation, the past and the future, or switching between success

Table D.8 Scenario elements

Descriptor	Description
Scenario element	A box with a description of something relevant to the scenario.
Essential element	Something that is necessary to the scenario, presented in a dark grey box.
Non-essential element	Something not necessary to the scenario but helps provide context or examples of possibilities within the scenario and is presented in a light grey box.
Issues and assumptions element	Things that are important to the scenario and understanding the other elements and relationships presented. These things are presented in a grey box.

and failure scenarios. At other times, respondents thought aloud about possible actors, dynamics, factors or situations that might be relevant. As such, transcripts contain discontinuous pieces of information, spread between statements and responses to different questions in some cases.

Most statements only addressed segments of chains, for example some critical node, a waypoint, an important change, or a significant event, and as such were akin to post-it notes marking out scenario pathways and storylines; which is a common approach used in scenario development workshops. The theory of change analysis was akin to taking post-it notes (i.e., quotes) based on what the respondents said and organising them into logical sequences (i.e., chains) to form success and failure scenarios.

From the theory of change analysis, 175 scenarios were compiled forming a sample of possible futures. These scenarios ranged from very simple to complex, in some cases consisting of a single element, and in other cases with multiple steps through time, or layers from local to international, branching or merging elements, parallel conditions and related assumptions. Figure D.3 is an example of a scenario with four steps and seven layers, compiled based on a theory of change analysis of a survey response.

Most scenarios plotted include chains of scenario elements through time, but in 2 cases the scenario plot took the form of a snapshot of a future international regime addressing climate change. These 2 scenario plots essentially took the form of organisation diagrams showing a hierarchy of international institutions with roles in quotes. Figure D.4 is one of the two scenarios plotted in the style of an organisation diagram.

Note: With regards to record keeping, for each survey response a theory of change analysis document was prepared consisting of: a short description of the respondent; a summary of the current climate change situation using quotes from the respondent, a summary of each scenario including scenario plots and accompanying quotes; a transcript of responses organised into sections with headings indicating the content and key points in each section.

D.3.2 Thematic analysis and coding of scenarios

Thematic analysis and coding of scenarios broadly followed the steps from Braun and Clarke (2006) which included familiarisation, initial codes, searching for themes, reviewing themes, naming and defining themes, and reporting of themes.

Familiarisation with the data collected happened over several phases including the editing and checking of transcripts to produce clean transcripts, the review of clean transcripts taking notes and highlighting "if... then" relations and the plotting of scenarios. The generation of initial codes included making concise notes on each scenario as well as concise notes on each of the scenario elements within a scenario. While generating themes (i.e., codes), it became apparent that notes could be identified as regarding outcomes, types of responses, actors and interests, issues or options, or regarded assumptions. In addition to these things, scenarios were coded according to their complexity. Any other notes were recoded under "Notes" (Table D.9).

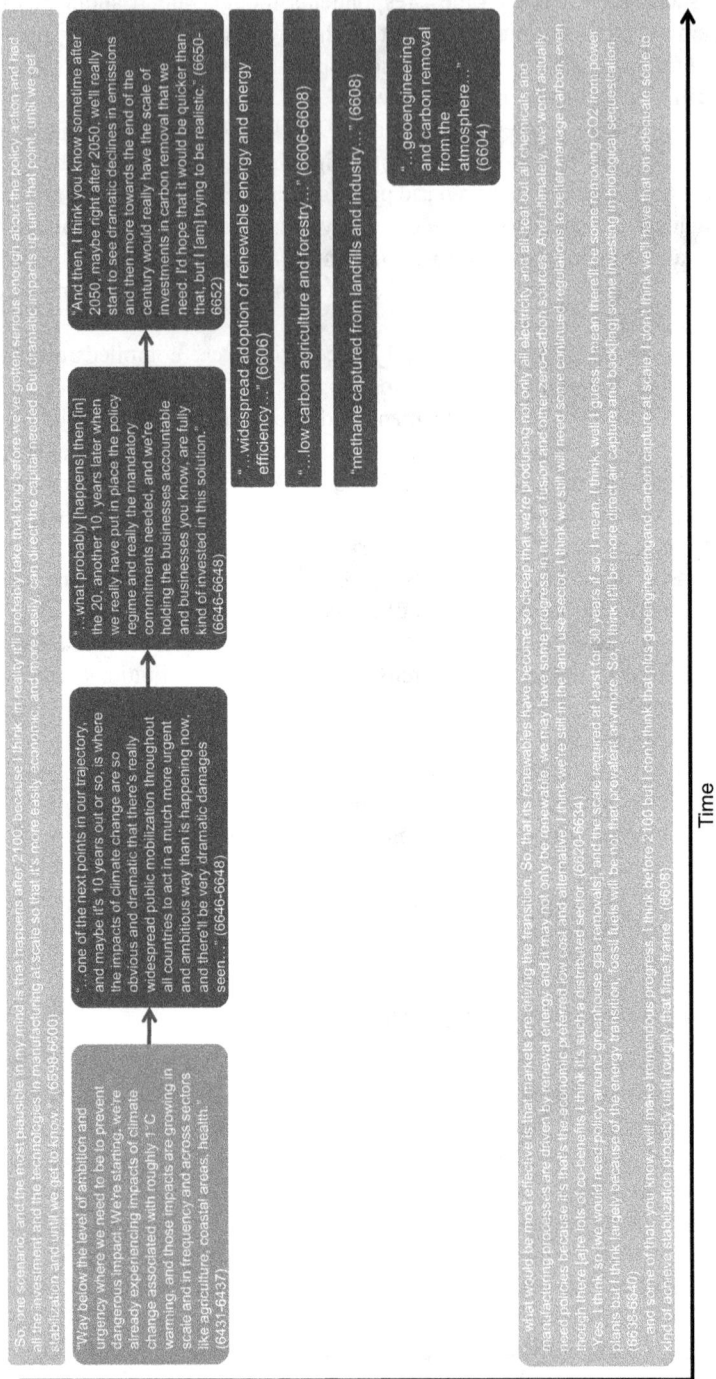

Figure D.3 An example of a scenario compiled using theory of change, with four steps and seven layers, including non-essential (grey) and essential (dark grey) elements as well as assumptions (light grey). The numbers in brackets refer to rows in the database of transcribed responses

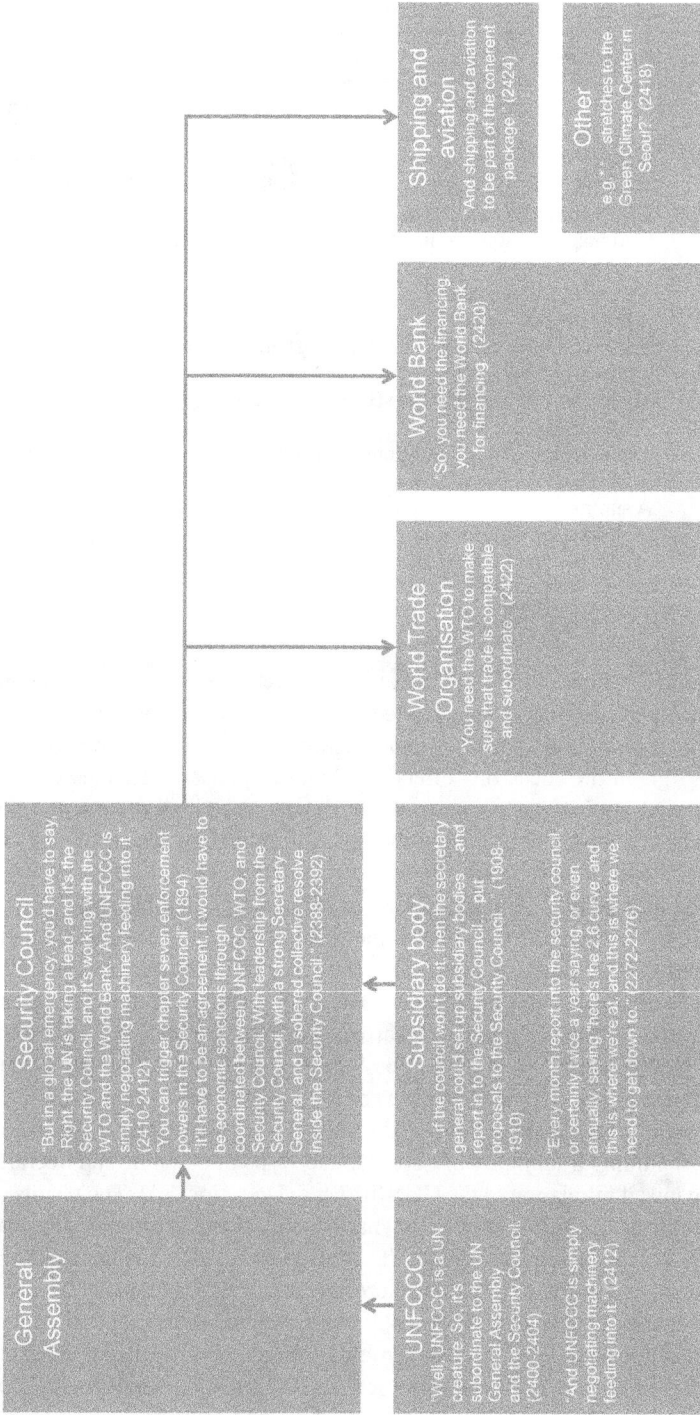

Figure D.4 Example of a plot showing the international regime and organisation of international institutions at a point in time.

Table D.9 Fields for which notes were recorded and themes developed

Fields	Definition
Complexity	A simple description of the extent to which different parts of the scenario are shown to influence each other based on the form of scenario plots.
Assumptions	Stated (i.e., explicit) or unstated (i.e., implicit) beliefs about the system that has explanatory power with regards to the scenario.
Issues and options	Issues are concerns or problems in the scenario.[a] Options are actions, or inactions, that can be decided upon by actors in the scenario.
Actors and interests	Actors are individuals or groups with the ability to make decisions and influence the scenario. Interests are the issues and options that an actor cares about in the scenario.
Response	The actions of actors in the scenario (i.e., options taken by actors in the scenario).
Outcome	The extent to which conditions at the end of a scenario are said to fulfil the UNFCCC objective.
Notes	A summary of any other information important to the interpretation or understanding of the scenario.
Coding	Short (comma separated) list of the most important themes in the scenario.

[a]Adapted from Merriam-Webster (2019).

The search for themes that could meaningfully reflect common and encompassing ideas with explanatory power started with a comparison of notes (i.e., initial codes) from across all the scenarios, specifically notes related to "outcomes" and "responses" given the focus on "effective global responses." Themes were iteratively reviewed until it was considered that themes described the most important characteristics of possible futures, at the level of individual scenarios and collectively when it comes to the entire sample of possible futures. At this stage themes were named and defined. Scenarios were then coded (in the "Coding" field) according to the most important themes in the scenario (i.e., the most salient themes).

Principles guiding the selection of words to name and define themes included using simple English while at the same time ensuring terms are identifiably grounded in survey responses and semantically correct. The first sentence of each definition describes the concept in the broadest possible and most essential terms while subsequent sentences may include examples or provide additional context.

Reporting of overarching themes and results of the thematic analysis is done in Chapter 4, including a summary of notes for each type of outcome and overarching themes organised in relation to the CCNIIC Model (Section 4.4). The results of the thematic analysis, including themes, scenarios and salient quotes are presented in Chapters 5–9. It is important to note: the study presented in this book did not attempt to limit the sample of possible futures to four or five reference scenarios. Instead all scenarios, and scenario elements, were accepted as being possible futures. As such, this book provides a broad sample of the multiverse of possible futures. Chapters 5–9 present multiple possible scenarios, even for a single theme, and include many headings reflecting the diversity of themes identified.

D.4 Data analysis

Data analysis involved identifying important themes and conditions, multiple possible pathways between these conditions, and identifying the preconditions that contribute towards effective global responses. The point of departure for data analysis was the overarching themes from Chapter 4 identified in Section 4.4. Importantly, the searchable sample of possible futures included searchable text in scenario titles, scenario elements and quotes, notes and themes. As such, keywords were used to search for relevant success, failure or other scenarios.

Having a searchable sample of possible futures made it possible to conduct "thematic chain analysis" of keywords. Thematic chain analysis involved identifying a theme of interest such as social unrest, making a search for relevant scenarios, which were then reviewed and a table created, with the theme of interest in the central column, preconditions recorded in the left column and subsequent conditions (i.e., following conditions) recorded to the right (Table D.10). Themes were highlighted in bold and quotes from respondents were recorded in quotation marks.

From thematic chain analyses, and the analyses in Chapters 5–9, it was possible to construct maps with multiple possible pathways, including possible preconditions and possible responses for example related to social change and behaviour, political will and policy, business and economic activity, GHG removals, and international cooperation as well as possible responses and subsequent conditions (Chapter 10). These maps highlight interventions and conditions that could contribute towards effective global responses to climate change.

The analysis, including thematic chain analysis, was organised into five parts. The first part addressed the influence climate change might have on actors and the global response to climate change from Chapter 5. From the analysis of these scenarios and themes, a typology of climate change signals and responses was created along with an analysis of response criteria (Section 10.2).

The second part of the analysis addressed scenarios and themes regarding actors and their interests from Chapter 6. The analyses of overarching actor interest

Table D.10 Example of "thematic chain analysis" of the theme "social unrest"

Preconditions	Social unrest	Following conditions	Scenario ID
Insecurity: "So, first, you feel the bite just in terms of heat, cost of living, food availability" **Defence**: "when food is not available, or when temperatures become unbearable or when wealthy people are perceived to be protecting themselves behind high walls"	**Social unrest**: "social disruption" "And I think that that could happen in our cities. That could happen in different places... the risk of breakdown of law and order and of social systems"	**Institutional failure**: "So if you've got a breakdown in law and order your ability to create a new order in an intentional way is compromised"	80919b

themes included social change and behaviour (Section 10.3.1), political will and policy (Section 10.3.2), as well as business and economic activity (Section 10.3.3). The third part of the analysis addressed response option scenarios and themes from Chapter 7, including an analysis of GHG removal options (Section 10.4). The fourth part of the analysis addressed international cooperation themes from Chapter 8, including coalitions and issues of power and capacity (Section 10.5).

The fifth part of the analysis is the discussion in Chapter 11 which addresses the findings from Chapter 10 and the literature reviewed in Chapters 2 and 3. An important part of the analysis was the qualitative assessment of response types and the likelihood of these responses fulfilling the UNFCCC objective. If a response type is consistent with dramatic reductions in GHG emissions described in Section 2.2.4, then it was considered more likely to be effective. If a response was consistent with business as usual or limited reductions in GHG emissions, then it was considered unlikely to be effective. However, as noted in Section 2.2.3, there is a small chance that global warming might be limited to below 2°C even if atmospheric GHG concentrations are above 450 or even 500 ppm (CO_2-eq), which complicates the analysis, requiring consideration of serendipity.

Note: Failure scenarios were identified and thematically analysed but due to space constraints, these scenarios do not feature in the analysis. It is unclear what influence additional analysis and discussion of failure scenarios might have had on the results presented in this book.

Note

1   If the respondent had responded "yes" the global response is on path towards fulfilling the UNFCCC objective, then the next question would have regarded success scenarios (Figure D.2).

Appendix E
Respondent sentiment

Respondent sentiment was sampled to help get a sense of possible respondent biases. Of the 27 people surveyed, respondent sentiment was collected from 24 respondents.[1] The results below are presented based on the sentiment of the 24 respondents from whom data was collected (Figure E.1).

With regards to whether the world is a better place to live today than it was in the past, nearly 40% either agreed or strongly agreed, while just over 40% disagreed or strongly disagreed. Just over 20% of the sample were neutral. With regards to the world being a better place to live in the future, over 50% of the sample were neutral, nearly 30% agreed or strongly agreed and 17% disagreed. As such, there was a spread of sentiments.

There were only a few obviously optimistic or pessimistic individuals in the sample. For example, there were two optimistic respondents that agreed or strongly agreed that the world is a better place today and will be better in the future, meanwhile there were three pessimistic respondents who disagreed the world is better today than it was in the past and disagreed that the world will be a better place to live in the future.

If neutral responses are ignored,[2] then one respondent can be said to have indicated the world is peaking, as they agreed the world is a better place than it was in the past but disagreed that the future would be better. Meanwhile, three respondents indicated the world is as bad as it might get, as they disagree that the world is better today than it was in the past and agreed the world will be a better place in the future.

Interestingly, regardless of respondent sentiment being generally optimistic, pessimistic or something else, respondents universally indicated that the global response to climate change is failing during their interviews (Section D.2.1.3).

With regards to whether the UNFCCC should be abandoned, 96% of respondents disagreed or strongly disagreed with the statement. Only one respondent indicated that they strongly agreed that the UNFCCC should be abandoned, however this may have been a mistake filling in the response sheet, as they also strongly agreed that the UNFCCC is essential for an effective global response to climate change. Regardless, it can be said, respondents overwhelmingly disagreed with the idea of abandoning the UNFCCC.

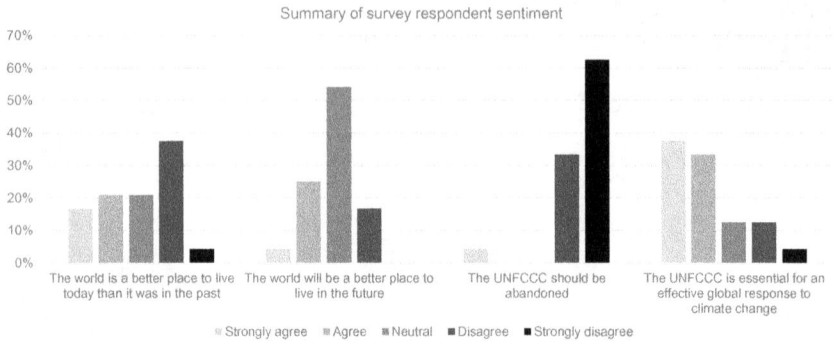

Figure E.1 Summary of survey respondent sentiment.

A large majority (just over 70%) of respondents agreed or strongly agreed that the UNFCCC is essential for an effective global response to climate change. However, nearly 30% were neutral, disagreed or strongly disagreed. As such, the sample included a limited number of respondents who were sceptical of the UNFCCC's contribution to the global response.

From the respondent sentiment data, it can be said that there is a relatively even spread of optimists and pessimists in the sample, however, the sample lacked respondents who believe the UNFCCC should be abandoned and generally included people who considered the UNFCCC to be essential for an effective global response to climate change, although nearly 30% didn't necessarily hold this view and as such may have been somewhat sceptical of the UNFCCC.

From Section D.2.2.1 the "survey sample" is biased but it was an open question as to what extent the survey sample biases the "sample of possible futures." Sentiment analysis suggests the survey sample includes respondents with mixed levels of optimism and faith in the UNFCCC and its role in the global response to climate change. These results are encouraging as the purpose of the survey was to collect a broad sample of possible futures.

Notes

1 The first respondent was not sampled as the sentiment question had not yet been included in the semi-structured interview, another respondent failed to press the submit button on the electronic form, hence their response was not collected, and lastly, one of the respondents from the COP24 was an impromptu interview, starting in the queue for coffee, hence the sentiment form was not filled in.
2 Neutral responses are ignored because it's not possible to tell if the respondent is unsure of what they think, or thinks life is the same as in the past or will be the same in the future.

Appendix F
Summary of scenario types and complexity

Each scenario was analysed and coded starting with outcomes. It was anticipated that scenario outcomes would be classified as being either, "success" "failure", or "other" However, some scenario plots had branching success or failure outcomes and as such a new category of scenario needed to be included.

Collectively, the scenarios compiled constitute a sample of possible futures. In total, 175 scenarios were compiled and plotted graphically. Of these, 8 were branching success or failure scenarios, 48 were failure scenarios, 68 were success scenarios, and 51 were other scenarios (Webb 2021). Of the 48 failure scenarios, 21 featured parties withdrawing from international agreements and 6 featured international climate agreements being amended in response to follow-up questions. Of the 68 success scenarios, 24 featured greenhouse gas removals and 22 featured stringent enforced agreements mostly in response to follow-up questions, but some of these scenarios were unprompted. Table F.1 provides descriptions of scenario types according to outcome codes.

Each scenario had a level of complexity reflected in the form (i.e., shape) of the scenario plot. Information recorded from scenario plots included the number of steps, whether the scenario was linear, had merging elements or branching

Table F.1 Scenario type by outcome

Scenario type	Description
Branching success or failure scenario	A scenario that shows diverging chains of scenario elements, one of which goes towards fulfilling the UNFCCC objective and the other goes towards not fulfilling the UNFCCC objective. Branching success or failure scenarios include critical nodes.
Failure scenario	A scenario that does not fulfil the UNFCCC objective or contributes towards not fulfilling the UNFCCC objective.
Success scenario	A scenario that results in fulfilling the UNFCCC objective or contributes towards fulfilling the UNFCCC objective.
Other scenario	A scenario that shows other conditions or events that could affect the climate or the global response to climate change, as well as other issues or dynamics related to the global response to climate change without necessarily contributing directly to either fulfilling or not fulfilling the UNFCCC objective, including extreme events.

elements, the number of layers and whether there were interlinkages or parallel conditions including assumptions (Table F.2 and Figure F.1).

With regards to the plotted form these scenarios took, the maximum number of steps in time in a scenario was 7 and the maximum number of layers (from more local to more global) was 14. Meanwhile, 15 scenarios were very simple consisting of one element only (i.e., 1 step and 1 layer). The most common form of scenario had 3 steps and 2 layers accounting for 25 scenarios. Most scenarios had 5 or less steps and 8 or less layers. Given the number of things that could happen for example between now and 2100, and from local to international levels, it is clear from the limited number of steps and layers, these scenarios are very broad and sweeping descriptions of possible futures. However, this can be a good thing when it comes to plausibility, given the conjunctive rule of probability theory (Yudkowsky 2008).

Interestingly, branching success and failure scenarios had an average of 3.1 steps and 3.4 layers, while success scenarios were very similar with an average of 2.9 steps and 3.7 layers. Using the number of steps and layers as a proxy for complexity, failure scenarios were generally simpler with an average of 2.9 steps and 2.4 layers while other scenarios were generally the simplest with an average of 2.6 steps and 2.0 layers.

Table F.2 Scenario complexity descriptors

Descriptor	Description
Steps	The maximum number of scenario elements in a chain from left to right, i.e., steps in time.
Linear	A simple chain of scenario elements through time, represented by horizontal arrows connecting these elements from left to right.
Merging	Two or more scenario elements joining a single scenario element in the future (to the right), represented with connecting arrows.
Branching	A scenario element that is connected to two or more scenario elements in the future (to the right), represented with connecting arrows.
Layers	The maximum number of scenario elements (including issues and assumptions boxes as well) from bottom to top (i.e., from more local to more global).
Interrelations	Scenario elements that are parallel in time and are directly connected, represented with a vertical arrow.
Parallel conditions	A condition that is parallel in time with other conditions, but not directly interrelated with other elements. In practice, this is presented as two boxes parallel in time but without connecting vertical arrows. Parallel conditions only refer to dark grey or grey boxes, not issues and assumptions in light grey boxes.
Assumptions	Stated (i.e., explicit) or unstated (i.e., implicit) beliefs about the system that has explanatory power with regards to the scenario.

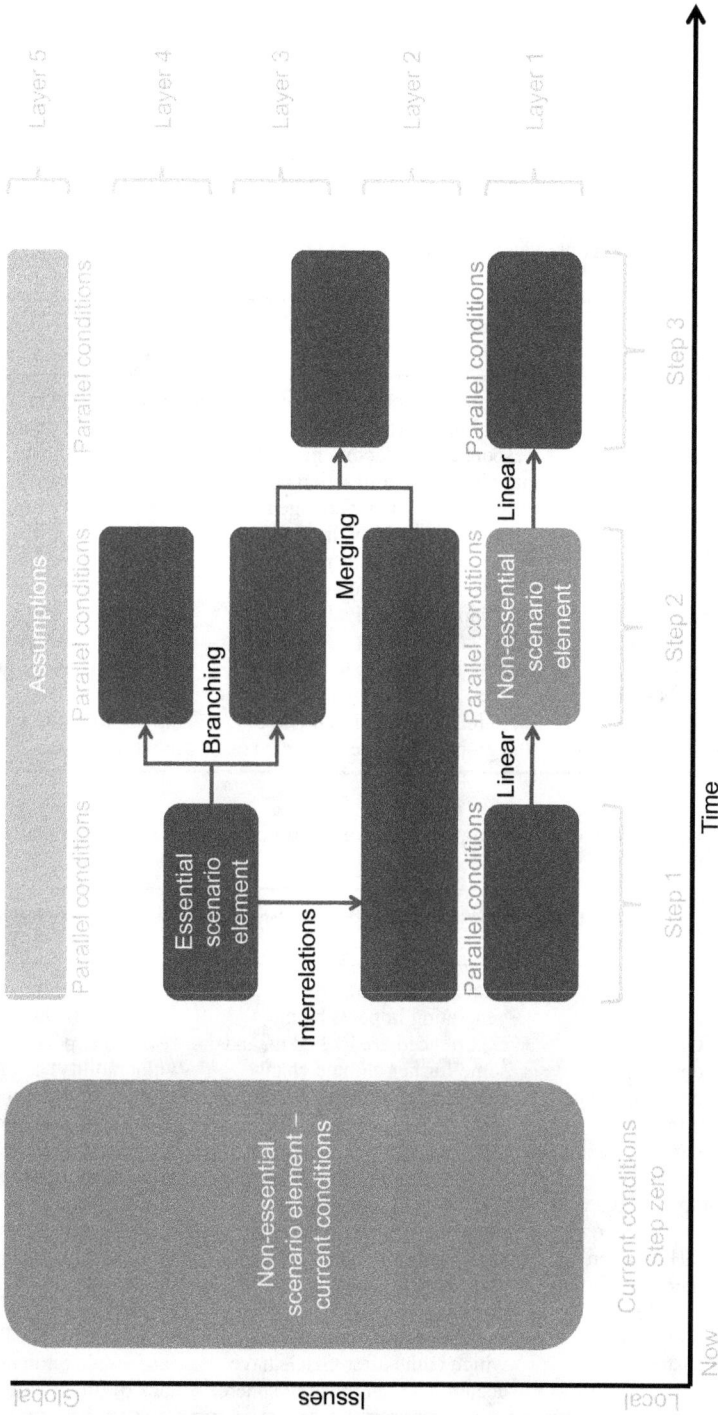

Figure F.1 Schematic illustration of a scenario plot including descriptors used to describe scenario complexity.

Appendix G
Timeliness and scale of responses

Table G.1 Cynical and non-responses and implications for the timeliness and scale of global respons‹ climate change

Trigger or driver:	Contribution to the global response	Timeliness of global response	Scale of global response
Cynical response	Some actors push their own interests increasing GHG emissions.	Due to the deliberate negative contributions to the global response, other actors need to act quicker on mitigation and removals if climate change and its impacts are to be limited.	Other actors need to increase the scale of mitigation and removals if climate change its impacts are to be limite‹
Non-response	Actors contribute to GHG emissions or contribute to adaptation, mitigation or removals without considering climate change.	Negative contributions mean other actors need to act quicker on adaptation, mitigation and removals if climate change and its impacts are to be limited. Positive contributions help improve the timeliness of the global response.	Negative contributions mean other actors need to do mo on adaptation, mitigation a‹ removals if climate change and its impacts are to be limited. Positive contribut‹ help improve the scale of t global response.

Table G.2 Impact and risk responses and implications for the timeliness and scale of global respons‹ climate change

Response trigger/driver	Contribution to the global response	Timeliness of global response	Scale of global response
Risk response	Actors mitigate and remove GHGs based on the risk of climate change impacts on human systems.	Climate change hazards generating impacts being experienced are locked in, and some further climate change risks may also be locked in.	Adaptation and improveme in climate resilience (i.e. reduced exposure and vulnerability) can limit impacts on human systen There will likely be a ne‹ for removals but at a sm‹ scale due to timely mitig
Impact response	Actors mitigate and remove GHGs when there are climate change impacts on human systems.	The hazards generating the impacts being experienced are locked in, as are further climate change and related physical hazards due to the lag in climate and related changes which come after GHGs have accumulated in the atmosphere.	Adaptation and improveme in climate resilience can impacts on human systen but will need to be scale‹ to address future locked– hazards. There will need large-scale removals due lack of timely mitigation

Table G.3 Response triggers and drivers and implications for the timeliness and scale of global responses to climate change

Response trigger/ driver	Contribution to the global response	Timeliness of global response	Scale of global response
Cost-benefit response	Actors mitigate and remove GHGs based on an assessment of costs and benefits, and because of discounting rates applied, immediate impacts and costs of climate action are more strongly weighted in decisions than risks in the future.	Cost-benefit approaches are more likely to respond to impacts than risk of unprecedented hazards (see impact response)	Decisions to act on climate change depend on cost-benefit analyses and only those actions with a positive net present value taking into account the social cost of carbon, are acted upon. Most likely, only a subset of what needs to be done to limit climate change will actually be done and the scale is unlikely to be adequate until a cost-benefit analysis supports large-scale removals of GHGs.
Enlightened response	There is urgency and ambition, in part due to climate change considerations but possibly due to other considerations. As such, enlightened responses are likely the be leading responses.	Enlightened responses may come ahead of impacts being realised, and as such make timely contributions to the global response, although any impacts already experienced will likely be locked in already.	The scale of an enlightened responses depends on the extent to which awareness creates change, including the breadth of actors that are enlightened.
Cooperation response (variation of enlightened response)	States decide to cooperate on climate change because they have seen the benefits of international cooperation in addressing other challenges	The timing of the responses is not due to any particular climate impact or risk, but rather has arisen due to unrelated experiences (e.g., pandemic and international response)	The scale of the global response is dependent on the ambition of states cooperating on climate change.

Table G.3 Continued

Response trigger/ driver	Contribution to the global response	Timeliness of global response	Scale of global response
Emergency risk response	There is urgency and ambition in the response to risks.	An emergency response to climate change and related hazards could help ensure timely adaptation and mitigation efforts. However, it could be that the risks generating the emergency response are already locked in unless there are GHG removals.	Resilience is essential to limit the impacts of climate change on actors and their interests. Mitigation efforts can limit long-term climate changes and impacts but will have limited influence on the identified emergency risks. GHG removals at massive scales will be required to lower atmospheric GHG concentrations and may be adequate to limit the emergency risk. Other geoengineering interventions might be necessary to address the emergency risk.
Emergency impact response	There is urgency and ambition in the response to current impacts.	The range of hazards generating the emergency response are already locked in and is part of the new normal for the foreseeable future.	Adaptation efforts will be essential to limit the vulnerability and exposure of actors and their interests. Mitigation efforts can limit long-term changes but will have no impact on the emergency. GHG removals at massive scales will be required to lower atmospheric GHG concentrations but the influence of this intervention will take time. Other geoengineering interventions might be necessary to address the emergency.

Response attitude	Contribution to the global response	Timeliness of global response	Scale of global response
Cooperative	Cooperation is by its nature limited to common interests, which in most cases are a subset of the problem. So, cooperation can lead to positive contributions, cooperation alone is unlikely to resolve the climate change problem.	Waiting for an agreement to cooperate can delay timely interventions or innovation.	Non-participation and non-compliance can limit the scale of cooperative interventions and initiatives.
Practice focused	Can be positive or negative depending on the characteristics of practices developed and deployed.	Research, development and piloting of practices (e.g., management options and policies) can help ensure there are practice options for addressing climate change in the future, when they may be especially needed.	Can lead to the creation of practice options (e.g., management options and policies) that might be scaled up. Can also lead to support of practice options that already exist, or have been piloted already, helping these options scale up. Management options can have unintended consequences which may be problematic at scale, depending on the characteristics of the unintended consequences.
Technological	Can be positive or negative depending on the characteristics of technologies developed and deployed.	The development of technologies can help ensure there are new technological options for addressing climate change in the future, when they may be especially needed.	Can lead to the creation of technological options that might be scaled up for example through competition. Can also lead to support of existing or close-to-market options that already exist, helping these options scale up. Technologies require inputs to production and related wastes, all of which could be problematic at scale, depending on the characteristics of inputs and wastes.
Competitive	Can be positive or negative depending on the incentives (including prices, taxes, subsidies, regulations and laws) and the actions being incentivised.	Competitive attitudes can lead to timely actions for example when actors believe "the early bird gets the worm." Competitive attitudes can lead to a situation where actors wait for options to be more viable, e.g., "the second mouse gets the cheese."	With the right incentives, competition can scale up interventions, especially technology and practices including business models.
Defensive	Local and domestic adaptation to the extent that is possible given available resources, but little or no mitigation or GHG removals.	Efforts to improve resilience may be enhanced but mitigation and GHG removals will not even be initiated, due to the belief other actors will emit GHGs.	Difficult to achieve mitigation or GHG removals at scale as defensive attitudes lead to each actor acting in isolation on their own interests. Adaptation may be achieved at scale if the actor has sufficient resources.

Appendix H

Effective responses and serendipity

Table H.1 Effective responses, including preconditions, that could conceivably fulfil the UNFCCC objective

Response	Preconditions for effectively fulfilling the UNFCCC objective
Responses relying on ambition	**Responses with high or very high ambition leading to leadership, enlightenment or emergency action on the climate change problem. These scenarios rely on actors taking action and forming effective coalitions to address climate change.**
Emergency response	Emergency responses involve very high ambition and urgency. Emergency responses could be considered an extreme end of the spectrum of enlightened responses, consisting of "involuntary enlightenment" (Beck 2006) where climate change impacts and risks force awareness and change.
to impacts	Emergency responses to impacts need climate change and related impacts to be reversible or at least able to be stabilised rapidly. However, given the lags in climate change and related impacts, this may or may not be possible. Furthermore, an emergency response will require technologies and practices capable of limiting climate change and related impacts.
to risks	Emergency responses to climate change and related risks need climate change and related risks to be reversible (i.e., lock-in of risks to be avoidable). A high degree of serendipity is also required, in the form of an unambiguous future climate risk signal that is able to generate an emergency response.
Enlightened response	Enlightened responses include an awareness of impacts and risks to human and managed systems as well as other considerations for example ethical or justice-related. Some level of serendipity is likely, and this could include some other factor influencing people to respond to climate change (e.g., responding to some other international crisis such as a pandemic and then pushing for international cooperation on climate change).
Leadership response	Acting with high to very high motivation due to concerns based on climate science, climate risks or other considerations. Leadership responses from social, political, and business actors can help create the preconditions for effective responses for example helping generate and guide social change, create political will and policies, or invest in technologies and practices when most other actors don't.

Responses relying on serendipity	**Responses with high, medium, low, zero or even negative ambition, but through very good or even blind luck manage to fulfil the UNFCCC objective. These scenarios could include much lower than expected climate sensitivity, or the prices of climate resilient low emissions technologies and practices falling much quicker than expected.**
Very lucky cost-benefit-based response	Cost-benefit-based responses are based cost-benefit analyses and as such, discount medium to long-term climate change impacts and risks to low values, unless a very low social discount rate is used. A high degree of serendipity is required in the form of either: A. climate resilient low emissions technologies and practices being available at a lower cost much quicker than expected, making responding cost-effective using a standard discount rate and a given social cost of carbon (depending on the level of ambition); or B. Climate sensitivity is much lower than expected hence the scale of climate change and related impacts is much lower than anticipated, allowing time for technological and changes in practices to become available following established learning curves.
Blind luck non-response	It is conceivable that non-responses could still result in an effective global response to climate change, but this would be through blind luck and an extremely high degree of serendipity rather than anything else. Such serendipity could be in the form of either: (1) climate resilient low emissions technologies and practices being available at lower cost much quicker than expected, making responding cost effective using a standard discount rate; or (2) Climate sensitivity is much lower than expected hence the scale of climate change and related impacts is much lower than anticipated.
Failed cynical response	It is conceivable that if cynical responses fail, and there is a large amount of blind luck, then the global response to climate change could fulfil the UNFCCC objective assuming either climate resilient low emissions technologies and practices become cheaper much quicker than expected or climate sensitivity is much lower than expected.

Appendix I

Preconditions and the role of serendipity

Table I.1 Preconditions and the role of serendipity

Tier 1 preconditions	The role of serendipity
Climate change related	
Scale of climate change can be addressed by technologies and practices	The scale of climate change and related impacts depends upon climate sensitivity, atmospheric concentrations of greenhouse gases and other climate forcers. In a high serendipity scenario, climate sensitivity would be lower than anticipated. The concentration of greenhouse gases in the atmosphere depends upon anthropogenic greenhouse gas emissions as well as the balance of GHG absorption and releases by the ecosystems and the rest of the environment through unmanaged processes. In a low serendipity scenario, there could be increased releases, or even net releases, of GHG from the environment into the atmosphere through unmanaged processes.
Climate change and related impacts are reversible or can be quickly stabilised	Given that many scenarios include overshoot with GHG removals from the atmosphere used to eventually stabilise atmospheric concentrations of GHGs, in a low serendipity scenario a threshold could be passed where the feedback mechanism kicks in causing the release of GHGs from the environment, or cascade effects creating impacts that are irreversible. In a high serendipity scenario, such tipping points and thresholds are not breached.
Response related	
The global response (i.e., transformation) is timely and at scale	In a high serendipity scenario, climate-resilient low emissions technologies and practices quickly become cheaper, or are easier to apply, than expected, and are applied at scale much earlier than expected.
Contingencies are available for addressing extreme climate change or other scenarios	In a high serendipity scenario, technologies and practices that can be applied in case of extreme climate change or other scenario, are available with little extra investment or research, but in a low serendipity, a lot of investment in research and development would be required, and still contingencies are inadequate.
Other scenario related	
Other things (e.g., events or dynamics) don't negatively influence scale, climate, or the global response to climate change	The global response to climate change is not the only thing that can influence climate change and related impacts, and furthermore, other things might also influence the global response to climate change. In a low serendipity scenario, other things could happen, such as manmade or natural catastrophes including war, an economic depression, large volcanic eruptions, or even cosmic events that would profoundly affect climate or the global response to climate change.

Index

Note: **Bold** page numbers refer to tables, *italic* page numbers refer to figures and page numbers followed by "n" refer to end notes.

abrupt irreversible changes 14–15, 177, 237, 247, 256–257, 298; *see also* irreversibility
accountability 82–**83**, 88, 155–*156*
acidification *see* ocean acidification
actions *see* climate action *and* options
actors 2, 4; *see also* actors and interests
actors and interests 79, 138, *141*, *153–154*, 159, *161*, 181; *see also* social change and behaviour, political will and policy, *and* business and economic activity
adaptation 237; capacities 34–35, 85, 89, 106; defensive 76; ecosystems 33–34, 61, 63, 134–135; financing 82, 128; litigation 112; local government 128; lock-in 72; options 22–23, 106, *41–42*; overshoot scenario 56; pathways 177–*179*; preconditions for effective global responses 185, 189–192, 196–199; rates of 68, 93, 98, 113
afforestation *see* forests
Africa 52, 64, 124, 143, 276–279
agriculture mitigation 25, 93, 102, 107, 172; production **61**, 98, 112, 133–134; technologies and practices 92, 120; vulnerability 67, 118
aid 125
ambition 19, 44–46, 237; scenarios 69–71, 73, 75, **86**, 108–109, 123; pathways 139–140, 144–146, *149–150*, **161–162**, *165–166*, *167–168*; preconditions for effective global responses *178–180*, *185*, **188–192**, 197, **293–294**, **296–297**
Antarctica 3, 14, 61, 132, 143, 177
apathy *see* non-response
Arctic 3, 63, 143, 176

ASEAN 120, 122
atmospheric greenhouse gas removals 103–108
Australia 16, 91, 118, 121, 126, 213
aviation 115, 178

backcasting 49–53, 57, 267, 270–271
behaviour *see* social change and behaviour
biodiversity 61, 125, 177
bioenergy carbon capture and storage 27, 104, 106–107, 113
biological systems 12, 14, 139, 238, 257, 259; scenarios 56–58, 61–62, 65
black swan 14, 53, 238, 273; scenarios 130–135
Bodandsky and Diringer 2010 40, 152
Borio et al. 2014 50, 58n1
Bostrom 2013 16–17, 163
Bostrom and Cirkovic 2008 2, 16, 19, 21–22
Brazil 82, 115, 118, 120–121
BRICS 116, 120–121
buildings 23, 25, 76, 151, 259
business *see* business and economic activity
business and economic activity: scenarios *56–57*, 94–98; related pathways **139**, **153**–*161*, 164, 181

Canada 64, 128
capacity **34**–*35*; pathways *146–147*, *149*, *157–158*, 177; preconditions for effective global responses **181–183**, **188–189**, 191–192, 198, 200, 202–203; scenarios 104, 111, 128, 130
carbon budget 13, 19–20, 24–25, 126–127, 173

carbon capture and storage *see* atmospheric greenhouse gas removals
carbon dioxide removals 20, 25, 27, 67, 104, 173, 183, 238
carbon dioxide 10, 12–15, 20, 27–28, 105
carbon price 89, 95, 111, 119, 127, 159–*160*, *166*–167
carbon tax *see* taxation
Carney 2015 12, 18–19, 21, 50
cascade effects 177, 187, 189, 198, 298
catastrophe 2, 16, 49, 239; scenarios 56, 61–63, 70–71, 74–75, 119, 132–133; pathways 148, *156*–*158*, **162**–**163**, 167, 177
catastrophic risk *see* risk
China: climate impacts 61, 64, 200; consumption 83; defensive response 94; geopolitical groupings 31; geopolitical power 65, 121, 124, 132; global leadership 109, 120–121; international cooperation 116, 117, 182; international relations 122, 124; disasters 213; national interest 118, 161; social licence 109
cities 3, 24, 36; scenarios 78, 84, 91–92, 120, 127–128, 135; pathways and preconditions 143, 208, 261, 263, 285
civil society: behavioural responses 82–83; coalitions 102; leadership 108, **153**; military crowd control 94; non-state actors 32, 38, 40, 52
civilisation 3, 16, 133, 207
climate action: pathways 71, 138, 145, **150**, 161; preconditions for effective global responses 171, 184, 188–189, 202; *see also* options
climate change 3, 10–11; *see also* climate signal, impacts, risks
Climate Change National Interests International Cooperation Model 37–43, 56–57, 138–139
climate change signal *see* climate signal
climate crisis 16; pathways 148, 155–156, 158, 175, 177; preconditions for effective global responses 196–197, 200; scenarios **61**, 63, 65, 108, 119, 124; *see also* crisis
climate forcing *see* radiative forcing
climate impact *see* impacts
climate policy *see* policy
climate policy architecture 32, 36, 128
climate regime *see* international climate regime

climate regime complex 36, 42, 260–265; *see also* international regime
climate response options *see* options
climate response system *see* Climate Change National Interests International Cooperation Model
climate risk *see* physical risk
climate signal 144, 177–178
coal 26, 28, 31, 70, 90, 106
coalitions 40–41, 43; pathways *159*–*161*, *165*–*168*, 171, 174; preconditions for effective global responses **181**–**183**, **186**–**190**, *192*–*193*, 197–200, 202; scenarios 102, 121–123, 127
collapse: socio-economic related 61, 114, 132, 134, 175; environment related 3, 63, 177
commercial viability 159, 161–162, 165, 178, 192, 199
compensation: claims 18; land use 93, 107; carbon prices 111, 167; enforcement mechanisms 125; international cooperation 126
competitive response 77–78, 82, 152, **178**–**180**, **185**–**186**, 241, 295; *see also* response attitudes
compliance (international agreements) *39*–41, 44–47, 124, 126–127, 152, *166*–*168*
Conference of Parties 33, 42–43, 109, 260–264
conflict 175, 184; scenarios 64, 95, 112–113, 115, 132–133
consumers 80, **84**, 92, 122, 144–145, 160
consumption 82, 126, 152, 155–*156*, 159, 161, 178; *see also* over-consumption 79–80, 82–84, 94, 101, 103
cooling 14, 67, 131, 133, 184
cooperative responses 152, **178**–**180**, **185**–**186**, 241, 295; scenarios 73, 75–76, 123; *see also* response attitudes
coral 15–16, 108, 201
corporations 36, 92, 97, 116, 120
cost benefit response 146, *149*–**150**, **152**, *168*, *179*, *185*; *see also* response triggers and drivers
countries *see* national interests
COVID-19 203, 206–208, 208–213
crisis 74–75, 131; *see also* climate crisis
crystal ball 4, 48, 52, 54
cynical response 241; pathways 144, 146, *148*–**150**, 161, 177; preconditions for effective global responses *179*–**180**,

182, *185*–**186**, 192, 197, 199; scenarios
69–**70**, 78, 91, 95; *see also* response
triggers

decisions 4, 5, 40–41; scenarios 63, 69,
71–72, **83**, 86
decision making 49–50; pathways
145–147, 149, *153–154*, *180*;
preconditions for effective global
responses 199, 203; scenarios 93, 115
defensive response 75–76, 152, 178, 185,
192, 241
deforestation *see* forests
developed countries 95, 111, 118,
128–129, 200
development 10–11, 24, **34**–**35**, 190, 198;
see also research and development,
sustainable development *and* sustainable
development goals
developing countries 111, 120, 125,
128, 202
diplomacy 43, 117, 127, 167
direct air capture *see* atmospheric
greenhouse gas removals
disruption 14; of human systems (social,
political and economic) 18, 35, 84, 92,
132; of natural systems 63, 118, 133;
risk conception 22, 146, 148
distribution: of impacts and risks 15,
65, 72, *140–142*, *149*, *155–156*; of
income, wealth, costs and benefits 81,
167, 175; of power 92, 96; of food
98, 113; preconditions for effective
global responses 175–177, 181–182,
189–190, 200
disasters 62–63, 74, *156–158*, 162, 175, 177
drought 14, 62, 63, 68, 139, 140

earth 3, 11, 28; scenarios 56, 101, 106,
116, 132
ecosystems 11, 15, 33–35, 259, **268**–**269**;
scenarios 56, **61**–**63**, 77, **84**, 103,
134–135; pathways **139**–**140**, 177–178;
preconditions for effective global
responses 190, 194, 200–201, 298; *see
also* biological systems *and* physical
systems
economic development *see* development
economic activities 20–21, 55, *56–58*, 199;
scenarios 64, 83, 85, 94–98; pathways
139, 144, 153–161, *164*
effective global responses 3–6, 184–187,
249; criteria 51–52, 267–269;

preconditions for 66–68, 171–173,
184–191, 196–200; *see also* effective
international cooperation
effective international cooperation criteria
43–47
efficiency 24–25, 31, 34; scenarios 90, 97,
102, 106–107, 127–129
electric vehicles 24, 68, 84, 92, 98, 101
emergency response 242; pathways 146,
148–149, *165–166*, 168 *177–179*;
preconditions for effective global
responses *185*, 190, 192, 197, 294,
296; scenarios **72**–**74**, 87, 94; *see also*
response triggers
energy: atmospheric greenhouse gas
removals 67, 163–*164*, 183, 200; business
and economic activity 96–98, 111, 148,
161–162; earth's energy balance 11,
258; efficiency 82; geopolitics 31–32,
126, **262**; leadership 199; policy
90–91; preconditions for effective global
responses 173–174, 178, 188, *190*; sector
24–**27**; technology 101–102, 105–107,
110–111
enforcement 44, 115, 122, 124–125,
166–167, 242
enlightened responses 131, 148–150,
185–186, 197, 242, **293**; pathways
155–156, 159–160, 165–166, 178–180;
scenarios 71–72, 74, 81, **84**, 119–120;
see also response triggers
Environment and National Interests Model
258–259
equity 34–**35**, 98, *172*, 175, 190
eruptions 11, 14, 131, 133–134, 184
Europe 14, 18, 143; scenarios 67, 89, 94,
115, **118**, 123
exposure 17–18, 85, **140**–**141**, 242,
258–**259**; *see also* risk
extreme heat 10, 18, 23, 61, 143

failure criteria *see* success and failure
criteria
farms 3, 68, 105, 135, 145
feedbacks 12–13, 17, 22, 28, 239, *258;*
pathways 163, 177; scenarios 61–63, 66,
76, 103, 132, 134–135
Figueres et al. 2017 19–20, 24, 173
finance *41–42*, 243, 260–263; Paris
Agreement *34–35*; pathways *153–154*,
159–160, 165–167; preconditions for
effective responses **172**–**174**, 188, 190;
scenarios 68, 89, 97, 111, 125–126, 128

finite game 2, 194, 201
flexibility 45–47, *166*, 243
flooding 18, 39, **61–62**, 93, **139–140, 142–143**
food 15, **33–35**, 243; scenarios **61–62**, 64, 68, 98, 113, **118**; pathways **143**, *160*; preconditions for effective global responses 189–190, 192
forestry *see* forests
forests 14, 24–25, 27; pathways *164*; preconditions for effective global responses 183, 192; scenarios 63, 68, 77, 93, 102, 105, 107, 122; *see also* landuse
forcing agents (and radiative forcing) 11, 13, 184, *258–259*
foresight methods 48–51, 243; methods used 51–55; *see also* backcasting; reverse stress test, stress test, thematic chain analysis; and, theory of change
fossil fuel 21, **26**; pathways *149*; preconditions for effective global response to climate change **173–174**, 177, 188, **190**; scenarios 70–71, 90–91, 95, 97, 110–111, 126
France 182, 200
funding 18, 70, 89, 128

G7 116, 121
game theory 2, 8n1; *see also* finite game, infinite game *and* two-level game
geoengineering 76, 107, 132, 243, 249; *see also* greenhouse gas removals *and* solar radiation management
geopolitical power *see* power
global response system 36–43
global stocktake 268
global warming *see* climate change
globalism 115
green economy 26, 68, 95, 113, 160; energy and technologies 110, 120; Green New Deal 81, 88–89; growth 77, 129; *see also* re-industrial revolution
greenhouse gas removals *see* atmospheric greenhouse gas removals
greenhouse gases 11–14, 17–23, 25–28, 33–35, *37–39*, 244; scenarios 62–63, 66–68, 71–72, 131, 134; pathways 145, *157–158, 163–165*, 172–174, *178–179*; preconditions for effective global responses *184–185*, **187–192**, 196–203; *see also* atmospheric greenhouse gas removals *and* forcing agents

Greenland 14, 61, 132, 177
Grubb et at. 2014 19, 21–22, 146, 148, 150–151, 177

Hawken 2017 173
hazards 17–18, **37–39**, 60–62, 72, 140–141, **258–259;** *see also* risk
health 11, 15, 27, 143, 256, 282; scenarios **61–62**, 67, 87, 91, 132; *see also* COVID-19
heat *see* extreme heat
hope 22, 31, 204; scenarios 67, 82, 107, 118, 133, 134
hopelessness *see* non-responses
human and managed systems 3, 245; climate impacts and risks to 11, 13, 56, **139–144**, 175, 198; responses *147*, *156–157, 159–160, 168*, 176, 181–182; scenarios 58, 63, 65, 87, 89
hypocritical *see* non-response

ice sheet 3, 14, 61, 63, 132, 177
impacts 11–12, *139, 142*–144, 149, 154, 245; pathways 155–*160*, 165–166, 168 175–177, *179*; preconditions for effective global responses 185–186, **189–190**, 196–204; scenarios 60–64; *see also* climate signal
incentives 78, 148, 199, 203; landuse 68, 105, 164; political will and policy 44, 90, 102, 107, 157–160; geopolitical 124, 126, 162
India 82, 89, 91, 117–118, 120–121, 124
industry 11, 23–26; pathways 153, *159–160, **162***; scenarios 76, 89–92, 95, 97–98, 106–107, 128
inequity 85, 184
infinite game 2, 171, 194, 196
infrastructure 91, 97, 143, 157–158, 165, 259
interests *see* actors and interests
international cooperation 246; effectiveness criteria 43–47; geopolitical leadership 120; geopolitical power and influence 120–122; global response system 37–43, 138–139; options 126–129; pathways 159–160, 165–169; preconditions for effective response 181–183, 199–200; stringent enforced agreement 122–126; triggers and drivers 117–120; *see also* international regime; international regime complex; *and*, models

International Monetary Fund 116
international regime *56–58*, 114–117, 127, 165, 246; *see also* climate regime complex *and* international cooperation
internationalism 165; *see also* globalism
investment 4, *41*–42, 111, 161; accountability 83; atmospheric greenhouse gas removals 67–68; cities 92; defensive 76; ethical 81; in infrastructure 91, 97; international regime 36, 117, 118, 128, **262**; law 112; pathways *156*–159, **162**, 165; preconditions for effective global responses 173–**174**, 190; in renewable energy 26, 97; research and development 76, 96, 107–108; scale and timing 25, 77, 105, 107; signals 90; social 24
irreversibility 14–15, 177, 238, 247

Jacinda Ardern 109, 120, 208, **211**
Japan 91, 94, 118, 123, 128
Jellyfish Model of international cooperation 260–265

Keohane and Victor 2011 2, 36, 38, 42, 260–265
Kingdon 1995 10, 110, 176, 191, 203
Kyoto Protocol 33, 35, 43, 126, 161, 192

law: international 122, 131; localisation 92, 97; legal precedent and contracts 84, 108, 112–113, 154, 201; *see also* regulation *and* liability risk
law making 39, 41–42, 44, 120
land use 11, 18, 27, 33, 247; pathways *157–158*, 162; preconditions for effective global responses 177–178, 192, 199; scenarios 68, 93, 96, 106–107, 110
large scale singular event 15, 247; pathways *142*, *149*, *158*, 177, 198; *see also* catastrophe
leadership 108–110, 153, 185–**186**, 188–**189**, 197–199, **296**; business and economic activity **83**, 95, 159–*160*; geopolitical 74, 115, 117, 120–121, 123, 166; political will and policy 25, **84**, 86; social change and behaviour 81, 82, **84**, 109, 155–*156*
legal precedent *see* law
legislation 20, 33, *39*, *41*–42, *157–159*; *see also* law making

liability risk *see* risk
litigation *see* risk
lobbying **22**, 109, 157–160
local government 32, 38, 40, 43; pathways 153, 57, 167; scenarios 86, 92; *see also* subnational
localisation 91–93, 97, 126–127, 157–160, 166–167
long lived greenhouse gases 11–12, 27
luck *see* serendipity

maladaptation 23, 191
marine 11, 62, **101**, **139**, 178
measurement, reporting and verification *see* transparency reporting and verification
meteorite 133, 134
methane 12, 14, 28; scenarios 63, 107, 134
migration 64, 89, 112, 118, 125, 175
military 94, 115, 132
minerals 31, 122, 126
mitigation *22–27*, *41*–42, 248; pathways 145–146, 148, **150–152**, 161, *163–164*; preconditions for effective global responses **173–174**, *177–179*, 182–183, *185*, **187–190**, 196–201; scenarios 66–68, 72, 97, 103–104, 112, 118–119; *see also* atmospheric greenhouse gas removals
Montreal Protocol 36, 125, 245
movement *see* social movement
multilateral development banks 36, 38, 116, 165
multiverse 2, 54, 138, 284

national interests: model 36–43; scenarios 57, **117–118**; pathways 138–139, 163, *165–166*, 181, 188
nationalism 115, 165
nationally determined contributions 19–20, 29n3, 31–32, 44–45, **264**; pathways *166–167*; scenarios 66, 89, 107, 123, 126, 128
natural resources 21, 23, 112–113, 122
New Zealand 64, 109, 208–209
no-trigger response 69–71; *see also* non-response
non-government organisations *see* civil society
non-responses 69–71, 150, 177, **180**, **182**, **186**; apathy and hopelessness 131, 201, 237; scenarios 96, 110
non-state actors 32, *36–43*; scenarios 91, 127; pathways **153**, **163**, 181, 199

ocean acidification 10, 14–15, 18, 28, 39,
   259; scenarios 61, 108
oceans 11, 134
oil and gas **25–26**, 31, 88, 102, 173
options *see* response options
over-consumption 79–80, 82–84, 94,
   101, 103

Pacific 143
pandemic 74–75, 131, 184, 192; *see also*
   COVID-19
Paris Agreement 19–20, 32–36, 40, 42–46,
   201, 268–269; pathways *166*–167;
   preconditions for effective responses
   171–175, 187–188, 190–191, 198;
   scenarios 88, 123, 125–128, 202
participation 21, *39–42*, 44–47, 248–
   249, 270; pathways 152, *166, 168*;
   preconditions for effective global
   responses 181, 188, 200; scenarios 75,
   122, 126–127, 295
pathways 20; to 1.5°C 25–26; business
   driven 157, 159–160; global warming
   13–14; greenhouse gas removals
   163–164; international cooperation
   165–168; methods 49–51, 281,
   285; politically driven 157–158;
   socially driven 85, 153, 155–*156*;
   towards effective global responses
   172–174, 187; *see also* Representative
   Concentration Pathways *and* Shared
   Socio-economic Pathways
penalties 125–126, *165–166*
permafrost 12–13, **61–63**, 75, 134, 139
physical risk *see* risk
physical systems 12, 139, 249, 258–259;
   scenarios 56–*57*, 61–*62*, 87
planet 64, 74, 82, 102, 131–135, 203
policy *see* political will and policy
political will and policy: scenarios *56–57*,
   84, 86–94; related pathways **139, 153**–
   161, 164, 181
population 131–133, 155
power 65–66, 249; coalitions 40, 174,
   181–183, 188–190, 195; business
   and economic activity 88, 94, 116,
   190; defence industry 76; emergency
   powers 73; geopolitical 117–118,
   120–122, 182; local empowerment
   91–93, 127; pathways 146–*147*,
   149, 165–167, 192; political will and
   policy 88; preconditions for effective
   global responses 194–195, 198–203;

social change and behaviour 80, 84,
   130; United Nations Security Council
   115, 118
practice focused attitude 76–77, 146
practices *see* technology and practices
precipitation 10, 13, 18, 28
preconditions for effective global responses
   see effective global responses
private sector 24, 42, **153**; scenarios 90, 96,
   102, 107, 110, 112
Project Drawdown 24, 173
the public 77, 88, 105, 108–109, 191
Putnam 1988 21, 39–41, 181
puzzle 2, 4, 9, 48, 149, 196; *see also* riddle

radiative forcing *see* forcing agents
re-industrial revolution 97, 159–160
reasons for concern 15, 64, 140–143,
   149, 168
regime complex *see* climate regime
   complex
regulation 88–89, 159–161; enforcement of
   laws 126; *see also* stringency
Representative Concentration Pathways 13,
   25, 123, 250
representative key risks *see* risk
research and development 27, 264;
   pathways **162**–*164*, **166–167**;
   preconditions for effective global
   responses **188–189**, 199; scenarios 76,
   89–90, 96, 107, 111, 126
resilience 11, **34**, 42; pathways 145,
   152, *157–158, 179*; preconditions for
   effective global responses **190–192**,
   198; scenarios 68–69, 76, 92–93
response attitudes 75–78, 251, **295**;
   pathways 146, **149**, 152, 178
response options 22, *145*, **162–163**; conflict
   112–113; finance and investment 111;
   greenhouse gas removals 103–108; law
   and contracts 112; leadership 108–110;
   migration 112; preconditions for
   effective global responses 174, 180–
   181, **186–189**, 191, 195, 199; prices
   110–111; technology and practices
   100–103; win set 40–*41*; *see also*
   adaptation; atmospheric greenhouse gas
   removals; mitigation; *and* solar radiation
   management
response triggers and drivers 69–75, 146,
   149, 293; *see also* cost-benefit response;
   cynical response; emergency response;
   enlightened response; non-response

responsibilities **34;** scenarios 73, 77,
82–83, 109, 112, 128; preconditions for
effective global responses 161, 172, **190**
reverse stress test 50–53, 267, 270–271
riddle 2, 196, 206; *see also* puzzle
risk 252; catastrophic risk 2, 16, 132, **163,**
167; liability risk 18–19, 50, *83–84,* 112,
247; physical risk 12, 16–19, 21–22,
140, 249, 258–259; Representative Key
Risks 15, 247, 250, 251; transition risk
18–19, 21, 140, 144–145, 199, 254; *see
also* exposure, hazard, *and* vulnerability
Russia 91, 118, 126, 128, 182, 200

Saleemul Huq 74
sample of possible futures 48, 52, 55, 57
sanctions 38, 115, 121–123, 125–126,
165–166
Saudi Arabia 91, 106, 126
scenarios: from the literature 22, 24–**26,** 28;
methods 48–55; from this study 55–135;
see also foresight methods
sea level rise 3, 10, 13, 18, 22, 39;
pathways **143,** 177, 194; scenarios 61,
63, 65, 85, 91, 93
Secretary General *see* United Nations
Secretary General
security **22,** *39*; scenarios **61,** 73–76,
115–119, 123–124; pathways 146, 148,
151, 181–182
serendipity *168,* 171; preconditions for
effective global responses 184–187,
191–192, 194, 197–198, 202
Shared Socio-economic Pathways 13,
21, 252
shareholders: scenarios 81, **94,** 97, 112;
pathways **153,** *155–156, 159–160*
shipping 115
Short-Lived Climate Pollutants 11,
27–28, 36
singular large scale event *see* large scale
singular event
Small Island Developing States 31, **162–
163,** 194, 201
social change and behaviour: scenarios
55–56, **80–86;** related pathways **139,**
153–161, 164, 181, 200
social contract: pathways 155–157, 159,
161, 169; preconditions for effective
global responses 181, **186,** 200, 253;
scenarios 79–80, 85–86; *see also* social
permissions
social licence *see* social contract

social movement: pathways *155–156*;
scenarios 82, 84, 92, 102, 109, 115
social permissions 85; pathways 155, 167,
171; preconditions for effective global
responses 181, 185–**186,** 188–**189,** 198,
200, 207
soil: greenhouse gas removals 27, 103, 183;
management 77, 93, 105
solar radiation management 28, 184, 253;
scenarios 89, 101, 103, 132, 135
South Africa 82, 120–121
states *see* national interests
Stavins et al. 2014 36, 38, **45–47,** 114,
262–263, 269–270
storms 10, 13, 140
stress test 93, 157–158, 202, 253; *see also*
reverse stress test
stringency 44–47, 123, 152, 166, 253
subnational 18, 32, 35, *262–263*; pathways
157–158, 166; scenarios 126–128
subsidies 24, *39,* 42, 159, 162; pathways
160, 164; scenarios 89, 96, 107, 120
success and failure criteria **33–34, 189**
sustainability *see* sustainable development
sustainable development 27, *34–35,* 110;
see also sustainable development goals
sustainable development goals 25, 35,
83, 173
systems *see* biological systems; climate
response system; climate regime
complex; ecosystems; human and
managed systems; and, physical systems

tariffs 95, 123
Task Force on Climate Related Financial
Disclosures 18, 140
tax 24, *39,* 42, 111, 120, 295; green carbon
tax 89, 116, 120, 123, 127
TCFD 2017 18–19, 140
technological attitude *see* response attitudes
technology 76–77, 89–90, 96–97, 100–103;
see technology and practices
technology and practices 139, *154, 158,*
162; scenarios 56, 86, 90, 100–101, 104
terrestrial 15, **27, 139,** 200
theory of change 50–51, 54, 254, 279–282
thematic analysis 54–55, *267,* 279, 281,
284, 286
thematic chain analysis 51, 55, 153,
279, 285
thresholds 15, 22, 50, 163; preconditions
for effective global responses 187,
189, 198

tipping point 15, 22, 254; environment scenarios **61**, 63–64, 66, 70–72, 108, 119; technology and practice scenarios 78, 90; preconditions for effective global responses 145, 177, 187, **189**, 198
trade 18, 20, 31, 33, 42–43, 260; scenarios 92, 95, 97–98, 117, 121–125; pathways *159–160*, 165
transition 24, 85, 96, 144–145
transition risk 18–19, 21, 140, 144–145, 199, 254; *see also* risk
transformation 23, 26, 31, 145, 172, 187–**190**, 198; scenarios 74, 81–82, 126
transformation risk 145, 254
transparency, reporting and verification 124, *165–167*
transport 23–26, 97, 101, **143**, 161–162, 199
trigger responses 71–75; *see also* response triggers and drivers
Trump 84, 85, 120, 199
trust 81, 90, 133; distrust 76, 85, 124
Tuvalu 65, 121
two level game 40–41, 181

United Kingdom 82, 115, 182, 200, 208–209, 273
United Nations 16, 37, 42–43, 261–262, 283; scenarios 64, 115, 125, 131, 133
United Nations Food and Agriculture Organisation 116
United Nations Framework Convention on Climate Change 32–47, 51–52; scenarios 66, 83, 105, 115–116, 123, 126–129; pathways 139, 165; preconditions for effective global responses 171, 184, 189, 194, 206, 260–265, 268–269, 283
United Nations Paris Agreement *see* Paris Agreement
United Nations Secretary General 115–116, 283
United Nations Security Council 122–124, 129, 132, 165, 182, 200
United States of America 16, 40; impacts and risk 64, 118; social movement 81–82; leadership 88–89, 109; cynical response 91; green economy 97, 157; nationalism 115; geopolitical power 121–123, 200; non-participation 128; solar radiation management 132; competition 161, 199; United Nations Security Council 182

volcano *see* eruptions
vulnerability 17–18, 85, 140–**141**, 255, **258**–259; *see also* risk

war 131–133, 184; *see also* conflict
waste 103, 152, 167, 178, 295
water: groundwater 93; potable 25, 42, 91, 106; security 15, **61**, 64, 113, 143; vapour 11–12; *see also* ocean acidification
wind energy 92, 101
World Bank 115, 123, 165, 283
World Trade Organisation 36, 38, 165; *see also* trade

youth 82, **84**, 109, *156*
Youba Sokona 23

For Product Safety Concerns and Information please contact our EU
representative GPSR@taylorandfrancis.com
Taylor & Francis Verlag GmbH, Kaufingerstraße 24, 80331 München, Germany

9 7 8 1 0 3 2 7 3 7 8 3 6